AF664374

DISASTER
ECOLOGY AND ENVIRONMENT

DISASTER
ECOLOGY AND ENVIRONMENT

Arvind Kumar

2024

Daya Publishing House®

A Division of

Astral International Pvt. Ltd.

New Delhi – 110 002

First Publishes, 2007

Reprinted, 2024

ISBN: 978-81-70354-87-1 (HB)

Disclaimer:

Every possible effort has been made to ensure that the information contained in this book is accurate at the time of going to press, and the publisher and author cannot accept responsibility for any errors or omissions, however caused. No responsibility for loss or damage occasioned to any person acting, or refraining from action, as a result of the material in this publication can be accepted by the editor, the publisher or the author. The Publisher is not associated with any product or vendor mentioned in the book. The contents of this work are intended to further general scientific research, understanding and discussion only. Readers should consult with a specialist where appropriate.

Every effort has been made to trace the owners of copyright material used in this book, if any. The author and the publisher will be grateful for any omission brought to their notice for acknowledgment in the future editions of the book.

Published by : **Daya Publishing House®**
A Division of
Astral International Pvt. Ltd.
– ISO 9001:2015 Certified Company –
4736/23, Ansari Road, Darya Ganj,
New Delhi-110 002
Ph. 011-43549197, 23278134
E-mail: info@astralint.com
Website: www.astralint.com

Printed at : **Replika Press Pvt. Ltd.**

Preface

The watery planet (earth) was formed over 4,600 million years ago from hot gases and cosmic dust. Primitive life probably evolved about 3,500 million years ago in the form of a single cell organism and bacteria; 400 million years ago life forms began to colonize the dry land. The human species evolved about 100,000 years ago. It was not until about 10,000 years ago that modern man emerged with the development of village, agriculture and the domestication of animal and plant species. From this perspective, the human species may seem to be an insignificant episode in the planet's history. However, the global population is growing by 75-80 million a year and is expected to reach approximately 8 billion by the year 2025. This enormous population requires food, water, clothes, house, medical care, transport, education, employment and recreation. The world we live in has to provide the raw materials for all these activities. The problem we face today is managing the earth's resources by utilizing man's ingenious technology, to support a huge population while not destroying the planet we live on.

The awareness of environmental disaster of the 1990s has enabled us to look at our activities and examine the effect that we are having on the planet earth. Now, we are facing major man-made global environmental problems, such as acid rain, greenhouse effect, smog, ozone depletion and eutrophication. Therefore, the crying need of the time is to proper environmental management and mass awareness, because environmental disaster has complex linkage with economic, social and ecological factors and processes. It is necessary not only to halt further degradation but to active restore and rehabilitate degraded ecosystems. The management of environmental disaster demands a synergy between ecology, sociology, economics, culture and technology, together with appropriate policy and institutions.

This book is the compilation of esteemed articles of internationally acknowledged experts in the field of aquatic biology with the intention of providing a sufficient depth of the subject to satisfy the needs at a level which will be comprehensive and interesting. The present book will be useful to the students, research scholars, scientists in the field of Environmental management and ecoplanners, politicians and other people with similar interest.

My special thanks and appreciation go to the scientists whose contributions have enriched this volume. I wish to express my sincere gratitude to Dr. P.C. Hembram, Hon'ble Vice Chancellor, S.K.M. University, Dumka, Professor M.C. Dash, Hon'ble former Vice Chancellor of Sambalpur University, Professor N.C. Datta of Calcutta University, Professor K.C. Pandey of Lucknow University, Professor S.K. Konar of Kalyani University, Professor D.K. Belsare of Bhopal University, Professor Professor U.C. Goswami of Gauhati University, Professor P.C. Mishra of Sambalpur University, Professor P. Natarajan of Kerala University, Professor P. S. Murthy of Bangalore University, Professor Ajit Varma of JNU, Professor A.L. Bhatia of Jaipur University, Professor A.K. Mittal of BHU, Professor S.P. Hosmani of Mysore University, Professor K. Kapoor of Udaipur University, Professor K.B. Reddy of Nagarjuna University, Professor M. Vikram Reddy of Pondicherry University, Professor B.K. Tiwari of NEHU, Professor G. Tripathi of Jodhpur University, Professor R. Ramalingam of Annamalai University, Professor Sharif U. Ahmad of Nagaland University, Professor G.K. Kulkarni of Aurangabad University, Professor S.U. Mehram of Nagpur University, Professor S.K. Battish of PAU, Professor B.M. Sharma of Manipur University, Professor B.D. Joshi of Hardwar University, Professor G.C. Pandey of Faizabad University, Professor K. C. Sharma of Ajmer University, Professor M. Raziuddin of Hazaribag University, Professor U.S. Bagde of Mumbai University, Professor Gurdeep Singh of I.S.M., Dhanbad, Dr. P.K. Goel of Karad, Professor S.P. Roy and Shri Tribhuwan Poddar of Bhagalpur University, Bhagalpur for encouragements.

I also express my deep sense of gratitude to my parents whose blessings have always prompted me to pursue academic activities deeply. I am also thankful to my wife, *Kumari Bimla* and my two lovely sons, *Kumar Pallav Shivshankaran* and *Kumar Prasun Ramakrishnan* whose natural smiles extended to me relief all through this tiresome endeavour.

Last but not the least, I am also thankful to Mr. Anil Mittal, Proprietor, Daya Publishing House, New Delhi for taking keen interest in bringing out of this book. Finally, I will always remain a debtor to all my well-wishers for their blessings, without which this book would not have come into existence.

Professor A. Kumar

Contents

Chapter 1

Contamination Impact of Chemicals Produced by Moulds

Kailash Kumar Sah & Arvind Kumar
Environmental Science Research Unit, Post Graduate Department of Zoology
S.K.M. University, Dumka, India

The term "mycotoxin" is used to designate a group of compounds produced by some strains of fungi belonging to the genera *Aspergillus, Pencillium, Fusarium, Alterrnaria, Helminthosporium* and *Cladosporium*. The disease syndromes that result from the intake of mycotoxins are called mycotoxicoses, which should not be confused with mycoses in which actual mould growth in the host is involved. Toxic metabolites from moulds have been known since long e.g. ergot poisoning is the most common and amongst the known earlier examples of mould poisonings. The possibility of moulds contaminating feed stuffs and producing highly toxic substances under favourable conditions was largely ignored with the outbreaks of Turkey 'x' disease in England in 1960 when at least 100,000 turkey poults died after consuming groundnut meal which contained highly toxic metabolites known as aflatoxins. To due over 350 mycotoxins have been identified and characterized. However, it is significant that only a small percentage of the known mycotoxins have actually been implicated in field outbreaks, which include, aflatoxins, ochratoxins, trichothecenes, zearalenone, citrinine, vomitoxin, patulin, sterigmatoxystin etc. The role of remaining toxins in human and animal health is yet to be explored.

The Food and Agricultural Organization (FAO) of the United Nations recently estimated that 25 per cent of the world's grain supplies are contaminated with mycotoxins. They pose a serious threat to food safety. Food losses caused by toxigenic moulds and consequent contamination affect food and animal feed availability for effective use. The mycotoxin problem is of grave concern for the farmers, traders and the feed manufacturers resulting in huge economic losses in terms of downgraded value of feeds, rejection of feeds, their marketability and losses in productivity and health hazards in the livestock. Mycotoxins are also important from public health point of view since these toxins and their metabolites are the common source of contamination in milk, meat and eggs.

Mycotoxin contamination of foods and feeds has gained global significance because of the significant economic losses associated with their impact on human health, livestock and poultry productivity and domestic and international trade in various agricultural crops and commodities. The mycotoxins are the potential source of environmental contamination leading to a serious quality and safety problems of foods and feed confronting human population and livestock including poultry in both developed and developing countries. In terms of their economic implications, mycotoxin are one of the most important contaminants of the feed chain. According to recent FAO estimates world losses of foodstuffs due to mycotoxins are in the range of 1000 million tonnes annually.

A number of cereal and oilseed based feeds are currently known to be contaminated with various mycotoxins. Among cereal crops, the most affected is maize. Maize an important ingredient in the diet of many poor people and also common ingredient in livestock feeds. Groundnut and cottonseed meals that form important ingredients of animal feeds are also contaminated with high levels of several mycotoxins. The main mycotoxins that have been detected in cereal-based feeds in regular occurrence are the aflatoxins, ochratoxin A, zearalenone, T-2 toxin and sterigmatoxystine. Mycotoxin contamination in feeds is of much concern in dairy and feed lot cattle, pigs and poultry because of their high intake of stored concentrated foods. In practice, the major impact of mycotoxins remains beneath the surface because the affects of continued low-to-moderate level consumption are variable undramatic and not easily recognized.

Mycotoxins can enter the human and dietary system by indirect or direct contamination. Indirect contamination of food or feed can occur when an ingredient of a process has previously been contaminated with toxin producing moulds and, although the moulds may be killed or removed during processing, mycotoxin may remain in the final products. Such contamination of different ingredients represents the main point of entry of many mycotoxins into food and feed chains. In direct contamination, the product later becomes infected with a toxigenic mould with subsequent toxin formation. Almost all foods and feeds are susceptible of mould growth at some stage during their production, processing, transport and storage.

Mycotoxin-contamination of dairy products is of great public health significance because of their toxic, carcinogenic, mutagenic, and/or teratogenic in nature. Dairy products can be contaminated directly, by mould growth during handling processing, transportation storage, and distribution of milk and its products, resulting in the production of aflatoxins. Indirectly, aflatoxins occurring in animal feed stuffs can be carried over to the milk. Aflatoxins have been detected in almost all the parts of the world.

Mycotoxins are relatively stable and remain in the feed even after the death of the mould responsible for toxin production. Chemically these are relatively of low molecular weight, non-antigenic, capable of producing toxic effects. These can enter into the body through ingestion, inhalation or direct skin contact and even minute quantities are responsible for significant health hazards in exposed individuals/animals/birds. At high concentration many mycotoxins can produce immediate acute disease syndromes, while at lower levels they can be carcinogenic, mutagenic, teratogenic or estrogenic. These can also affect the growth rate of younger ones and may interfere with native mechanism of resistance and immunogenic responsive thus increasing their susceptibility to infectious agents.

The level of mycotoxin contamination varies from country to country and within countries and depends on a wide range of environmental factors, agricultural practice and food hygiene. Many types of toxin producing fungi as mentioned earlier, have shown to have toxin potential to man and animals, depending upon the nature of toxigenic mould, suitable substrate for their growth and a suitable environment, *e.g.* moisture and temperature. Toxin production occurs only, when all the conditions are fulfilled.

Many countries have set stringent mycotoxin levels for foodstuffs ranging from zero to 50 $\mu g kg^{-1}$. This necessitates effective monitoring and control of mycotoxins in foodstuffs. The control of mycotoxins involves surveillance procedures to identify the nature and extent of the mycotoxin, together with quality control.

Controlling Measures

Highly concerted efforts are required to adopt integrated approaches to prevent or greatly reduce the problems related to mycotoxins. This is important the mycotoxin producing moulds can invade various foods and feeds at different stages of food production, storage, transport, and processing of foods and feeds.

Effective crop management practices are extremely important to prevent or control microbial pathogens of food crops for producing quality food and feed products. Proper harvesting, drying of crop produce to safe moisture levels, and storage environments can aid in restricting microbial contaminants. Hygienic conditions and pest-free environment in warehouses and storage buildings are extremely important to control microbial spoilage in stored products. Some non-hazardous insecticides / fumigants can be applied to control insect pests.

Hygienic conditions in food processing plants and feed mills are of utmost importance in manufacturing various quality products. Storage of feeds under dry, hygienic conditions can help check the development of moulds and bacterial infections. Application of certain anti-mould products (e.g. copper sulfate, propionic acid) to feeds can prove useful to prevent and reduce the contaminants.

At the farm level, effective management in rearing houses, hygienic designs of buildings and equipment, and personnel hygiene are of great significance in reducing the microbial infections in animal and poultry birds, and thereby minimizing the infection risks in humans.

In order to maintain the quality of food and feed during transportation, containers and packaging material play a vital role in restricting microbial products, particularly perishable and semi-perishable products, can be of great help. Polythene wrapping for fresh and frozen meat is the most effective method. Under the HACCP, it is imperative to adopt regular schedules of clearing and disinfecting of equipments and processing plants for effectively reducing microbial contamination.

Different systems of Hazard Analysis Critical Control Point (HACCP) to identify, evaluate and control, microbial hazards should be integrated at different production levels wherein each link of the food chain should be carefully analyzed to elucidate where and how specific hazards are introduced. Apart from respecting the legislations for permissible levels of mycotoxins, products and the food and feed industries must share responsibilities for the quality and safety of food and feeds.

Chapter 2

Impact of Heavy Metals on Algal Flora in Effluent Channel Balanagar, Hyderabad

P. Manikya Reddy and D. Venugopal Rao
Department of Botany, Osmania University, Hyderabad – 500 007

ABSTRACT

Impact of heavy metals on algae was studied over a period of one year. Water and algal samples were collected and analyzed at monthly intervals at three stations along the course of a channel. Lead, Manganese, Zinc, Iron and Copper were estimated at all the stations. In the channel all the metals were recorded in high concentrations and showed a decrease from stations I to III. Diatoms were constituted the dominant group of algae. Multiple Regression Analysis was employed to discover the relative importance of various metals on the abundance and distribution of algae. Mathematical equations were derived involving the metals for better prediction of algal number.

***Keywords**: Heavy metals, Algae, Pollution, MRA.*

Introduction

The most common pollutants in water are high concentrations of organic matter, plant nutrients, suspended mineral particles, deoxygenating substances and small quantities of heavy metals. Mercury, Cadmium, Chromium, Manganese, Iron, Copper, Zinc, Cobalt, Nickel and Lead have attracted the attention of many workers. Some of them are required as trace elements by living organisms though others (Hg, Cd and Pb) do not appear to be important in the metabolism of the organisms. Organisms are capable of extracting these elements from water in which the metals are present in very low concentrations.

Municipal sewage and industrial wastes introduce metals directly into surface water. In many rivers and lakes the human input of trace elements is many times greater than the natural input (Stumm and Baccini, 1978 and Nriagu *et al.*, 1979).

Because algae serve as the basis of all food chains and food webs, accumulation of heavy metals by them may adversely affect the entire ecosystem. Mercury and Cadmium are grouped under "black list" because of their toxicity and persistence whereas Zn, Cu, Ni, Cr, Pb, Co were grouped under gray list. Which are regarded as less dangerous than those of black list by "European Economic Community". The sensitivity of algae to metals has been demonstrated at various levels by Say and Whitton (1977), Whitton (1984), Prahalad and Seenayya (1989), Sudhakar *et al.* (1991), and Sudha Rani and Manikya Reddy (2003).

The effect of heavy metals on the microbiota in aquatic ecosystems are influenced to a large extent by the physico-chemical factors. Metal toxicity is influenced by pH (Richter and Theis, 1980), Carbonate ions (Babich and Stotzky, 1979), Phosphates (Rana and Kumar, 1974) and hardness. Organic matter is also greatly influenced the mobility, bioavailability and toxicity of heavy metals to algae.

The present article deals with certain heavy metals, their distribution, seasonal fluctuations and their impact on algae.

Materials and Methods

The effluents channel originates near Jeedimetla industrial area and passes through Balanagar after which it merges with Kukatpally nala at Fathenagar and finally enters into Hussainsagar lake. It receives various industrial wastes at the first phase of its flow later it receives domestic wastes from various localities along its stretch and serves as the main feeding channel to the lake Hussainsagar.

Three sampling stations were selected along the course of the channel. Station-I located at Balanagar, where it receives industrial effluents. Station-II at Fathenagar, which receives industrial effluents and domestic sewage. Station-III located at Begumpet flyover bridge.

Water and algal samples were collected from all the stations at monthly intervals. For heavy metals, sampling, preservation, digestion and preparation of samples the procedures given in APHA (1985) were followed. The final estimation was done with the help of Atomic Absorption Spectrophotometer (Perkin-Elmer-272). For the collection and analysis of algae the technique of Blum (1957) and Venkateswarlu (1969) was followed. The data were statistically analyzed by following the methods of Nageswara Rao (1983).

Results

The average values of heavy metals are given in Table 2.1. It is evident from the table that station-I recorded very high concentrations of metals whereas station-III recorded comparatively low concentrations of metals. Iron was recorded in very high concentration at all the stations.

Lead is a serious cumulative poison. It ranged between 1.9 and 7.5 mg/l at station-I, 1.5 to 5.2 mg/l and 1.5 to 8.0 mg/l at stations-II and III respectively. Manganese in surface waters present both in suspension in the quadrivalent state and trivalent state in a relatively stable soluble complex. At station-I it varied widely between 16.52 and 28.0 mg/l. At station-II it was recorded in between 2.19–14.59 mg/l, thus showing a 116 per cent decrease from station-I. At station-III it was recorded 2.7 to 8.7 mg/l, showing 44 per cent decrease in its concentration from station-II. The highest values were recorded in March and May. Zinc is an important micronutrient for growth and metabolism of algae.

Zinc varied between 7.08–21.8 mg/l at station-I and 1.0–15.9 mg/l at station-II, showing 116 per cent decrease from station-I. At station-III, it was recorded between 0.7 and 10.4 mg/l showing 65 per cent decrease in its concentration from station-II. Maximum values were recorded during summer. Iron was the most dominant heavy metal in the present study at all the stations. At station-I it was in the range of 9623–20547 mg/l and at station-II it fluctuated between 16.4 and 383.0 mg/l, showing 10000 per cent decrease from station-I. At station-III it ranges between 8.4 and 183.1 mg/l, a decrease of 40 per cent from station-II. At station-I copper fluctuated between 0.63 and 6.8 mg/l. At station-II in the range of 0.5–5.0 mg/l, showing a decrease of 102 per cent from station-I. At station-III it was recorded in the range of 0.08–2.28 mg/l, showing 41 per cent decrease from station-II. High values were recorded during summer.

Table 2.1: Average Values of Heavy Metals and Certain Chemical Parameters (All values in mg/l, except pH)

Metals	*Station-I*	*Station-II*	*Station-III*
Lead	4.57	3.4	4.5
Manganese	20.13	8.2	6.16
Zinc	16.01	8.1	4.9
Iron	12,995	129.6	98.1
Copper	2.96	1.47	1.0
pH	4.8	6.7	7.3
Phosphates	1.8	1.32	1.36
Hardness	709	412	456
Organic matter	23.3	24.3	19.1

In the effluent channel four groups of algae were recorded. Bacillariophyceae were dominant at all the stations followed by cyanophyceae and euglenophyceae. Chlorophycean members occupied the last position at all the stations (Table 2.2).

Table 2.2: Percentage of Different Groups of Algae

Groups	*Station-I*	*Station-II*	*Station-III*
Bacillariophyceae	45.5	35.24	42.62
Cyanophyceae	20.06	18.12	17.26
Chlorophyceae	16.68	16.13	15.9
Euglenophyceae	17.60	30.12	24.31

Discussion

At station-I lead, manganese, zinc and copper together accounts for 60 per cent of algal variance in a statistically significant manner. Lead accounts for 12 per cent algal variance and exhibits negative relationship with algal growth. Sudhakar *et al.* (1991) reported that lead influence the growth of diatoms to the positive side. The diatom number is influenced by copper to the extent of 18 per cent in stepwise regression analysis. Manganese contributes 15 per cent algal variance and exhibits negative relationship with algal number. Sudhakar *et al.* (1991) also observed negative effect of Mn on algae. A unit change in its concentration brings about 19 units change in algal number. Patrick *et al.* (1969) reported that high concentrations of Mn favoured the development of diatom communities. This is obvious in the present study. The model derived is as follows:

Diatoms number = 51.83 – 28.53 × lead + 19.94 × Manganese – 9.62 × Zinc – 22.64 × Copper

$$(R^2 = 0.60, F = 2.71, df = 4, 7, n = 12)$$

At station-II, Pb, Zn, Fe and Cu together account for 45 per cent of algal variance in a statistically significant manner. Cu and Zn account for 10 per cent and 18 per cent algal variation with negative relationship. Sudhakar *et al.* (1991) reported that high concentration of zinc proves to be beneficial to the growth of diatoms. In the present study decrease in Zn concentration from station-I to III showed the decrease in algal number. The proposed equation is as follows:

Diatoms number = 135.9 + 111.05 × Pb – 24.8 × Zn + 1.37 × Fe + 25.8 × Cu

$$(R^2 = 0.45, F = 1.44, df = 4, 7, n = 12)$$

At station-III except Pb remaining factors influence the diatoms to the extent of 71 per cent variation in algal number. Zinc accounts for 17 per cent of algal variation with negative side. A change of one unit in its concentration brings 114 units change in algal number. Iron contributes 30 per cent of variation in algal number. Manikya Reddy and Venkateswarlu (1985) and Sudhakar *et al.* (1991) observed an inverse relationship between Iron and diatoms. Cu and Mn together account for 24 per cent of variation in algal number with negative relationship. Cu alone contributing 20 per cent of algal variation, while manganese brings about 4 per cent of algal variation. The contribution of copper in influencing the growth is considerably large as one unit change in its concentration results in 251 units change in algal number. The model derived is as follows:

Diatoms number = 468.7 = 37.88 × Mn – 114.94 × Zn + 4.01 × Fe – 251.7 × Cu

$$(R^2 = 0.71, f = 4.45, df = 4, 7, n = 12)$$

Cyanophyceae

At station-I the contribution of Pb, Mn, Fe and Cu appears to be less significant on the algal growth. From MRA it is clear that Zn is playing significant role, accounts for 16 per cent of algal variation with negative influence. A unit change in its concentration brings about 100 units change in algal growth. Manikya Reddy and Venkateswarlu (1985) and Sudhakar *et al.* (1991) reported that Zinc and organic matter are related positively with algal number. This indicates an increase in organic matter decrease the Zinc uptake.

At station-II Mn, Cu and Zn together account for 60 per cent of variation in algal number. Among these Cu brings about 51 per cent of algal variance with positive relation on algal growth. One unit change in its concentration brings about 59 units in algal number. Zinc contributes 3 per cent algal variation with positive influence. Mn alone explains 3 per cent of algal variation with negative side. From Stepwise Regression Analysis it is evident that the Pb and Fe contribution appears to be less significant. The proposed equation is as follows:

Cynophyceae number = 84.6 – 15.81 × Mn + 7.55 × Zn + 59.3 × Cu

$$(R^2 = 0.60, F = 4.05, df = 3, 8, n = 12)$$

At station-III Mn and Cu together account for 26 per cent of algal variations in statistically significant manner. Among these two, copper alone accounts for 19 per cent of variation in algal number. Mn accounts for 7 per cent of algal variance with positive side. The model derived is as follows:

Cynophyceae number = 22.61 + 11.33 × Mn + 48.55 × Cu

$$(R^2 = 0.26, F = 1.63, df = 9, 2, n = 12)$$

Chlorophyceae

At station-I Pb, Mn, Fe, Cu together account for 39 per cent of variation. Iron and Cu together account for 20 per cent variation in algal number. At station-II Cu and Zn together account for 58 per cent of algal variation. Among these Cu individually influences to the extent of 51 per cent with positive manner. This is in agreement with the findings of Sudhakar *et al.* (1991). A unit change in its concentration brings about 46 units change in algal number. The proposed equation is as follows:

Chlorophyceae number = 48.38 + 7.6 × Zn + 46.6 × Cu

$$(R^2 = 0.58, f = 6.39, df = 9, 2, n = 12)$$

At station-III except Pb remaining metals Fe, Mn, Zn and Cu together account for 52 per cent of variation in algal number of these Mn alone responsible for 29 per cent variation with positive side. Cu proves to be beneficial to the growth of green alga, because it accounts for 6 per cent of variation with positive relationship, as a unit change in its concentration results in 87 units change in algal number. The model derived is as follows:

Chlorophyceae number = 197.46 + 25.8 × Mn + 26.28 × Zn – 0.9 × Fe + 87.5 × Cu

$$(R^2 = 0.51, f = 1.8, df = 4, 7, n = 12)$$

Euglenophyceae

At station-I Pb, Mn, Zn and Cu collectively influenced to the extent of 81 per cent of algal variation. Copper alone contributes 46 per cent of algal variants with positive manner. A unit change its concentration brings about 13 units change in algal number. Mn contributes 11 per cent of variation in algal number with negative side. Negative effect of Mn on algae could be result of its non availability due to antagonistic role of Pb, thereby increasing the levels of Mn in water (Upitis *et al.*, 1973). The proposed equation is as follows:

Euglenophyceae number = 26.8 + 10.0 × Pb – 7.63 × Mn + 4.24 × Zn + 13.17 × Cu

$$(R^2\ 0.81, F = 7.89, df = 4, 7, n = 12)$$

At station-II Cu and Zn together responsible for 54 per cent of algal variation. Among these Cu alone contributes 48 per cent of algal variation with positive side. The influence of Cu is so significant that one unit change in its concentration brings about 81 units change in algal number. Pb contributes 25 per cent of variation with negative side. This may be due to the presence of higher amounts of organic matter, total hardness and phosphates decrease the uptake of heavy metals. Similar observations were also made by Manikya Reddy and Venkateswarlu (1985). The model derived is as follows:

Euglenophyceae number = 57.8 + 11.2 × Zn + 81.7 × Cu

$$(R^2 = 0.54, f = 5.33, df = 9, 2, n = 12)$$

At station-III Cu and Pb are influenced to the extent of 31 per cent of algal variation. Among these Cu alone contributing 16 per cent of variance in algal number with positive side. Whereas Pb contributes 15 per cent of variation with negative side. The proposed equation is as follows:

Euglenophyceae number = 122.5 – 33.6 × Pb + 15.3 × Cu

$$(R^2 = 0.31, f = 2.02, df = 9, 2, n = 12)$$

From the foregoing account it is clear that diatoms are directly related with Mn and Fe and inversely with Zn and Cu at all the stations. Sudhakar *et al.* (1991) and Manikya Reddy and Venkateswarlu (1985) found an inverse relationship between Cu and Zn with diatoms. Such a relationship is also evident in the present study. At high concentrations of Zn and Cu associated with good growth of chlorophyceae was observed by Manikya Reddy and Venkateswarlu (1985). In the present study also Zn and Cu exhibited direct relationship with green algae. Zn, Cu and Mn influence the growth of cyanophyceae. Pb, Cu and Zn are significant for the growth of euglenophyceae. Copper shows direct and Manganese shows an inverse relation with euglenoid flagellates. The bioavailability and toxicity of metals are influenced to a large extent by physico-chemical variables. In the present study pH, hardness, organic matter and phosphate content in water are found to be more significant in influencing the metal availability to algae. Further the interactions among multiple heavy metals, influence the response of algae to stress by these metals.

Several workers have pointed out the tolerance of algal species to certain metals and non-tolerance to others and use them as metal pollution indicator organisms. Patrick (1978) mentioned that low concentration of manganese favoured the growth of diatoms. Butcher (1955) observed *Chroococcus, Stigeoclonium tenue* as resistant to copper than other species. Manikya Reddy and Venkateswarlu (1985) reported that growth of *Stigeoclonium tenue* was in thick masses in high concentrations of metals. In the present study *S. tenue* was shown luxuriant growth at all the stations. This clearly indicates that these can be used as good indicator of metal pollution. Among diatoms the species of *Nitzchia* dominated over others whereas euglenoid flagellates mainly represented by *Euglena* species and which can be used as good indicators of metal pollution.

References

APHA, AWWA and WPCF, 1985. *Standard Methods for the Examination of Water and Wastewater*, 13th ed. Washington, USA.

Babich, H. and Stotzky, G., 1979. Abiotic factors affecting the toxicity of lead to fungi. *Appl. Environ. Microbiol.*, 38: 506–514.

Blum, J.L., 1957. An Ecological study of the algae of the Saline River, Michigon, *Hydrobiologia* 9: 361–408.

Butcher, R.W., 1955. Relation between the biology and the polluted condition of the Trent. *Vern. Int. Ver. Limnol.*, 12: 827–832.

Manikya Reddy, P. and Venkateswarlu, V., 1985. Ecological studies in the paper mill effluents and their impact on the river Tungabhadra: Heavy metals and algae. In: *Proc. Indian Acad. Sci. (Plant Sci.)*, 95: 139–146.

Nageswara Rao, G., 1983. *Statistics for Agricultural Sciences.* Oxford and IBH Publishing Co., New Delhi.

Nriagu, J.O., Kemp, A.L.W., Wong, H.K.T. and Harper, N., 1979. Sedimentary record of heavy metal pollution in lake Erie. *Geochem. Cosmochim. Acta*, 43: 247–258.

Prahalad, A.K. and Seenayya, G., 1989. Physico-chemical interactions and bio-concentration of zinc and lead in the industrially polluted Hussain Sagar lake, Hyderabad, India. *Environmental Pollution*, 58: 139–154.

Rana, B.C. and Kumar, H.D., 1974. Ecological studies on the uptake of pollutants copper, zinc and phosphates by certain algae. *Indian J. Ecol.*, 1: 1–11.

Richter, R.O. and Theis, T.L., 1980. Nickel speciation in soil water system. In: *Nickel in the Environment*, (Ed.) J.O. Nriage. John Wiley and Sons, New York, pp. 189–202.

Say, P.J. and Whitton, B.A., 1977. Influence of zinc on lotic plants. II. Environmental effects on toxicity of zinc. *Hormidium rivulare. Fresh Water Biol.*, 7: 377–384.

Sudhakar, G., Jyothi, B. and Venkateswarlu, V., 1991. Metal pollution and its impact on algae in flowing waters in India. *Arch. Environ. Contam. Toxicol.*, 21: 556–566.

Sudha Rani and Manikya Reddy, P., 2003. Preliminary studies on metal concentration on Hussain Sagar lake. *Poll. Res.*, 22: 377–380.

Stumm, W. and Baccini, P., 1978. Man made perturbations of lakes. In: *Lakes Chemistry, Geology and Physics*, (Ed.) A. Lerman. Springer-Verlag, New York, pp. 91–126.

Upitis, V.V., Pakatue, D.S. and Nollendorf, 1973. The dosage of trace elements in the nutrient medium as a factor in increasing the resistance of *Chlorella* to unfavourable conditions of culturing. *Microbiology*, 42: 758– 762.

Venkateswarlu, V., 1969. An ecological study of algae of the river Moosi, Hyderabad, India with special reference to water pollution. I. Physico chemical complexes. *Hydrobiologia*, 33: 117–143.

Whitton, B.A., 1984. Algae as monitors of heavy metals in fresh waters. In: *Algae as Ecological Indicators*, (Ed.) L.E. Shubert. Academic Press, London, pp. 257–280.

Chapter 3

Studies on Some of the Commercially Important Fishes Caught in the Hobt Rigged with Different Categories of Bridle Lengths

Jaya Naik[1], B. Hanumanthappa[1] and R.M. Prabhu[2]
[1]Department of Fishery Engineering, College of Fisheries, Mangalore
[2]Karnataka Veterinary, Animal and Fisheries Sciences University, Bidar, India

ABSTRACT

Investigations were carried out to know the effect of bridle lengths in HOIBT on commercially available fish catch along Mangalore coast. The 19–21 m bridle lengths yielded higher catch. Perches, Ribbon fishes, Shrimps and Cephalopods were caught in large quantities during the fishing season lasting from September 2002 to May 2003 which may due to the availability of fish according to their habitat. Herding effect played an important role for catching larger quantity of fish and shellfishes in case of 19–21 m bridle length.

Keywords: Bottom trawls, Trawl catchability, Trawl efficiency, Trawl selectivity.

Introduction

Bottom trawl study supports the Fishery management process by providing stock assessment models with fishery-independent estimates of relative abundance based on catch per unit effort (CPUE) as well as age and size composition information. Precision of these CPUE estimates is partly affected by variability in the sampling efficiency of the trawl. To minimize this variability scientists pay close attention to the maintenance of strict fishing procedures such as gear deployment, retrievel and towing speed. However, study must also be conducted with uncontrollable environmental factors such as currents that can impact trawl geometry (Fridman, 1986) and perhaps those which influence the capture efficiency of the trawl.

Bottom trawl efficiency is a function of three processes *viz.*,

1. Horizontal herding wherein the movement of fish is towards the center of the trawl path in response to stimuli provided by sand, mud and bridles (Engas and Godo, 1989a; Ramm and Xaio, 1995; Somerton and Munro, 2001).
2. Vertical herding where the movement of fish is in and out of the trawl path relative to head rope height in response to stimuli such as head rope or vessel noise (Ona and Godo, 1990; Nunnallee, 1991; Mitson, 1995; Aglen, 1996; Godo *et al.*, 1999).
3. Escapement of fish which results in the loss of fish through the trawl mesh or underneath the footrope (Engas and Godo, 1989b; Walsh, 1992; Ona, 1999; Dham, 2000; Munro and Somerton, 2001). Since bottom trawls typically use large bridle lengths most escapement occurs due to passing of fish underneath the foot rope, especially near its center where fish gets concentrated by herding (Walsh, 1992; Ona, 1999). Variation in footrope which has contact with the bottom in this area could impact the capture efficiency of the trawl.

A recommended study was established with a gear experiment conducted in 1999 by Somerton and Weinberg (2001) which reported that the bottom tending performance of the footrope bridle varied with speed through water.

In marine sciences, biologists have long known that the presence of a vessel and of various types of fishing gears may influence the behaviour of the fish. Fish reaction to trawl and other gears have been identified in several works (Aglen, 1985; Misund and Aglen, 1992; Surronen *et al.*, 1997). Such knowledge of influences on fish behaviour has also been used to develop fishing gears, which take advantages of the fishes reaction to the gear (Ferno and Olsen, 1994).

The status of marine fishery in India is in a crucial phase now. The production has progressively increased by nearly 6 times during the past fifty years. However, there are a few warning signals in the characteristics of fish landings which suggest that the resources in the inshore waters are being fully exploited and the scope for increasing production from the present level is limited. During the earlier phase of marine capture fisheries development, the fisheries resources remained rather under utilized, whereas in the latter phases (especially in the 1990's) most of the resources have been either fully exploited or as feared over exploited consequently. The present status of marine fisheries causes for quick implementation of appropriated management measure to sustain production. However, the marine fisheries sector, which has thus enjoyed free access to the resources, is not prepared to face strength restrictive management measure.

Correct rigging of the trawl to the foot rope and head rope arrangements of trawl door is very important for smooth operation of the trawl, the trawl gear is towed by means of two warps, these warps are not directly attached to the trawl doors but by means intermediate pieces. The trawl boards are connected to the bridles, the diameter of the bridles are selected according to the type of trawlers and size of the trawl. The arrangement of bridle length is an important factors which influences the mouth opening of the trawl gear. As the mouth opening is increased the amount of catch will also increase, because of its swept area and herding effect. Keeping all these views, an attempt has been made to know the efficiency of the commercially operated HOBT rigged with varying bridle lengths.

Materials and Methods

The experiment has been conducted to know the commercially important fish catch in the varying categories of bridle lengths namely 15–18 m, 19–21 m, 22–24 m, 25–27 m, 28–30 m and 31–33 m. The stern trawler operated by the College of Fisheries, Mangalore has been used. The commercially important fishes such as Silver bellies, Sciaenids. Perches, Lactarius, Carangids, Ribbonfishes, Pomfrets, Anchovies, Shrimps and Cephalopods have been taken into consideration. For all the species caught the lengths and weights have been measured and their total catch and percentages were tabulated in the Tables.

Results and Discussion

The commercially important species obtained with different bridle lengths are Silver bellies (Table 3.1), Sciaenids (Table 3.2), Perches (Table 3.3), Lactarius (Table 3.4), Carangids (Table 3.5), Ribbonfishes (Table 3.6), Pomfrets (Table 3.7), Anchovies (Table 3.8), Shrimps (Table 3.9) and Cephalopods (Table 3.10). The total catch obtained in case of Silverbellies was 2491 kg for Sciaenids 1363 kg, for Perches 32,400 kg, for Lactarius 1448 kg, for Carangids 1542 kg. In case of Ribbon fishes 34,046 kg, Pomfrets (2628 kg), Anchovies (653 kg), Shrimps (10,397 kg) and Cephalopods (22,867 kg). During experimental sampling using all different categories of bridle lengths HOBT. The different size ranges of fish caught in the different categories of bridle lengths are also presented in the Tables. Highest catch of all the 10 species mentioned above were caught in a single bridle length of 19–21 m. This is due to the optimum mouth opening of the net for all the species. The herding effect in case of 19–21 m bridle lengths helped in catching large quantities of all species when compared to the other categories of bridle lengths. Similar results were obtained by Naik *et al.* (2004), while conducting the experiments on month wise basis for all the species caught in the commercially High Opening Bottom Trawl (HOBT).

Table 3.1: Size Range and Percentage of Total Catch of Silver Bellies for Different Categories of Bridle Lengths

Average Catch Rate Obtained for Different Categories of Bridle Lengths	*< 10 cm*		*Total Catch (Kg)*
	Weight (Kg)	*% of Total Catch*	
15–18 m bridle lengths	408	16.37	408
19–21 m bridle lengths	645	25.89	645
22–24 m bridle lengths	295	11.84	295
25–27 m bridle lengths	545	21.89	545
28–30 m bridle lengths	328	13.16	328
31–33 m bridle lengths	270	10.85	270
Total	**2491**	**100.0**	**2491**

The highest catch was observed in case of Perches, Ribbon fishes, Shrimps and Cephalopods, which may be due to their availability with variation in seasonal conditions.

Table 3.2: Size Range and Percentage of Total Catch of Sciaenids for Different Categories of Bridle Lengths

Average Catch Rate Obtained for Different Categories of Bridle Lengths	*< 10 cm*		*10–15 cm*		*> 15 cm*		*Total Catch (Kg)*
	Weight (Kg)	*% of Total Catch*	*Weight (Kg)*	*% of Total Catch*	*Weight (Kg)*	*% of Total Catch*	
15–18 m bridle lengths	144	18.39	24	6.50	36	14.36	204
19–21 m bridle lengths	210	26.84	90	24.39	30	11.95	330
22–24 m bridle lengths	135	17.24	60	16.26	40	15.93	235
25–27 m bridle lengths	140	17.87	75	20.32	75	29.88	250
28–30 m bridle lengths	54	6.89	50	13.56	30	11.95	134
31–33 m bridle lengths	100	12.77	70	18.97	40	15.93	210
Total	**783**	**100.0**	**369**	**100.0**	**251**	**100.0**	**1363**

Table 3.3: Size Range and Percentage of Total Catch of Perches for Different Categories of Bridle Lengths

Average Catch Rate Obtained for Different Categories of Bridle Lengths	*< 10 cm*		*10–15 cm*		*> 15 cm*		*Total Catch (Kg)*
	Weight (Kg)	*% of Total Catch*	*Weight (Kg)*	*% of Total Catch*	*Weight (Kg)*	*% of Total Catch*	
15–18 m bridle lengths	480	11.18	4800	19.43	240	7.03	5520
19–21 m bridle lengths	1050	24.47	5400	21.86	510	14.95	6960
22–24 m bridle lengths	900	20.97	2500	10.14	500	14.66	3900
25–27 m bridle lengths	800	18.66	5000	20.24	1000	29.36	6800
28–30 m bridle lengths	660	15.38	4000	16.19	700	20.52	5360
31–33 m bridle lengths	400	9.32	3000	12.14	460	13.48	3860
Total	**4290**	**100.0**	**24700**	**100.0**	**3410**	**100.0**	**32400**

Table 3.4: Size Range and Percentage of Total Catch of Lactarius for Different Categories of Bridle Lengths

Average Catch Rate Obtained for Different Categories of Bridle Lengths	*< 10 cm*		*10–15 cm*		*> 15 cm*		*Total Catch (Kg)*
	Weight (Kg)	*% of Total Catch*	*Weight (Kg)*	*% of Total Catch*	*Weight (Kg)*	*% of Total Catch*	
15–18 m bridle lengths	29	8.84	120	13.33	–	–	149
19–21 m bridle lengths	75	22.86	300	33.34	30	13.63	405
22–24 m bridle lengths	80	24.39	120	13.34	80	36.38	280
25–27 m bridle lengths	65	19.81	170	18.88	50	22.72	285
28–30 m bridle lengths	29	8.86	30	3.34	20	9.09	79
31–33 m bridle lengths	50	15.24	160	17.77	40	18.18	250
Total	**328**	**100.0**	**900**	**100.0**	**220**	**100.0**	**1448**

Table 3.5: Size Range and Percentage of Total Catch of Carangids for Different Categories of Bridle Lengths

Average Catch Rate Obtained for Different Categories of Bridle Lengths	*< 10 cm*		*10–15 cm*		*> 15 cm*		*Total Catch (Kg)*
	Weight (Kg)	*% of Total Catch*	*Weight (Kg)*	*% of Total Catch*	*Weight (Kg)*	*% of Total Catch*	
15–18 m bridle lengths	36	11.18	240	23.09	–	–	276
19–21 m bridle lengths	75	23.29	330	31.73	30	16.66	435
22–24 m bridle lengths	35	10.86	80	7.09	40	22.23	155
25–27 m bridle lengths	130	40.37	160	15.38	20	11.11	310
28–30 m bridle lengths	26	8.07	130	12.50	70	38.89	226
31–33 m bridle lengths	20	6.23	100	9.61	20	11.11	140
Total	**322**	**100.0**	**1040**	**100.0**	**180**	**100.0**	**1542**

Table 3.6: Size Range and Percentage of Total Catch of Ribbonfishes for Different Categories of Bridle Lengths

Average Catch Rate Obtained for Different Categories of Bridle Lengths	*10–15 cm*		*> 15 cm*		*Total Catch (Kg)*
	Weight (Kg)	*% of Total Catch*	*Weight (Kg)*	*% of Total Catch*	
15–18 m bridle lengths	588	7.79	4800	17.45	5388
19–21 m bridle lengths	1425	18.89	6600	24.00	8025
22–24 m bridle lengths	1780	23.59	2000	7.27	3780
25–27 m bridle lengths	1885	24.99	6000	21.83	7885
28–30 m bridle lengths	1228	16.26	5000	18.18	5228
31–33 m bridle lengths	640	8.48	3100	11.27	3740
Total	**7543**	**100.0**	**27500**	**100.0**	**34046**

Table 3.7: Size Range and Percentage of Total Catch of Pomfrets for Different Categories of Bridle Lengths

Average Catch Rate Obtained for Different Categories or Bridle Lengths	*< 10–15 cm*		*> 15 cm*		*Total Catch (Kg)*
	Weight (Kg)	*% of Total Catch*	*Weight (Kg)*	*% of Total Catch*	
15–18 m bridle lengths	72	14.46	420	19.73	492
19–21 m bridle lengths	42	8.43	390	18.30	432
22–24 m bridle lengths	106	21.28	300	14.08	406
25–27 m bridle lengths	52	10.46	200	9.38	252
28–30 m bridle lengths	180	36.14	500	23.47	680
31–33 m bridle lengths	46	9.23	320	15.04	366
Total	**498**	**100.0**	**2130**	**100.0**	**2628**

Table 3.8: Size Range and Percentage of Total Catch of Anchovies for Different Categories of Bridle Lengths

Average Catch Rate Obtained for Different Categories of Bridle Lengths	10–15 cm		Total Catch (Kg)
	Weight (Kg)	% of Total Catch	
15–18 m bridle lengths	108	16.53	108
19–21 m bridle lengths	105	16.07	105
22–24 m bridle lengths	175	26.79	175
25–27 m bridle lengths	55	8.44	55
28–30 m bridle lengths	60	9.18	60
31–33 m bridle lengths	150	22.99	150
Total	**653**	**100.0**	**653**

Table 3.9: Size Range and Percentage of Total Catch of Shrimps for Different Categories of Bridle Lengths

Average Catch Rate Obtained for Different Categories of Bridle Lengths	< 10 cm		10–15 cm		Total Catch (Kg)
	Weight (Kg)	% of Total Catch	Weight (Kg)	% of Total Catch	
15–18 m bridle lengths	1200	15.78	336	12.03	1536
19–21 m bridle lengths	1275	16.78	1050	37.62	2325
22–24 m bridle lengths	1200	15.77	205	7.37	1405
25–27 m bridle lengths	1755	23.07	500	17.91	2255
28–30 m bridle lengths	1316	17.30	200	7.16	1516
31–33 m bridle lengths	860	11.30	500	17.91	1360
Total	**3335**	**100.0**	**2791**	**100.0**	**10397**

Table 3.10: Size Range and Percentage of Total Catch of Cephalopods for Different Categories of Bridle Lengths

Average Catch Rate Obtained for Different Categories of Bridle Lengths	< 10 cm		10–15 cm		> 15 cm		Total Catch (Kg)
	Weight (Kg)	% of Total Catch	Weight (Kg)	% of Total Catch	Weight (Kg)	% of Total Catch	
15–18 m bridle lengths	384	8.91	2400	18.89	960	16.39	3744
19–21 m bridle lengths	330	7.68	3000	23.62	1800	30.72	5130
22–24 m bridle lengths	725	16.83	1200	9.46	800	13.65	2725
25–27 m bridle lengths	1040	24.14	3000	23.62	1000	17.06	5040
28–30 m bridle lengths	1548	35.94	1500	11.82	500	8.53	3548
31–33 m bridle lengths	280	6.50	1600	12.59	800	13.65	2680
Total	**3313**	**100.0**	**7000**	**100.0**	**5860**	**100.0**	**22867**

Conclusion

By this it has come to known that for the optimum mouth opening of the trawl net will have the better catch in all the groups of fishes and shellfishes caught in the experimental gear. In 19–21 m category of bridle lengths will have a good herding effect when compared to all other categories, hence it has been considered that well suited category for catching demersal fishes, because of its swept area and less time for the escapement of species.

References

Aglen, A., 1985. Sonar observations of the behavior of the herring schools in relation to fishing vessels. *ICES Fisheries Acoustic Science and Technology Working Group*, Tromoso, May, 1985.

Aglen, A., 1996. Impact of fish distribution and species composition on the relationship between acoustic and swept area estimates of fish density. *ICES J. Mar. Sci.*, 53: 501–505.

Dahm, E., 2000. Changes in length composition of some fish species as a consequence of alterations in the ground gear of GOV–Trawl. *Fish. Res.*, 49: 39–50.

Engas, A. and Godo, O.R., 1989a. The effects of different sweep lengths on the length composition of bottom: Sampling trawl catches. *J. Cons. Int. Explor. Mer.*, 45: 263–268.

Engas, A. and Godo, O.R., 1989b. Escape of fish under the fishing line of Norwegian sampling trawl and its influence on survey results. *J. Cons. Int. Explor. Mer.*, 45: 269–276.

Ferno, A. and Olsen, S. (Eds.), 1994. *Marine Fish Behavior in Capture and Abundance Estimation*. Fishing News Books, Blackwell Science Ltd., Oxford, 216 pp.

Fridman, A.L., 1986. *Calculation for Fishing Gear Designs*. Fishing News Books Ltd., Farnham, UK.

Godo, O.R., Somerton, D.A. and Totland, A., 1999. Fish behavior during sampling observed from free floating buoys: Application for bottom trawl survey assessment. Council for the exploration of the sea. CM 1999/J: 10, 14 pp.

Jaya Naik, Hanumanthappa, B., Prabhu, R.M. and Varadaraju, S., 2004. Conducted studies to know the effect of bridle lengths. *Environ. and Ecol.*, 22(4): 787–790.

Mitson, R.B., 1995. Underwater noise of research vessels: review and recommendations. *International Council for the Exploration of the Sea. Coop. Res. Rep.*, 29: 61.

Misund, O.A. and Aglen, A., 1992. Swimming behavior of fish schools in the North Sea during acoustic surveying and pelagic trawl sampling. *ICES J. Mar. Sci.*, 49: 325–334.

Munro, P.T. and Somerton, D.A., 2001. Maximum likelihood and nonparametric methods for estimating trwal footrope selectivity. *ICES J. Mar. Sci.*, 58: 220–229.

Nunnallee, E., 1991. An investigation of the avoidance reaction of pacific whiting (Merluccius products) for demersal and mid water trawls. *International Council for the Exploration of the Sea*. CM 1991B: 5, 21 pp.

Ona, E., 1999. Determining the entrance position of the fish trawls. *Internationul Council for the Exploration of the Sea*. CM 1999/J: 13, 10 pp.

Ona, E. and Godo, O.R., 1990. Fish reaction to trawling noice: the significance for trawl sampling. *Rapp. P.V. Reun. Cons. Int. Eplor. Mer.*, 189: 159–166.

Ramm, D.C. and Xaio, Y., 1995. Herding in ground fish and effective path width of trawls. *Fish. Res.*, 24: 243–259.

Somerton, D.A. and Munro, P.T., 2001. Bridle efficiency of a survey trawl for flat fish. *Fish. Bull.*, 99: 641–652.

Somerton, D.A. and Weinberg, K.L., 2001. The effect of speed through the water on the footrope contact of a survey trawl. *Fish. Res.*, 53: 17–24.

Surronen, P., Lehtonen, E. and Wallece, J., 1997. Avoidance and escape behavior by herring encountering mid water trawls. *Fish. Res.*, 29: 13–24.

Walsh, S.J., 1992. Size dependent selection at the footgear of a ground fish survey trawl. *N. Am. J. Fish. Manage.*, 12: 625–633.

Chapter 4

Efficiency of Micro-irrigation in Cane Culture: An Agroclimatic Study for Pune

S. Venkataraman*
Retired Director, India Meteorological Department and
Former WMO/FAO Agromet. Expert
59/19, Navsahyadri Society, Pune – 411 052, India

ABSTRACT

The agrometeorology of evaporation from bare and cropped soil surfaces is briefly dealt with. Methodology for determining monthly transpiratory need and possible evaporative depletion for Eksali and Dedsali cane crops at Pune mainly from pan evaporation data are detailed. Format for entry of relevant data and procedure for assessment from the data generated of (*i*) net micro-irrigation requirement (*ii*) quantum of water saved and (*iii*) use-efficiency of micro-irrigation under the two cropping systems are presented.

Keywords: Evaporative loss, Transpiratory need, Net micro-irrigation, Water use-efficiency.

Introduction

The soil moisture needs for optimal transpiration and hence yield of a crop is its minimum, essential water need. When soil moisture is provided through surface irrigation there is the inevitable loss of irrigation water as evaporation from the soil surface. The use of micro-irrigation has been mooted to save irrigation water used in cane culture. In micro-irrigation water is delivered directly to

* E-mail: svr@sridhar-v.com

the root zone to avoid evaporative losses. Savings in water of 60 per cent is claimed for micro-irrigation. There are two main systems of cane culture namely, the Spring-planted crop of about 12 months duration and the near-monsoon crop of about 18 months duration. To gauge the likely savings in irrigation water in the above two systems of cane culture, an understanding of the process of evaporation via the soil surface in bare soil and from soil surfaces of (*i*) ground shading crop and (*ii*) crop with incomplete crop cover is necessary and the same is detailed below.

Evaporation from Soil Surfaces

Immediately after irrigation of a soil column to field capacity, evaporation from the top surface of the soil proceeds at rate of the Evaporative Power of Air, EPA, which is the potential capacity of the overlying air to desiccate substances and is the resultant of a number of factors like solar radiation, vapour pressure deficit of air and wind speed. This phase of evaporation, called the 'Constant Rate Phase,' lasts till a fixed quantum of moisture appropriate to the soil type is lost. Subsequently, evaporation occurs from a sub-surface layer till a quantum of water equal to that in the surface layer is lost. In this phase, controlled by the hydraulic conductivity of the soil, daily evaporative loss goes on diminishing and is hence called the "Declining Phase." Further evaporative loss from the soil occurs as vapour transport due to thermal gradients in the soil. Vapour transport is not of relevance in periodically irrigated soils. Evaporative loss in the declining phase can be obtained (Ritchie, 1972) from the relation:

$$E = Ct^{1/2} - C\,(t-1)^{1/2}$$

In the above C is a soil specific constant and "t" is time in days after commencement of the declining phase. With crop's growth, the soil surface gets increasingly shaded and EPA over the soil surface, EPA, progressively decreases to a steady value of 0.2 times EPA at time of ground-coverage by the crop (Ritchie and Brunett, 1971; Venkataraman *et al.*, 1976). Thus as the crop grows the constant rate phase gets lengthened with a corresponding reduction in duration of the declining phase.

Materials and Methods

It was decided to carry out a preliminary agroclimatic study for Pune on efficiency of micro-irrigation with reference to (*i*) an Eksali crop planted in mid-February and (*ii*) a Dedsali crop planted in Mid-June. For both the crops attainment of ground covering stage was taken as 3 months from planting. Allen *et al.* (1992) have assumed that maturity period in Sugarcane will be the last 1/3rd of the crop's life for crops. This apparently is not correct. Thus for both the Eksali and Dedsali crops maturity period was taken as the last 4 months of the crop's life. The agricultural rainy period (Venkataraman, 2001) at Pune in which irrigation requirement will be minimal, is from mid-June to end September (IMD, 1995).

For the type of soils in which sugarcane is grown in and around Pune the amount of evaporable water from the surface and sub-surface layers will be 10 mm each (Allen *et al.*, 1992) and value of 'C' will be 5.0 (Ritchie, 1972). Evaporation measured by a mesh-covered Standard Class A pan, EP can be related to peak evapotranspiration, PET of crops. The ratio PET/EP shows variations in time and place and at Pune, PET of tall crop will (*i*) equal to EP in the rainy season, (*ii*) 5 per cent higher than EP in summer, (*iii*) equal to 1.33 times EP in winter and (*iv*) 10 per cent higher than EP at other times (Venkataraman *et al.*, 1984).

In the cane growing season at Pune, EPA can range from 10 mm/day during the planting of the Eksali crop to 1 mm/day from ground covering stage of both crops in the rainy and winter months. The daily evaporative loss for a period of 10 days following field-capacity irrigation for the EPA range mentioned above is given in Table 4.1.

Table 4.1: Daily Evaporation Via the Soil Surface in Cane Fields at Pune

EPAs mm/day	Evaporation, mm										
	Days										
	1	2	3	4	5	6	7	8	9	10	Total
10.0	10.0	2.1	1.6	1.3	1.2	1.1	1.0	0.90	0.85	0.00	20.0
5.0	5.0	5.0	2.1	1.6	1.3	1.2	1.1	1.0	0.90	0.85	20.0
4.0	4.0	4.0	3.1	1.9	1.5	1.3	1.1	1.1	0.95	0.90	19.8
2.5	2.5	2.5	2.5	2.5	2.1	1.6	1.3	1.2	1.1	1.0	18.3
2.0	2.0	2.0	2.0	2.0	2.0	2.0	1.6	1.3	1.2	1.1	17.2
1.0	1.0	1.0	1.0	1.0	1.0	1.0	1.0	1.0	1.0	1.0	10.0

It can be seen from Table 4.1 that in the range of EPAs of 10.0 to 2.0 mm/day the cumulative evaporative loss from the soil surface is little influenced by level of EPAs.

The establishment phase, from planting to ground-coverage, of a spring planted Eksali crop will occur in the hot weather period in which evaporative loss will be a significant fraction of the crop's irrigation need. For a near-monsoon planted Dedsali crop (*i*) the establishment will be in rainy weather and (*ii*) the summer period will occur next year with the crop fully grown. Sugarcane exhibits a tendency to curtail its water uptake during maturity to around 60 per cent of the potential (Allen *et al.*, 1992) which will be the last 4 months of the crop's life (Venkataraman, *et al.*, 1976). Thus, micro-irrigation can be expected to be more efficient in the early stages of a spring planted Eksali crop and in the maturity period of both Eksali and Dedsali crops.

Results and Discussion

Climatological means of EP at Pune for all the 52 standard weeks were taken. To simplify computations, mean values of EP for the corresponding monthly crop periods from sowing to harvest were worked out. The EP values were converted into values of peak ET of Sugarcane, PETsc, using the ratios of PET tall crop to EP mentioned above. Peak transpiratory need of sugarcane from ground coverage to commencement of maturity was taken as equal to 0.8 times PETsc. In the maturity period transpiratory need of the cane crop was taken as 0.7, 0.5, 0.4 and 0.4 times PETsc in the 1st, 2nd, 3rd and 4th months of maturity respectively. Evaporative Potential (EP) at the soil surface, Es, was taken as equal to 0.8, 0.55 and 0.3 times PETsc in the 1st, 2nd and 3rd months of crop growth respectively and as equal to 0.2 times PETsc from 4th month of crop growth. Transpiratory need was taken as (*i*) 1.0 minus Es from ground-shading to commencement of crop maturity and (*ii*) as 0.7, 0.5, 0.4 and 0.4 in the 1st, 2nd, 3rd and 4th months of maturity. For the Eksali crop irrigation at weekly intervals in the first two months of crop growth, at 10-day intervals upto start of rains and 15 day intervals from October to mid-February were assumed. For the Dedsali irrigation intervals of 15 days from October to harvest was assumed. Likely evaporative loss from a surface irrigation was computed by procedures and formula detailed above.

The data-entry and its particulars for computation relating to efficiency of micro-irrigation are shown in Tables 4.2 and 4.3 respectively. In this the quantum of irrigation water saved by micro-irrigation will be the likely evaporative loss for surface irrigation. For irrigated crops passing through the rainy season determination of net irrigation requirement in the rainy period is a must. For this the adequacy of rainfall in meeting the crop's water need has to be assessed. It is seen from Table 4.4, that the transpiratory need in the second half of September will be (*i*) 55 mm for the Eksali and (*ii*) 55 and 35 mm respectively in the first and second rainy seasons respectively for the Dedsali crop. The Minimum Assured Rainfall, MAR at the 50 per cent probability level in the second fortnight of September at Pune is 52 mm (IMD, 1995). Thus only one irrigation is required in the period from mid-September to mid-October for the cane crops. In the period from mid-June to mid-September the transpiratory need of the Eksali crops will be 331 mm while MAR at 50 per cent probability is 236 mm. For the Eksali crop evaporative loss of rainfall will be a maximum of 80 mm. Thus rainfall will be inadequate to meet its water needs. Since the rainfall amounts relate to minimum assured rainfall the actual climatological need for supplementary micro-irrigation in the rainy season for the Eksali crop is 100 mm. For the Dedsali crop in the first-year rainy period MAR at 50 per cent probability of 236 mm exceeds the transpiratory need of 169. No supplementary irrigation is called for. However, in the rainy period from the second year the transpiratory need is 318 mm. Thus, a supplementary micro-irrigation of 100 mm needs to be given. If the supplementary irrigation has been one of surface irrigation, evaporative loss would have been one of 20 mm per irrigation.

Table 4.2: February 15 Planted Eksali Sugarcane Crop at Pune

Crop age, month	1	2	3	4	5	6	7	8	9	10	11	12
EP mm/day	6.7	8.9	10.2	8.9	5.4	4.0	4.1	4.4	4.4	3.9	3.8	4.6
PETsc mm/day	6.9	9.3	10.7	9.3	5.4	4.0	4.1	4.6	5.3	5.2	5.1	5.5
Tc	0.15	0.45	0.70	0.80	0.80	0.80	0.80	0.80	0.70	0.50	0.40	0.40
T mm/month	29	130	225	231	130	99	102	110	115	78	63	68
E mm/month	72	72	60	60	*	*	20	40	40	40	40	
Efficiency %	71	36	21	21	*	*	*	15	26	34	39	37

* Rainy period; E is 20 mm in rainy period and in 8th week.

Table 4.3: June 15 Planted Dedsali Sugarcane Crop at Pune

Crop age, month	1	2	3	4	5	6	7	8	9	10	11	12
EP mm/day	5.4	4.0	4.1	4.4	4.4	3.9	3.8	4.6	6.7	8.9	10.2	8.9
PETsc mm/day	5.4	4.0	4.1	4.6	5.3	5.2	5.1	5.5	6.9	9.3	10.7	9.3
Tc	0.15	0.45	0.70	0.80	0.80	0.80	0.80	0.80	0.80	0.80	0.80	0.80
T mm/month	24	56	89	110	131	125	126	136	155	231	257	231
E mm/month	*	*	*	16	36	36	35	36	40	40	40	40
Efficiency %	*	*	*	13	21	22	22	21	21	15	13	15

Table 4.4: June 15 Planted Dedsali Sugarcane Crop at Pune

Crop age, month	13	14	15	16	17	18
EP mm/day	5.4	4.0	4.1	4.4	4.4	3.9
PETsc mm/day	5.4	4.0	4.1	4.6	5.3	5.2
Tc	0.80	0.80	0.70	0.50	0.40	0.40
T mm/month	130	99	89	69	66	62
E mm/month	*	*	*	16	36	36
Efficiency %	*	*	*	23	21	22

* Rainy Period; E is Nil in 1st rainy year; 20 mm in 2nd rainy year and 16 mm in 4th and 16th month.

In view of the above micro-irrigation requirement was taken as (*i*) cumulative transpiratory need plus 100 mm minus 386 mm in case of Eksali crop and (*ii*) cumulative transpiratory need plus 100 mm minus 577 mm in case of Dedsali crop. Water saved by micro-irrigation was taken as cumulative evaporation. From the above and from data at Table 4.4, it emerges that for the Eksali and Dedsali canes (*i*) the micro-irrigation requirements are 1094 and 1709, (*ii*) water saved are 464 and 427 mm and (*iii*) micro-irrigation efficiency are 30 per cent and 20 per cent respectively. A Dedsali crop can be expected to give 50 per cent more yield than an Eksali one. Thus the water use efficiency of micro-irrigated Eksali crop will only be 5 per cent more than a Dedsali crop. In case of surface irrigation the Dedsali crop will have about 10 per cent better water use-efficiency than the Eksali crop. In Eksali crop despite a shorter field-life duration the quantum of water saved is the same as that of the Dedsali crop. Again, 55 per cent and 35 per cent of water are saved in the establishment and maturity phases of the Eksali crop. For the Dedsali crop the water savings would be about 20 per cent and evenly spread out in the non-rainy crop phase.

It can be concluded that (*i*) for a Spring-planted Eksali crop provision of water by micro-irrigation is a must and (*ii*) despite passing through two rainy seasons the water use-efficiency of the Dedsali crop is not better than that of the Eksali one. There are wide variations in planting dates and field-life durations in sugarcane culture in India. However, one may hazard a guess that the savings in irrigation water due to micro-irrigation will be about (*i*) 20 per cent for a ground covering vegetative crop, (*ii*) 30 per cent during maturity and (*iii*) about 40 per cent for cane planted in spring or warm weather period.

References

Allen, R.G., Periera, L.S., Raes, D. and Smith, M., 1992. *Crop Evapotranspiration: Guidelines for Computing Crop Water Requirements*. Irrigation and Drainage Paper No. 56, Food and Agriculture Organisation (FAO), 300 pp.

India Meteorological Department (IMD), 1995. *Rainfall Probability for Selected Stations in India*. 1: 5.72.

Ritchie, J.T., 1972. Model for predicting evaporation from a row crop with incomplete crop cover. *Water Resources Res.*, 8: 1204–1213.

Ritchie, J.T. and Burnett, E., 1971. Dryland evaporative flux in a sub-humid climate. II. *Plant Influences. Agronomy J.*, 63: 56–62.

Venkataraman, S., 2001. A simple and rational agroclimatic method for rainfall zonation in dryland areas. *Ind. J. Environment and Ecoplanning*, 5(1): 135–144.

Venkataraman, S., Subba Rao, K. and Jilani, Y., 1984. A comparative study of the climatological estimation of potential evapotranspiration. *Mausam*, 35: 171–174.

Venkataraman, S., Subba Rao, K. and Raghava Rao, P., 1976. A preliminary lysimetric study on the evapotranspiration of the sugarcane crop at Anakapalle. India Met. Deptt. Pre-published Scientific Report No. 76/13.

Chapter 5

Length-weight Relationship and Relative Condition Factor of Freshwater Fish *Glossogobius giuris* (Hamilton) from Sadatpur Lake Near Pravaranagar, Ahmednagar District, Maharashtra

B.A. Pawar[1] and U.H. Mane[2]

[1]Department of Zoology, P.V.P. College, Pravaranagar, Tq. Rahata, Dist. Ahmednagar – 413 713, Maharashtra

[2]Department of Zoology, Dr. Babasaheb Ambedkar Marathawada University, Aurangabad – 431 004, Maharashtra

ABSTRACT

Glossogobius giuris is a predator fish abundantly found in Sadatpur lake and it has high economic value and considerable fishery importance in the area. The length-weight relationship of *G. giuris* can be described by the equation Male: Log W = – 4.8910 + 2.9097 Log L, Female: Log W = – 5.1493 + 3.0416 Log L and combined males and females: Log W = – 4.75181 + 2.85180 Log L which indicated closely the cubic law for the isometric growth of the fish in its natural habitat. The peak Kn values with fish length at 164.5 mm in both the sexes coincided with 50 per cent level of maturity. Subsequent fall was due to spawning. The seasonal changes taking place in Kn values was mainly related to greatly influence by cyclical changes taking place in the gonads and feeding intensity. The mean Kn values for different months and length groups indicated good conditions of the fishes in the Sadatpur lake with much scope for culture of the fish.

***Keywords**: Glossogobius giuris, Length-weight relationship, Relative condition factor, Maturation, Spawning.*

Introduction

Determination of mathematical relationship between length and weight of the fish has numerous practical applications in fishery biology (Sinha, 1981). The length-weight relationship is not only helps in establishing the mathematical relationship between the two but also in converting one variable into the other. It also has considerable importance in the study of fish population dynamics and patterns of growth in fish stocks in a particular area. In fishes, generally the growth pattern follows the cube law (Allen, 1938). The growth in weight of fishes, in general, is directly proportional to the cube of its length, but sometimes values of the relationship may depart from this (Le Cren, 1951) either due to environmental factors or condition of fish. Information is available on the length-weight relationship (Raje, 2000; Sunil, 2000; Subba and Ghosh, 2000; Jaiswar *et al.*, 2004; Rajasegar and Rajkumar, 2004; Saravanakumar *et al.*, 2004; Rajasegar, 2005).

Study on the relative condition factor (Kn) can yield information on many interesting events in the life history of a fish. A scrutiny of the relative condition factor at different body lengths can give valuable information regarding the maturation and spawning in the fishes whereas the condition at different months may give definite clues regarding the breeding seasons. Relative condition factor of some fishes has been already discussed by many authors (Usha and Naif, 1996; Kasim *et al.*, 1996 and Vijayakumaran, 1997). No attempt has so far been made to study the length-weight relationship and relative condition factor of *G. giuris* from Sadatpur lake. With a view to supplementing this, the present study was undertaken.

Materials and Methods

Monthly samples of *G. giuris* were collected from Sadatpur lake during June 1999 to September 2000. A total of 71 males and 28 females ranging in length from 80 to 205 mm and 80 to 175 mm respectively were used. Total length was taken from the tip of lower jaw to the end of caudal fin. Length was measured up to 1 mm by scale and weight was recorded up to 1 mg with the help of digital electronic balance. The method suggested by Le Cren (1951) was followed to compute the length-weight relationship. The length-weight relationship was determined by using the equation $W = aL^b$ and its logarithmic transformation Log W = Log a + b Log L, where, W = Weight, L = Length, 'a' and 'b' are constant. The relationship was established for both male and female by linear regression of the natural logarithms of the length and weight data.

Relative condition factor Kn (Le Cren, 1951) was established for male and female by using the formula $Kn = W/\hat{W}$, where $\hat{W}$ = Observed weight and W is the expected weight derived from the length-weight relationship. Kn values for various length groups and for different months were calculated after obtaining the mean lengths and mean weights for the corresponding length groups and months.

Results and Discussion

The logarithmic regression equations in respect of length-weight relationship for male and female *G. giuris* obtained are as follows:

Male: Log W = – 4.8910 + 2.9097 Log L (r = 0.9965)

Female: Log W = – 5.1493 + 3.0416 Log L (r = 0.9802)

The corresponding exponential formula can be expressed as follows:

Male: $W = 0.00001285\ L^{2.29097}$

Female: $W = 0.00000709\ L^{3.0416}$

The ANACOVA (Snedecor and Cochran, 1967) revealed no significant difference in the regression coefficient between the sexes of the species (Table 5.1). The data of both the sexes were pooled and the equation obtained was Log W = – 4.75181 + 2.85180 Log L. These are in accordance with the results on the length-weight relationship observed in case of *Tachisurus tenuispinis* (Dan and Mojumder, 1973), *Plotosus conius* (Sinha, 1981), *Arius subrostratus* (Usha and Nair, 1996) and *Johnius glaucus* (Raje, 2000).

Table 5.1: Comparison of Regression Lines of Length-weight Relationship of Males and Females *G. giuris* by ANACOVA

Sl.No.		d.f.	Σx^2	Σxy	Σy^2	Regression Coefficient	Deviation from Regression		
							d.f.	S.S.	M.S.
L	Within								
1.	Male	70	0.028423	0.82707	0.2425	2.9096932	69	0.001834	0.0000266
2.	Female	27	0.015022	0.045678	0.144905	3.0416002	26	0.006010	0.0002312
3.							95	0.007844	0.0000826
4.	Pooled, W	97	0.043445	0.128385	0.387405	2.9551156	96	0.0080125	0.0000835
5.			Difference	between	slopes		1	0.0001685	0.0001685
6.	Between, B	1	0.009766	0.023415	0.05593		97	0.0102809	
7.	Total W+B	98	0.053211	0.1518	0.443335				
8.			Between	adjusted	means		1	0.0022684	0.0022684

Comparison of slopes: F = 0.0001685/0.0000826 = 2.0399516 (d.f. = 1, 95)

No significant (Table value at 5 per cent = 4.0).

Allen (1938) suggested that the value of 'b' in an ideal fish could be 3, *i.e.* it should agree with the 'cube law', showing the isometric growth. However, Hile, (1936) and Martin (1949) were opined that it might vary between 2.5 and 4.0. In the present study the regression value 2.85180 was not significantly different from the theoretical value of 3, indicate closely the 'cube law' for isometric growth of *G. giuris* in Sadatpur lake. Hence the cubic formula adheres a proper presentation of length-weight relationship in this fish. A slight deviation in the value from 3 may be attributed due to seasonal variations in food availability and gonad stages. Isometric growth was also reported in *priacanthus hamrur* (Philip and Mathew, 1996) and the grass and bighead carp (Pandey and Sharma, 1998).

The correlation coefficient (r) between log-length and log-weight were 0.9965, 0.9802 and 0.9885 for males, females and pooled data respectively. As evident from 't' test, the coefficients of correlation were highly significant and the body weight positively and significantly correlated with the length. Similar observations also observed in *Tachysurus tenuispines* (Dan and Mojumder, 1973), *Arius subrostratus* (Usha and Nair, 1996) and *Glyptothorax telchitta* (Subba and Ghosh, 2000).

The monthly means of Kn values have been plotted for the period of June 1999 to September 2000 in Figure 5.1(A). The fluctuation of Kn values of males and females varied during the study period. The high peaks of Kn values were recorded in males during September, December and April, while in females during November, February and April may be considered as indications of preponderance of ripe gonads and time of breeding and feeding. The low Kn values recorded during June, July, August and October in males and during June and July in females, indicative of the spawning which causes a drop in the weight of the gonads (as stated by Shamsul Hoda, 1976; Sathyashree *et al.*, 1981; Annigeri, 1982).

Length group-wise relative condition values (Figure 5.1B) showed that an increase in the Kn values from 134.5 to 144.5 mm in both the sexes were probably due to the intensive feeding activity and maturation of gonads. Increase in Kn value from 154.5 mm in both the sexes possibly indicated the onset of maturity and the peaks at 164.5 mm coincided with 50 per cent level of maturity. Subsequent fall was perhaps due to spawning and lysis of the gametes. Very low level of Kn at 194.5 mm in males may be due to intensive spawning and low feeding status.

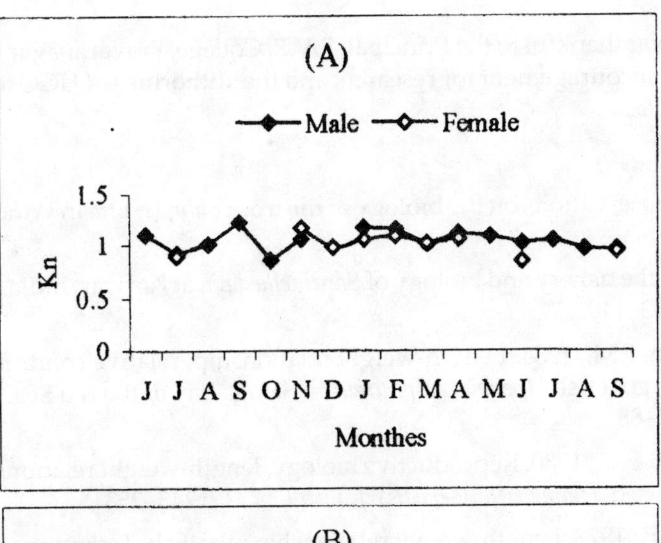

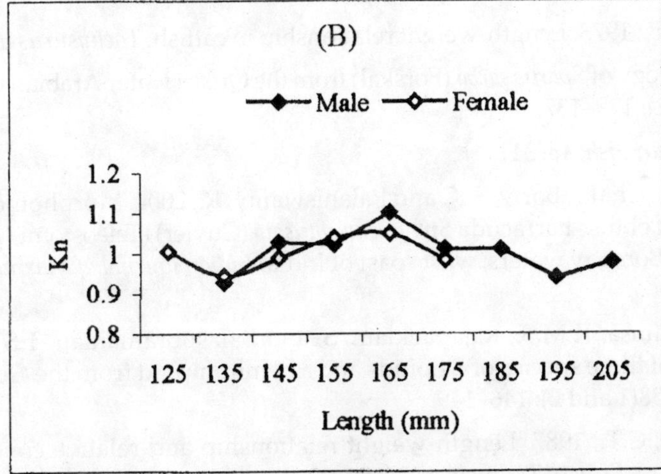

Figure 5.1: (A) Monthly Variation of Relative Condition Factor (Kn) of Male and Female *G. giuris*; (B) Relative Condition Factor (Kn) for Males and Females of *G. giuris* in Different Length Groups

The relative condition Kn could be used as an indicator to show well being of fish, its relative robustness and suitability of habitat. The estimated data showed that in *G. giuris* the mean Kn values for different months were 1.0445 and 1.0034 for males and females respectively and for different length groups were 1.00125 and 1.000633 for males and females respectively. This indicated good condition of the fish in the Sadatpur lake and also implies that food was not the major limiting factor. The Kn value 1 was also reported in *Tilapia zilli* (Dadzie and Wangila, 1980) and *Upeneus sulphureus* (Boraey and Soliman, 1984). It is thus, evident that the fluctuations in Kn value could be attributed to both the spawning cycle and feeding intensity. Similar reports were reported on *Johnius dussumieri* and *Johnius corutta* (Sriramachandramurty, 1979), *Daysciaena albida* (Kurup and Samuel, 1987) and *Sparsus sarba* (El-Agamy, 1989).

Acknowledgement

The author B.A. Pawar thankful to the Principal, P.V.P. College, Pravaranagar for providing the laboratory facilities and encouragement for research, and the authorities of UGC for award of UGC Teacher Fellowship.

References

Allen, K.R., 1938. Some observations on the biology of the trout *Salmo trutta* in Windermere. *J. Anim. Ecol.*, 7: 333–349.

Annigeri, G.G., 1982. On the fishery and biology of *Sardinella dayi*, at Karwar, India. *J. Mar. Biol. Ass., India*, 24(1 and 2): 133–140.

Boraey, F.A. and Soliman, F.M., 1984. Length-weight relationship, relative condition and food and feeding habits of the goat fish *Upeneus sulphureus* in Safaga bay of the red Sea. *J. Mar. Biol. Ass., India*, 26(1 and 2): 83–88.

Dadzie, S. and Wangila, B.C.C., 1980. Reproductive biology, length-weight relationship and relative condition of pond raised *Tilapia zilli (Gervais). J. Fish Biol.*, 17: 243–253.

Dan, S.S. and Mojumder, P., 1973. Length-weight relationship in catfish, *Tachysurus tenuispinis* (Day).

El-Agamy, A.E., 1989. Biology of *Sparus sarba* (Forskal) from the Quatri water Arabian Gulf. *J. Mar. Biol. Ass., India*, 31(1 and 2): 129–137.

Hile, R., 1936. *Bull. U.S. Bur. Fish*, 48: 311.

Jaiswar, A.K., Parida, P.K., Chakraborty, S.K. and Palaniswamy, R., 2004. Morphometry and length-weight relationship of obtuse barracuda *Sphyraena obtusata* (Cuvier) (Teleostomi / Actinopterygii / Sphyraenidae) from Bombay waters, west coast of India. *Indian Journal of Marine Sciences*, 33(3): 307–309.

Kashim, H.M., Ameer Hamsa, K.M.S., Rajapaekiam, S. and Balasubramanian, T.S., 1996. Length-weight relationship of three species of Belonids and a hemiramphid from the Gulf of Mannar. *J. Mar. Biol. Ass., India*, 38(1 and 2): 146–149.

Kurup, B.M. and Samuel, C.T., 1987. Length-weight relationship and relative condition factor in *Daysciaena albida* (Cuv.). *Fishery Technology*, 24: 86–92.

Le Cren, E.D., 1951. The length-weight relationship and seasonal cycle in gonad weight and condition in the Perch (*Perca fluviatilis*). *J. Anim. Ecol.*, 20: 201–219.

Martin, W.R., 1949. Univ. Toronto Stud. Biol. 58 Pub. Ont. Fish Res. Lab. 70, 1.

Pandey, A.C. and Sharma, M.K., 1998. Applicability of cube law in the length-weight relationship of the four species of exotic carps. *Proc. Nat. Acad. Sci., India*, 68(B): 1.

Philip, K.P. and Mathew, K., 1996. Length-weight relationship and relative condition factor in *Priacanthus hamrur* (Forskal). *Fishery Technology*, 33(2): 79–83.

Rajasegar, M., 2005. Length-weight relationship of *Nemipterus Japonicus* from Karaikkal coastal waters, Tamil Nadu. *J. Aqua. Biol.*, 20(1): 25–26.

Rajasegar, M. and Rajkumar, M., 2004. Length-weight relationship of catfish *Scatophagus argus* (Bloch) from Parangipettai coastal waters, Tamil Nadu. *J. Aqua. Biol.*, 19(2): 43–44.

Raje, S.G., 2000. Some aspects of the biology of *Otolithes cuvieri* and *Johnius glaucus* from Vera Val. *J. Ind. Fish. Ass.*, 27: 1–6.

Saravanakumar, K., Rajagopal, S., Vijayanand, P. and Yuvaraj, D., 2004. Length-weight relationships of the flying fish *Exocoetus volitans* from Cuddalore coastal waters, Tamil Nadu, India. *J. Aqua. Biol.*, 19(2): 45–47.

Sathyashree, P.K., Reddy, P.S. and Natarajan, R., 1981. Maturity and spawning of *Osteomugil speigleri* (Bleeker) in Porto Novo waters. *J. Mar. Biol. Ass., India*, 23 (1 and 2): 1–6.

Shamsul Hoda, S.M., 1976. Reproductive biology and length-weight relationship of *Thryssa dussumieri* (Val.) of the Pakistan coast. *J. Mar. Biol. Ass., India*, 18 (2): 272–287.

Sinha, M., 1981. Length-weight relationship and relative condition factor of the canine catfish–Eel *Plotosus canius* (Ham.). *J. Mar. Biol. Ass., India*, 23(1 and 2): 39–43.

Snedecor, G.W. and Cochran, W.G., 1967. *Statistical Methods*, 6th Edn. Oxford and IBH Pub. Co., New Delhi, p. 432–436.

Sriramachandramurty, V., 1979. Observations on some aspects of biology of the croakers *Johnius dussumieri* and *Johnius carutta* from Kakinada. *J. Mar. Biol. Ass., India*, 21 (1 and 2): 77–85.

Subba, B.R. and Ghosh, T. 2000. Length-weight relationship of a hill-stream fish, *Glyptothorax telchitta* (Ham.) from Saptakoshi river of Nepal. *J. Ind. Fish. Ass.*, 27: 79–82.

Sunil, M.S., 2000. Length-weight relationship in *Rasbora daniconius* (Ham.) from Achancoil River, Pathanamthitta, Kerala, India. *Inland J. Fish.*, 47 (3): 271–274.

Usha, P.G. and Nair, N.B., 1996. Length-weight relationship and relative condition factor in *Arius subrostratus* (Val.). *J. Anim. Morphol. Physiol.*, 43(2): 169–179.

Vijayakumaran, K., 1997. Growth and mortality parameters and some aspects of biology of striped eel-catfish, *Plotosus lineatus* (Thunberg) from North Andhra Pradesh Coast. *J. Mar. Biol. Ass., India*, 39(1 and 2): 108–112.

Chapter 6

Trypanosome Infection and Some Associated Physiological Changes in an Indian Cat-fish, *Clarias batrachus* (Linnaeus)

Jhuma Saha, Nimai Chandra Saha, Tushar Kanti Mukhopadhyay, Trilochan Midya and Nirmal Kumar Sarkar

Postgraduate Department of Zoology, Presidency College, Kolkata – 700 073, West Bengal, India

ABSTRACT

Trypanosome infection was observed in blood smears of a few specimens of an Indian cat-fish, *Clarias batrachus* (L.). The parasite was identified as *Trypanosoma danilewskyi*, which had not earlier been reported from the aforesaid fish species. Red blood cell count and haemoglobin content of blood were significantly lower in the infected fishes, as compared to the values observed in their uninfected counterparts. Lymphocyte and monocyte counts were significantly higher in the infected fishes than in the uninfected ones. Blood glucose concentration as well as hepatic glycogen and glucose concentrations were significantly lower in the infected fishes than in their uninfected counterparts. Plasma protein concentration was considerably but not very significantly decreased in the infected fishes. The overall findings are in conformity with the existing information on the effects of trypanosome infection in fishes. Moreover, some recommendations can be put forward from our findings for the investigators interested to work in this line in future.

Keywords: *Clarias batrachus, Trypanosoma danilewskyi, Blood.*

Introduction

Trypanosome infection has occasionally been reported in the blood of Indian freshwater fishes. For instance, *Trypanosoma danilewskyi* has been described in *Heteropneustes fossilis, Wallago attu, Notopterus notopterus* and *Mastocembelus armatus* (Qadri, 1962; Joshi, 1979); *T. mukundi* has been reported in *Heteropneustes fossilis* (Ray Chaudhury and Misra, 1973); *T. maguri, T. clariae, T batrachi* and *T. qadrii* have been reported in *Clarias batrachus* (deMello and Valles, 1936; Qadri, 1962; Tandon and Joshi, 1973; Narasimhamurti and Saratchandra, 1980; Gupta and Gupta, 1985); *T. aligaricus* has been described in *Channa punctatus* (Gupta and Gupta, 1985). Moreover, decreased blood haemoglobin and glucose levels, red blood cell count and serum cholesterol level have variously been reported in different fishes infected with *T. maguri, T. batrachi* and *T. aligaricus* (Tandon and Joshi, 1973, 1974; Tandon and Chandra, 1977; Gupta and Gupta, 1985).

In the present article, we describe trypanosome infection in the cat-fish, *Clarias batrachus* (Linnaeus); the parasite has been identified as *Trypanosoma danilewskyi*, which had not earlier been reported from the fish species under study. We also describe the blood picture (red blood cell count, haemoglobin content and differential counts of white blood cells), blood glucose and plasma protein concentrations as well as hepatic glycogen and glucose concentrations of the infected and the uninfected fishes. Our report is supposed to add to the existing information on trypanosome infection in fishes in general.

Materials and Methods

Twelve adult, male specimens of *Clarias batrachus* (20–22 cm in length and 95–110 g in weight) were collected from a freshwater pond of the Nadia district, West Bengal, India. Different non-piscivorous air-breathing fishes (*Clarias batrachus, Heteropneustes fossilis* and *Anabas testudineus*) are cultivated in that pond for commercial purpose. The fishes were actually brought for practical classes of postgraduate students specialising in Haematology. After one week of maintenance in departmental aquaria, two of the fishes were sacrificed for examination of blood smears. Coincidentally, trypanosome infection was observed in the smears of one specimen and we developed interest in examining blood smears as well as different biochemical parameters in the remaining ten of the fishes; the first two fishes could not be used for examination of all parameters. Methods used in the subsequent study are described below.

The fishes were sacrificed by striking their heads on a laboratory table and subjected to a puncture at the caudal peduncle. Blood samples were allowed to drip into clean centrifuge tubes from the caudal vein of the fishes. The samples were prevented from clotting by allowing a thin film of ethylenediamine tetracetate (0.2 ml of a 1.0 per cent solution) to dry on the inner wall of each tube 1 h before collecting the samples. The liver of each fish was then carefully dissected out, weighed readily and homogenised in presence of chilled phosphate buffer (pH 7.0; 2.0 ml of the buffer was added to 1.0 g of tissue).

Blood haemoglobin content was determined following the acid haematin method using a Sahli's haemometer (Mulchay, 1970). Total count of red blood cells was determined microscopically with the help of a haemocytometer having an improved-type Neubauer counting chamber, using a modified Dacie's fluid (1 ml of formalin, 3.13 g of sodium citrate and 0.1 g of brilliant cresyl blue dissolved in 100 ml of distilled water) as the diluting fluid (Blaxhall and Daisley, 1973). Blood smears were drawn on clean glass slides, air-dried, fixed in methanol for 15 min and stained with freshly prepared Giemsa's stain diluted with an equal volume of phosphate buffer (pH 7.0) for 40 min at room temperature. The smears were examined under the oil-immersion lens of a microscope for detection of parasites as well as determination of differential counts of white blood cells by the conventional longitudinal counting method (Dacie and Lewis, 1984). Cell measurements were carried out with the aid of an ocular micrometer. Glucose concentrations of blood and liver were estimated colorimetrically following a modified version of the well-known Somogyi's arsenomolybdate method (Nath and Nath, 1990). In the adopted method, 5.0 per cent zinc sulphate 0.5 N sodium hydroxide solutions have been used for deproteinization of blood or tissue homogenate while 5.0 per cent zinc sulphate and 0.3 N barium hydroxide solutions were used in its prototype. Glycogen concentration of liver was determined colorimetrically following the well-known alkali extraction-acid hydrolysis method (Hassid and Abraham, 1957). Plasma samples collected by centrifugation of blood samples at 3000 rpm for 30 min were used for colorimetric assay of protein concentration following the widely used Folin-phenol method of Lowry (Lowry *et al.*, 1951).

The differences between various results obtained from the infected and the uninfected fish specimens were analysed statistically by the help of Fisher's two-tailed t-test (Goon *et al.*, 1981).

Results and Discussion

Four out of the ten specimens of *Clarias batrachus* under study were found to harbour trypanosome infection in their blood. The parasite was monomorphic and appeared only in a slender form, which measured 38–44 μm in length (including flagellum) and 1.2–1.5 μm in width (at nuclear region). The nucleus was elongated and about 3 μm × 1 μm in size. A prominent kinetoplast, about 1.5 μm × 0.7 μm in size, lied near the posterior end of the cell-body and 10–12 μm away from the nucleus. Based on these characteristics, which are considered to be significant from taxonomic point of view (Laveran and Mesnil, 1904; Qadri, 1962; Woo, 1981, 1995), the parasitic was identified as *Trypanosoma danilewskyi*. The morphology of this parasite differs from that of different other trypanosome species reported to infect Indian cat-fishes like *Clarias batrachus* and *Heteropneustes fossilis*. For instance, *T. maguri* observed in *C. batrachus* (Tandon and Joshi, 1973) and *T. mukundi* observed in *H. fossilis* (Ray Chaudhury and Misra, 1973) are clearly dimorphic in appearance. *T. batrachi* and *T. qadrii* reported from *C. batrachus* are monomorphic but somewhat shorter and wider in appearance (Qadri, 1962; Narasimhamurti and Saratchandra, 1980; Gupta and Gupta, 1985). Again, *T. clariae* observed in *C. batrachus* is much shorter in size (deMello and Valles, 1936).

Till now, *T. danilewskyi* has been reported from different Indian freshwater fishes like *Heteropneustes fossilis, Wallago attu, Notopterus notopterus* and *Mastocembelus armatus* (Qadri, 1962; Joshi, 1979). Therefore, the present report may be considered as the first one on the occurrence of this trypanosome species in *Clarias batrachus*. The pond from which the fish specimens under study were collected was being used for cultivation of *C. batrachus, H. fossilis* and *Anabas testudineus*. Therefore, it might have so happened that a common leech vector like *Hemiclepsis marginata* (Qadri, 1962) transmitted the parasite to both the scaleless silurid fishes (*C. batrachus* and *H. fossilis*) simultaneously or to *C. batrachus* from *H. fossilis*, which has long been known to be a host of *T. danilewskyi* (Qadri, 1962; Joshi, 1979).

In both the infected and the uninfected fishes, the red blood cells were normochromic and normocytic in appearance, but the total count of these cells was significantly lower in the infected fishes (Table 6.1). Blood haemoglobin content was also significantly lower in the infected fishes (Table 6.1). *T. danilewskyi* has been reported to secrete some haemolysins capable of lysing red blood cells in infected goldfishes (Islam and Woo, 1991). Our findings are, therefore, in conformity with the aforesaid report. Our findings are also in agreement with a few more reports on the causation of anaemia by different species of trypanosomes in different species of Indian freshwater fishes (Tandon and Joshi, 1973; Gupta and Gupta, 1985).

Differential counts of the white blood cells revealed that the percentages of lymphocytes and monocytes were significantly higher while the percentage of neutrophils was significantly lower in the infected fishes than in their uninfected counterparts (Table 6.1). Our findings are consistent with that reported in *Clarius batrachus* and *Channa punctatus* infected by *Trypanosoma batrachi* and *T. aligaricus*, respectively (Gupta and Gupta, 1985). The increased lymphocyte count might be associated with production of antibodies by the hosts against the trypanosome parasites (Weir, 1983). On the other hand, the monocytes might have increased in number to participate in opsonin-mediated phagocytosis of the trypanosomes (Roitt *et al.*, 1996). Increased lymphocyte and monocyte counts in turn had caused a relative rather than a true form of neutropenia in the infected fishes.

Table 6.1: Some Haematological Findings in *Clarias batrachus* with and without Trypanosome Infection (the values represent means ± S.D.) (Red blood cell size is based on 50 cells in each group)

Parameter	*Uninfected Fish*	*Infected Fish*
Number of fish	6	4
Red blood cell count (million/mm^3)	2.52±0.16	2.05±0.11*
Staining of red blood cells	Normochromic	Normochromic
Red blood cell length (μm)	13.16±0.99	13.08±1.11
Red blood cell width (μm)	11.38±1.15	11.33±1.18
Haemoglobin content (g/100 ml)	13.08±0.25	10.77±0.27*
Neutrophil percentage	19.67±1.97	9.75±1.09*
Eosinophil percentage	2.00±0.58	2.00±0.71
Basophil percentage	0.67±0.74	0.75±0.22
Lymphocyte percentage	72.16±2.41	78.75±1.58*
Monocyte percentage	5.50±0.96	8.75±0.43*

*: Significantly different from the value in uninfected fishes at 1 per cent level.

Blood glucose concentration was significantly lower in the infected fishes than in their uninfected counterparts (Table 6.2). The finding reiterates an earlier report by others on decrease of blood glucose level in Indian freshwater fishes infected by trypanosomes (Tandon and Joshi, 1974). In the present study by us, the infected fishes were also observed to have significant reductions in hepatic glycogen and glucose concentrations, as compared to the uninfected fishes (Table 6.2). Therefore, it is likely that a rapid utilisation of glucose by the trypanosomes from the host blood (Tandon and Joshi, 1974) might have stimulated the hosts to enhance the rate of glycogenolysis in liver in order to combat their hypoglycaemic status. In the long run, the host fishes failed to maintain normal glycogen concentration in liver as well as normal glucose concentration in both liver and blood. Finally, plasma protein concentration was found to be considerably decreased in the infected fishes than in their uninfected counterparts (Table 6.2). However, the decrease was rather moderately significant from a statistical point ofview. It is, therefore, likely that the rate of utilisation of protein by the trypanosomes from the host blood was not so high as the rate of utilisation of glucose, resulting in a considerable but not any severe fall in plasma protein concentration of the hosts.

Based on the findings of the present study, some recommendations may also be put forward for the investigators interested to work in the same line in future. Firstly, it appears that one should examine blood smears of all fish species instead of any particular species available in a pond or tank, for detection of trypanosome infection. This will make it clear if a trypanosome species may be transmitted by a common vector to different species of fishes or not and also, if a trypanosome species has any host specificity or not. Secondly, while estimating blood glucose concentration in the infected fishes, one should also estimate hepatic glycogen concentration to understand if the hosts are really depleted of their carbohydrate resource or not owing to trypanosome infection. Finally, it seems that plasma protein concentration should be re-examined in as many fish species carrying trypanosome infection as possible to reach a more convincing conclusion on the effect of trypanosome infection on the plasma protein status of fishes in general.

Table 6.2: Some Biochemical Findings in *Clarias batrachus* with and without Trypanosome Infection (the values represent means ± S.D.)

Parameter	Uninfected Fish	Infected Fish
Number of fish	6	4
Blood glucose (mg/100 ml)	131.17±4.45	97.25±6.94**
Hepatic glycogen (mg/g)	21.57±0.95	17.62±0.72**
Hepatic glucose (mg/g)	17.63±0.79	13.85±0.64**
Plasma protein (g/100 ml)	2.82±0.11	2.66±0.05*

*: Significantly different from the value in uninfected fishes at 5 per cent level;
**: Significantly different from the value in uninfected fishes at 1 per cent level.

Acknowledgements

We are grateful to Dr. Prabir Kumar Bandyopadhyay, Reader, Parasitology Unit, Department of Zoology, University ofKalyani and Dr. Rupendu Ray, Reader (now, posted elsewhere), Parasitology Unit, Postgraduate Department of Zoology, Presidency College, for their kind assistance in our study.

References

Blaxhall, P.C. and Daisley, K.W., 1973. Routine haematological methods for use with fish blood. *J. Fish Biol.*, 5: 771–781.

Dacie, J.V. and Lewis, S.M., 1984. *Practical Haematology*, 6th Ed., Churchill Livingstone, Edinburgh, pp. 23–83.

deMello, F. and Valles, C., 1936. On the *Trypanosoma* found in the blood of the Indian freshwater fish *Clarias batrachus* (Linn.). *Proc. Indian Acad. Sci.*, 3(8): 120–124.

Goon, A.M., Gupta, M.K. and Dasgupta, B., 1981. *Basic Statistics*, 3rd ed. The World Press, Kolkata, pp. 306–310.

Gupta, N. and Gupta, D.K., 1985. Haematological changes due to *Trypanosoma batrachi* and *Trypanosoma aligaricus* infection in two freshwater teleosts. *Angewandte Parasitol.*, 26: 193–196.

Hassid, W.Z. and Abraham, S., 1957. Determination of glycogen and starch. In: *Methods in Enzymology*, Vol. 3 (Eds.) Colowick, S.P. and Kaplan, N.O. Academic Press, New York, pp. 34–36.

Islam, A.K.M.N. and Woo, P.T.K., 1991. Anaemia and its mechanism in goldfish *Carassius auratus* infected with *Trypanosoma danilewskyi*. *Dis. Aquatic Organ.*, 11: 37–43.

Joshi, B.D., 1979. On the occurrence of trypanosoma in the blood of some freshwater teleosts of Lucknow (U.P.), *India. Proc. Indian Acad. Sci.*, 88(B): 59–63.

Laveran, A. and Mesnil, 1904. *Trypanosomes and Trypanosomiases*. W.T. Keener and Co., Chicago, pp. 538–540.

Lowry, O.H., Rosenbrough, N.J., Farr, A.L. and Randall, R.J., 1951. Protein measurement with folin-phenol reagent. *J. Biol. Chem.*, 193: 265–275.

Mulchay, M.F., 1970. Blood values in the pike, *Esox lucius* (L.). *J. Fish Biol.*, 2: 203–209.

Narasimhamurti, C.C. and Saratchandra, B., 1980. Two new species of trypanosomes, *Trypanosoma channai* n.sp. from *Channa punctatus* and *Trypanosoma gadrii* n.sp. from *Clarias batrachus. Proc. Indian Acad. Sci.*, 89: 371–376.

Nath, R.L. and Nath, R.K., 1990. *Practical Biochemistry in Clinical Medicine*, 2nd ed. Academic Publishers, Kolkata, pp. 38–41.

Qadri, S.S., 1962. An experimental study of the life cycle of *Trypanosoma danilewskyi* in the leech, *Hemiclepsis marginata. J. Protozool.*, 9: 254–258.

Ray Chaudhury, S. and Misra, K.K., 1973. Two new trypanosomes from India. *Arch. Protistenk.*, 115: 10–17.

Roitt, I., Brostoff, J. and Male, D., 1996. *Immunology*, 4th ed., Times Mirror International Publishers Ltd., London, pp. 184–186.

Tandon, R.S. and Chandra, S., 1977. Studies on ecophysiology of fish parasites: Effects of trypanosome infection on the serum cholesterol levels of fishes. *Z. Parasiten*, 52: 195–198.

Tandon, R.S. and Joshi, B.D., 1973. Studies on physio-pathology of blood of freshwater fishes infected with two new forms of trypanosomes. *Z. Wissen. Zool.*, 185: 207–221.

Tandon, R.S. and Joshi, B.D., 1974. Effect of trypanosome infection on blood glucose levels of some freshwater teleosts. *J. Inland Fish. Soc. India*, 6: 81–82.

Weir, D.M. 1983. *Immunology*, 5th ed. Churchill Livingstone, Edinburgh, pp. 132–136.

Woo, P.T.K., 1981. *Trypanosoma danilewskyi*, a new multiplication process for *Trypanosoma* (Protozoa, Kinetoplastida). *J. Parasitol.*, 67: 522–526.

Woo, P.T.K., 1995. *Fish Diseases and Disorders*, Vol. 1. CAB International, Wallingford, UK, pp. 69–96.

Chapter 7

Effect of *Nerium odorum* Soland, on the Periodicity of Estrous Cycle in Female Albino Rats

Dixit Mukesh and A.O. Prakash*
SOS Pharmacy, Jiwaji University, Gwalior, M.P.

ABSTRACT

Nerium odorum which is known as to impair fertility in female rats has been studied for its effect on estrous cycle in mature rats. 50 per cent ethanolic extract of *Nerium odorum* at a dose of 250 mg per kg. body weight *p.o.* for 10, 20 and 30 days disturbed the vaginal cytology. In control rats the diestrus stage covered almost half of the cycle followed by estrus phase and then the proestrus and metaestrus. The administration of extracts did not alter the per cent duration of proestrus and metaestrus phases however shifting occurred from diestrus to estrus. Moreover the effect was dose dependent irrespective to duration of treatment. Prolonged estrus stage of the cycle was returned to normal after discontinuation of the extract.

Keywords: *Nerium odorum, Ethanolic extract, Estrous cycle.*

Introduction

The administration of antifertility agents of hormonal virtues *viz.* estrogenic or antiestrogenic are known to act on the reproductive tract in female rats. There are some anti-fertility agents which do not disturb the periodicity of the estrous cycle being non-hormonal and their *modus operandi* lies with the

* Corresponding Author: E-mail: krgc@rediffmail.com

toxic action at the level of development of blastocyst. Extract of *Nerium odorum* is known to impair fertility in female rats. Therefore, keeping in view the anti-fertility activity in the leaf extract of *Nerium odorum*, the present investigation has been undertaken to know whether the extract of *Nerium odorum* has any effect on the periodicity of estrous cycle in female rats.

Materials and Methods

Animals and their Maintenance

Virgin female rats of Wistar strain weighing 130±10 gm were collected from the animal colony of the department. These animals were maintained under uniform husbandry conditions of light and temperature (25–26 °C) and were fed with "Gold Mohar" rat pelleted diet and water *ad libitum*.

Preparation of Leaf Extract

Weighed dried leaves of *Nerium odorum* were grinded powdered and extracted with 50 per cent ethanol using dumping method. The powder form was dumped with 50 per cent ethanol in air tight glass jar and after about 15 days, the upper clear supernatent was separated and evaporated to dryness under low temperature and reduced pressure in a rotatory film vacuume evaporator. The oily black brown semi solid mass so obtained was considered as extract which was kept in sealed bottle and stored in refrigerator till its use. A small amount of this extract was taken out at the time of need.

Preparation of Standard Doses and its Administration

In earlier studies 25 mg/kg dose has shown significant antifertility activity and therefore for the present studies dose of 250 mg/kg body wt. was prepared by weighing 2.5 gm of dried extract which was macerated with gum acacia and finally suspended in mineral water up to 100 ml. 1.0 ml of this prepared extract was administered orally to a rat of 100 gm with the help of intragastric catheter.

Experimental Protocol

The selected female rats were examined for different stages of estrous cycle as described by Prakash (1979). Vaginal smear of each rat was examined daily at a definite time and the stages of estrous cycle were recorded (Figure 7.1). The female rats showing at least two regular cycles were selected for experimental purpose. The selected animals were distributed into different groups of control and experimental purpose. A dose of 250 mg/kg body weight was administered orally to all three groups of experimental rats for 10, 20 and 30 days and the control group received vehicle only. Vaginal smear of each rat was examined daily at definite time during the experimental period. After the last treatment the administration of extract was discontinued and the record of various stages of estrous cycle as revealed through the vaginal smear was examined and analysed statistically using student 't'-test.

Results

The effect of 50 per cent ethanolic extract of *Nerium odorum* on the periodicity of estrous cycle in adult cyclic rats (Table 7.1 and 7.2) revealed that in the control rats the diestrus stage covered almost half of the cycle followed by estrus of 30 per cent and then proestrus and metaestrus of short durations. When the extract of *Nerium odorum* was administered for 10 days there was an increase in the duration of estrus phase significantly. When the period of administration was increased to 20 and 30 days, there was further prolongation in the per cent duration of estrus stage, consequently the per cent duration of the other stage was reduced. The table also reveals that the administration of extract did not alter the per cent duration of proestrus and metaestrus, however, shifting occurred from diestrus to estrus. It is also interesting to note that the period of 10 days of administration of extract could

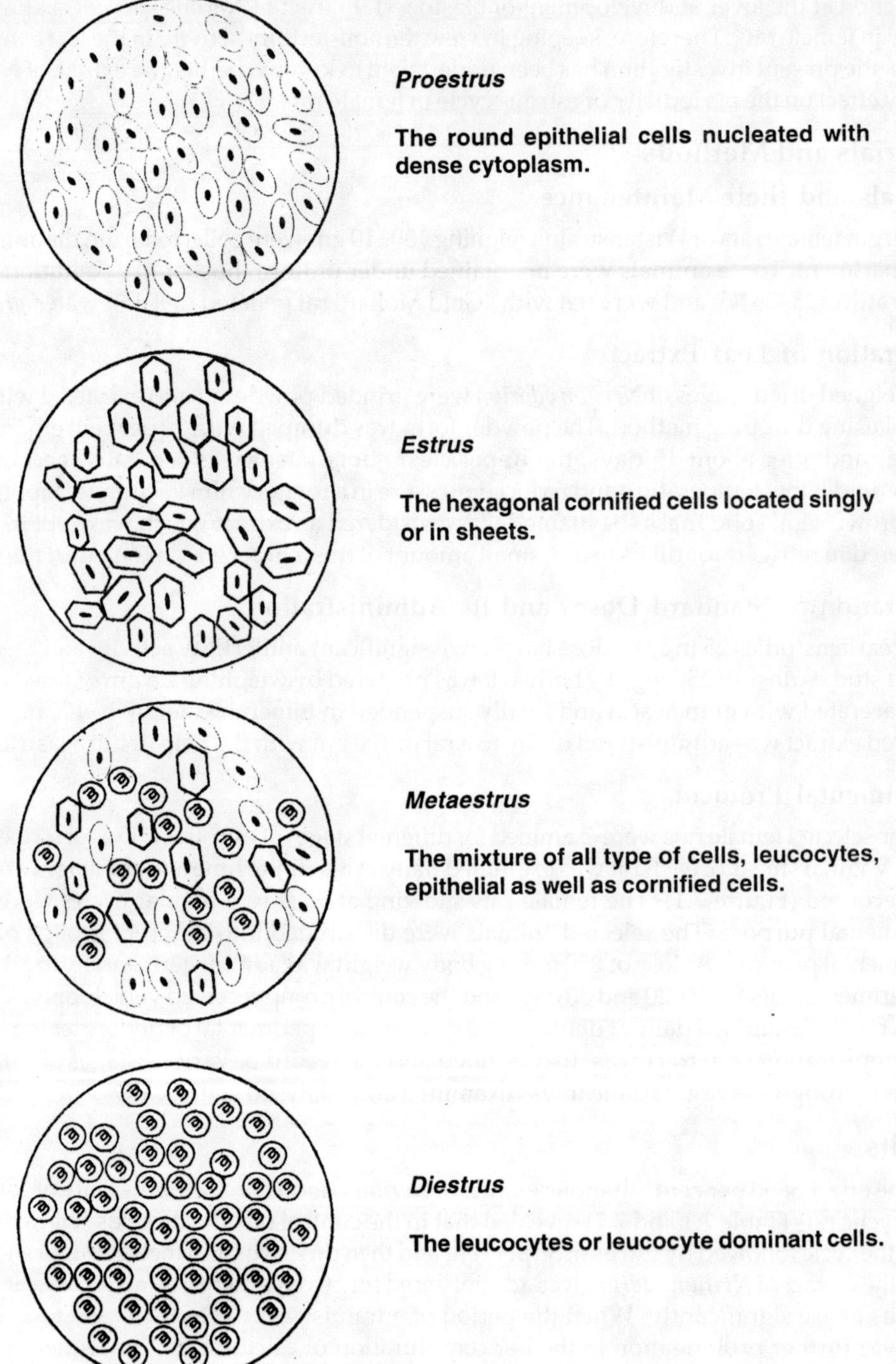

Figure 7.1: Various Stages of Estrus Cycle in Rats

significantly induce the changes, which persisted even up to 30 days. Therefore, the effect of extract of *Nerium odorum* on the per cent duration of estrous cycle clearly revealed that the effect was dose dependent irrespective to duration of treatment.

Table 7.1: Effect of 50 per cent Ethanolic Extract of *Nerium odorum* Soland on the Periodicity of Estrous Cycle in Adults Rats (in terms of per cent duration). 5 animals were used in each set. Values are mean ± SE.

Group No.	*Duration (Days)*	*Treatment*	*Per cent Duration of Stages of Estrous Cycle*				*Restoration Time for Normalcy in Cycle (Days)*
			Proestrus	*Estrus*	*Metaestrus*	*Diestrus*	
1	10	Control	12.0±0.6	28.0±1.0	14.0±0.7	46.0±2.1	6.2±0.3
		Treated	12.0±0.6	40.0±2.0*	10.0±0.5*	38.0±1.9*	
2	20	Control	12.0±0.6	29.0±1.5	15.0±0.7	44.0±2.2	8.3±0.5
		Treated	12.0±0.6	45.0±2.3*	13.0±0.6	30.0±1.5*	
3	30	Control	14.0±0.7	30.0±1.5	12.0±0.6	44.0±2.2	12.4±0.7
		Treated	12.0±0.6	48.0±2.4*	10.0±0.5	30.0±1.5*	

Statistical Analysis

*: P value vs. respective control < 0.05.

Table 7.2: Effect of 50 per cent Ethanolic Extract of *Nerium odorum* Soland on the Periodicity of Estrous Cycle in Adults Rats (in terms of duration in days). 5 animals were used in each set. Values are mean±SE

Group No.	*Duration (Days)*	*Treatment*	*Duration of Stages of Estrous Cycle*			
			Proestrus	*Estrus*	*Metaestrus*	*Diestrus*
1	10	Control	1.2±0.22	3.2±0.22	1.2±0.22	4.4±0.27
		Treated	1.2±0.22	3.6±0.44	1.2±0.22	4.0±0.35
2	20	Control	1.4±0.27	6.2±0.41	4.2±0.41	8.2±0.22
		Treated	2.0±0.5	9.2±0.22*	2.6±0.27	6.2±0.41
3	30	Control	6.2±0.41	9.2±0.22	2.6±0.27	12.0±0.35
		Treated	2.4±0.44	15.2±0.41*	3.2±0.22	9.2±0.22

Statistical Analysis

*: P value vs. respective control < 0.05.

Discussion

During a reproductive cycle, the ovary undergoes cyclic changes and so as the uterus. Consequently the vaginal epithelium induced remarkable changes morphologically and histologically. The cyclic changes in the vagina under the influence of female sex steroids estrogen and progesterone are known as estrous cycles. Turner and Bagnara (1975) have clearly shown that the duration of estrus cycle and also the conversion of one stage to another in a estrous cycle is hormonal dependent. The entire length of one estrous cycle in rats varies from 4–5 days and consists of proestrus, estrus, metaestrus and diestrus. The entire cycle is divided into follicular phase which contains proestrus and estrus whereas leucocytic phase contains metaestrus and diestrus.

The conversion of follicular phase to leucocytic phase is governed by estrogen and progesterone of ovarian origin which are further influenced by the pituitary gonadotropins (Lerner, 1969). Allen and Doisy (1923) discovered the estrous cycle in female rats and have shown the significant role of estrogen and progesterone in its inter or intra conversions. Thereafter, number of authors worked on the mechanism of estrogen and progesterone in relation to vaginal cornification. Estradiol-17β known as naturally occurring estrogen is also known to modify vaginal epithelial cells into cornified corn flake like cells (Turner and Bagnara, 1975 and Nalbandov 1970). The intra conversions of follicular phase to leukocytic phase is the inter play of estrogen and progesterone which are secreted by the ovary. Progesterone is also known to act as an antiestrogen and thus it also helps to regulate the estrous cycle. Lerner (1969) has studied the effect of steroidal estrogens on the periodicity of estrus cycle and has reported that administration of typical estrogen is known to induce vaginal cornification. However, when estrogenic agents are administered to the estrogen primed rats, the interruption of vaginal cornification occurs. Therefore, it is quite clear that a typical estrogen induces vaginal cornification and shows estrus stage where as the progesterone induces diestrus stage. Regulation of estrogen and progesterone is carried out through the feed back mechanism.

Medicinal plants which have shown potent antifertility activity in laboratory animals have been studied for their effect on the estrous cycle as these have shown significant hormonal properties in immature rats. Phyto estrogens and other estrogenic herbal preparations are known to alter the periodicity of estrus cycle in rats. Ethanolic extract of dried berries of *Embelia ribes* when administered to cyclic rats increased the duration of estrus phase (Prakash and Mathur, 1979). Similarly Kholkute *et al.* (1976) have reported that benzene extract of *Hibiscus rosa-sinensis* induced a prolong diestrus stage in adult rats. Barethia (1995) while working on the extract of *Apium graveolens* have shown a typical characteristics effect on the estrous cycle. When the dose was administered for 10 days there was an increase in the follicular phase of the estrous cycle, however, when the extract was administered for 20 and 30 days the same extract increased the leucocytic phase. This dual action of the extract has been explained on the estrogenic potentiality in the extract of *Apium graveolens*. In the present investigation author has observed that the administration of 50 per cent ethanolic extract of *Nerium odorum* to adult cyclic rats significantly increased the periodicity of estrus phase.

Therefore, on the basis of present finding it is concluded that the 50 per cent ethanolic extract of *Nerium odorum* which have shown promising antifertility activity in rats induced prolongation in follicular phase in the estrous cycle in adult rats. The estrus stage was persisted as long as the extract was administered which clearly indicated a synergistic effect. Findings also suggested that the estrogenic influence on the estrous cycle appeared is mild as feed back was not initiated. Moreover, studies also suggested that the prolonged estrus stage of the cycle was returned to normal after discontinuation of the extract although in the long period of treatment took more time to restitute. Therefore, it can safety be said that estrogenic effect of extract appeared temporary and the animals might resume the normal estrous cycle after discontinuation of the treatment.

References

Allen, E. and Doisy, E.A., 1923. An ovarian Hormone. Preliminary report on its localization, extraction and partial purification, and action in test animals. *J. Am. Med. Assoc., N.Y.*, 81: 819.

Barethia, R., 1995. Antifertility effect of *Apium graveolens* and *Crotalaria juncia* Linn. in female rats. *Ph.D. Thesis*, Jiwaji University, Gwalior.

Kholkute, S.D., Mudgal, V. and Deshpandey, P.J., 1976. Screening of indigenous medicinal plants for antifertility potentiality. *Planta Medica*, 29(2): 151.

Lerner, L.J., 1969. The biology of non-steroidal antifertility agents. In: *Contraception: The Chemical Control of Fertility*, (Ed.) Daniel Lednicer. Marcel Dekker Inc., N.Y., pp. 161.

Nalbandov, A.V., 1970. Reproduction in female mammals and birds. In: *Reprod. Physiol.*, (Ed.) A.V. Nalbandov. D.B. Taraporewala Sons and Co. Pvt. Ltd., Bombay, India, 2: 124-127.

Prakash, A.O, 1979. *Hibiscus rosa-sinensis* Linn. Effect on estrous cycle in rats. *Probe.*, 18: 79.

Prakash, A.O. and Mathur, R., 1979. Studies on estrous cycle of albino rats. Response to *Embelia ribes* Burn extracts. *Planta Medica*, 36: 134.

Turner, D.C. and Bagnara, J.T., 1975. *General Endocrinology*, 5th edition. W.B. Saunders Company, Tokyo, Japan.

Chapter 8

Transport and Transformation of Urea Based Fertilizers in an Inceptisol of Orissa

A.K. Dash, B.K. Mishra, M. Jena and B.B. Behera

Department of Soil Science, Agricultural Chemistry,

Orissa University of Agriculture and Technology, Bhubaneswar – 751 003, Orissa

ABSTRACT

A field experiment was conducted to study the mechanism of transport and transformation of prilled urea (PU), nimin coated prilled urea (NCPU) and Urea Super Granules (USG) in the field. Micro plots of size 506.25 cm^2 were enclosed with tin planket up to a depth of 20 cm. In each micro plot PU containing 387.5 mg N, NCPU containing 387.5 mg N with nimin (1 per cent) was applied on the soil surface where as USG containing 494.9 mg N was placed at 5 cm soil depth. The fertilizers were displaced vertically with rain water. The leachate samples collected from the surface as well as at 10 cm soil depths were analysed NH_2-N, NH_4-N and NO^-_3-N. No NO^-_3-N was detected in the leachate samples irrespective of source and depth of placement. In flood water sample, the amount of NH_2-N as well as NH^+_4-N was more with the application NCPU, intermediate with application of PU alone and least with deep placement of USG which was affected by depth of placement also. On the other hand, at 10cm soil depth the content of NH_2-N and NH^+_4-N was highest with deep placement of USG, intermediate with NCPU and the least in case of PU alone. The hydrolysis rate constant at 10 cm soil depth for USG, NCPU and PU alone was 0.0934 h^{-1}, 0.0552 h^{-1} and 0.0696 h^{-1} respectively due to variation in placement depth of nitrogenous fertilizer.

Keywords: Prilled urea, Nimin coated prilled urea, Urea super granules, Urea hydrolysis rate constant.

Introduction

Application of nitrogenous fertilizer in form of urea is a risky proposition for lowland rice. In lowland condition loss of nitrogen is manifold and that is the reason why nitrogen use efficiency is only 20–25 per cent. Many workers tried to increase the nitrogen use efficiency through placing USG in reduced soil zone. Application of USG in lowland rice increased grain yield than split application of PU as reported by Kapre *et al.* (1996) and Dash *et al.* (2003). Similarly apparent N recovery increased up to 51.4 per cent by using USG as reported by Singh *et al.* (1984). As USG was not available, nimin was used by Vyas *et al.* (1991) to increase the grain and straw yield of rice. Similarly Vimala and Subramanium (1994) reported the efficiency of nimin for increasing grain yield of lowland rice. The mechanism of transport and transformation of NCPU was not available. In the present investigation attempt has been made to study the transport and transformation mechanism of solid form of PU alone, NCPU and USG in the lowland rice field.

Materials and Methods

A field experiment was conducted in the central farm of OUAT, Bhubaneswar belonging to Inceptisol. Twenty-five days rice seedlings (cv. Lalat) were transplanted in the lowland with a common dose of 50 kg P_2O_5 and K_2O per hectare in form of single super phosphate (SSP) and muriate of Potash (MOP) respectively. As per the treatment plan, nitrogen was applied in form of PU and USG containing 46 per cent and 45 per cent N respectively and both of them marketed by IFFCO. The granule size of USG is 1 g approximately. The nimin, *i.e.* used in this study is marketed by M/s Godrej Agrovet Ltd. in liquid form. A micro plot size of 506.25 cm^2 was prepared in the center of each plot by putting tin planket to a depth of 20 cm. These plots contain T_1: surface application of N @ 387.5 mg as PU; T_2: surface application of N @ 387.5 mg as NCPU (nimin 1 per cent) and T_3: deep placement of N at 5m soil depth @ 494.9 mg as USG. In all these three treatments the nitrogen applied corresponds to 76 kg N ha^{-1}. Soil sampler at a depth of 0.1 m was placed in the centre of the micro plot by the help of soil auger. The PVC pipe of 0.2 m internal diameter and required length with ceramic cup at one end and a rubber cork at the other end (small holes on the side wall of tube for air entry) were used as soil solution sampler. After application of required fertilizer, that was displaced with rain water. Leachate samples at different soil depth were collected periodically. About 10 ml of leachate sample was pipetted out each time in an interval of 3 h in first day, of 6 h in second day, of 12 h in 3rd and 4th day, of 24 h in for next 6 days, of once in 4 days for 20 days, once in a fortnight for 30 days and once in a month till the harvest of the crop. The leachate samples were treated with a few drops of solution containing 5 mg L^{-1} of Phenyl Mercuric Acetate (PMA) soon after collection. Similarly flood water samples (about 50 ml) were collected simultaneously at each time from these micro plots. The leachate samples as well as flood water were analysed for determination of NH_2-N, NH_4^+-N and NO^-_3-N. Soil samples from 20 cm depth at the center of the microplot were collected before transplanting and after harvest of the crop. The samples were analysed for determination of water soluble NH_2-N calorimetrically (Mulvaney and Bermner, 1979), NH_4^+-N calorimetrically by Indophenol blue method (Keeney and Nelson, 1982).

Results and Discussion

Diffusion of NH_2-N in to Flood Water

From the table it was observed that NH_2-N increased with progress of time and reached the maximum value of 60.10 mg L^{-1} at 9 h in case of broadcasting of PU, of 51.11 mg L^{-1} at 18 h in case of broadcasting of NCPU and of 16.62 mg L^{-1} at 18 h in case of point placement of USG. The peak concentration of NH_2-N appeared latter in case of USG placement than that observed in case of PU as the diffusion of NH_2-N in soil is slow due to placement of USG at lower depth. Thereafter the concentration of NH_2-N decreased and attained the negligible value at about 32 h in case of PU, 76 h in case of NCPU and 64 h in case of USG. However, this value for USG was less than that observed in case of PU owing to deep placement of USG at 5 cm depth.

The content of NH_2-N diffused to flood water was computed from the following equation:

$$N = \frac{A\sum_{i=1}^{T} h_i c_i t_i}{1000 \times T} \qquad \text{......(1)}$$

where,

N: Amount (mg) of N diffused to flood water

A: Area (cm^2) of the planket

h_i: Depth (cm) of standing water on soil surface

c_i: Concentration (mg L^{-1}) or NH_2-N at time t_i (h)

T: Period (h) of collection of flood water sample, and $\sum_{i=1}^{T} c_i t_i$ will be area under BTC.

From the above equation it has been estimated that amount of applied N was diffused to flood water was highest (12 per cent) with NCPU, intermediate (11 per cent) with PU alone and least (2.7 per cent) with USG deep placement at 5 cm soil depth. Deep placement of USG reduced the diffusion of NH_2-N into flood water.

Leaching of NH_2-N at Soil Depth of 10 cm

The BTCs of NH_2-N in the leachate fractions in the treatment plot of T_1, T_2 and T_3 are presented in Figure 8.1. NH_2-N reached a soil depth of 10 cm soon after application of PU at the surface. The concentration of NH_2-N in the leachate fractions attained the maximum value of 6.99 mgL^{-1} at 46 h in case of T_1, 12.71 mg L^{-1} at 52 h in case of T_2 and 18.79 mg L^{-1} at 28 h in case of T_3. Thereafter, the concentration of NH_2N decreased and reached to a negligible amount at about 113 h, 160 h and 150 h in case of T_1, T_2, and T_3 respectively. The velocity of NH_2-N was calculated as per the following equation:

$$V_u = \frac{z}{t_m} \qquad \text{......(2)}$$

where,

V_u: Velocity (cm h^{-1}) of NH_2-N

z: Soil depth (cm) and

t_m: Mean resident time (h) at which the concentration of NH_2-N was maximum in the effluent samples.

The velocity of NH_2-N was calculated to be 0.21 cm h^{-1} in case of T_1, 0.189 cm h^{-1} in case of T_2 and 0.252 cm h^{-1} in case of T_3. The velocity of NH_2-N stemming from USG was higher than the PU.

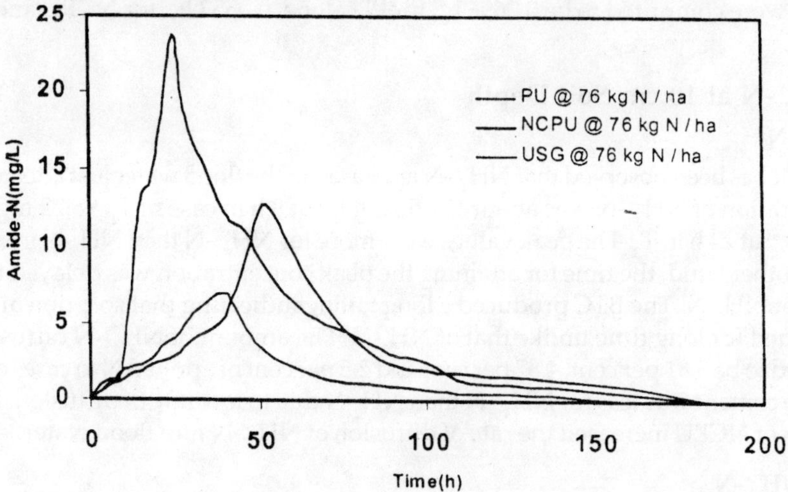

Figure 8.1: BTCs of Amide-N at Soil Depth of 10 cm Stemming from Placement of PU, NCPU and USG

The amount of N leached down the soil column was computed by using the equation given by Mishra *et al.*, 2000. The percentage of NH_2-N leached down the soil depth of 10 cm as calculated to be 3.75, 5.41 and 7.2 in case of T_1, T_2 and T_3 respectively indicating that in field soil almost all of the applied PU was hydrolysed irrespective of the nitrogen sources.

The mean resident time t_m was calculated by the following equation:

$$t_m = \frac{\sum_{i=1}^{T} t_i \cdot c_i \cdot \Delta t}{\sum_{i=1}^{T} c_i \cdot \Delta t} \qquad \text{......(3)}$$

where,

c_i: The concentration of NH_2-N at time t_i

T: Gross resident time and

Δt: Change in time.

The mean resident time (t_m) was calculated to be 47.3 h for PU alone, 52.8 h for NCPU and 28.1 h for USG. This value was small in case of USG since the displacement depth in case of USG was shallow (5 cm) as compared to displacement depth (10 cm) in case of PU. The rate constant for urea hydrolysis k was calculated as follows:

$$N = N_0 e^{-ktm} \qquad \text{......(4)}$$

where,

N_0: Amount (mg) of NH_2-N added to the soil and

k: Rate constant for urea hydrolysis.

The value of k were computed to be 0.0696 h^{-1} for PU alone, 0.0552 h^{-1} for NCPU and 0.0934 h^{-1} USG.

Leaching of NH_4^+-N at 10 cm Soil Depth

Diffusion of NH_4^+-N

From Table 8.1 it has been observed that NH_4^+-N appeared in the flood water just after application. Maximum concentration of NH_4^+-N was about 127.52 mg L^{-1} at 9 h in case of T_1, 98.72 mg L^{-1} in case of T_2 and 26.67 mg L^{-1} at 24 h in T_3. The peak values were more for NH_4^+-N than NH_2-N irrespective of treatments. On the other hand, the time for attaining the peak concentration was delayed for NH_4^+-N compared to that for NH_2-N. The BTC produced a long tailing indicating that sorption of NH_4^+-N in flood water continued for long time unlike that of NH_2-N. The amount of NH_4^+-N diffused to flood water was computed to be 3.81 per cent, 4.87 per cent and 2.5 per cent of applied N in cases of T_1, T_2 and T_3 respectively. The content was less for NH_4^+-N than NH_2-N due to low rate of diffusion. Like that of NH_2-N, application of NCPU increased the rate of diffusion of NH_4^+-N into flood water.

Displacement of NH_4^+-N

The NH_4^+-N content in the leachate fraction collected at a soil depth of 10 cm in different treatments are presented in Figure 8.2. NH_4^+-N in the leachate fraction applied soon after imposition of different treatments reached to a maximum value of 32.76 mg L^{-1} at 40 h for T_1, 30.5 mg L^{-1} at 52 h for T_2 and 172.9 mg L^{-1} at 96 h for T_3. After attaining the peak values, concentration of NH_4^+-N in the leachate fraction reduced drastically with progress of the time. In each case a long tail of BTC was observed which indicated that the sorption of NH_4^+-N in the soil was of high magnitude and absorbed in soil for a long period of time. The mean resident time, t_m, was 41.3 h, 42.6 h and 97.3 h for T_1, T_2 and T_3 respectively. It has been observed that the mean resident time is higher in case of USG than that of PU. Similarly velocity of NH_4^+-N in case of PU was slightly more than that of NCPU where as that value was lower with USG (Table 8.1). The amount of NH_4^+-N leached down the soil depth of 10 cm was

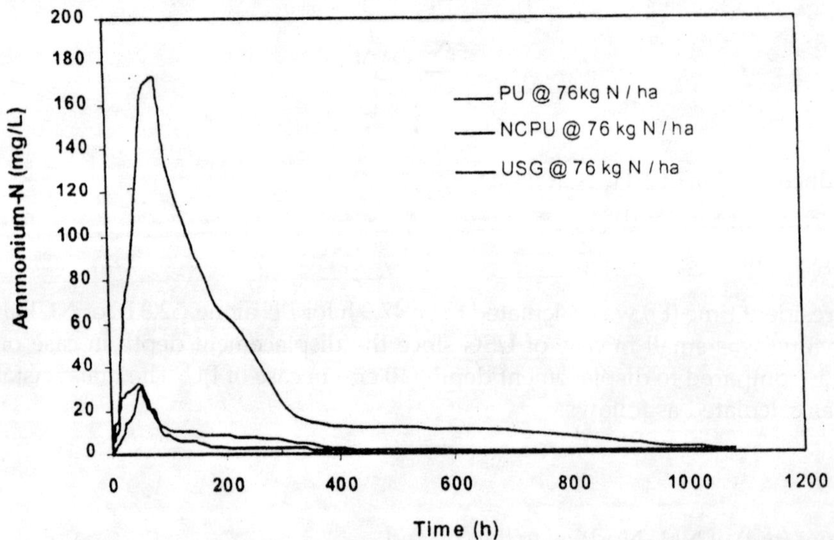

Figure 8.2: BTCs of Ammonium-N at Soil Depth of 10 cm Stemming from Placement of PU, NCPU and USG

computed to be 49.39 per cent in case of T_1, 33.46 per cent in case of T_2 and 131 per cent in case of T_3. On the other hand, leaching of NH_4^+-N was increased to 131 per cent when USG was point placed in the soil possibly some of NH_4^+-N from the organic pool of the soil might have been transformed and leached down the soil depth of 10 cm.

Table 8.1: Basic Information on Leaching of PU, NCPU and USG Under Field Condition

Parameters	*Treatments*		
	T_1	T_2	T_3
Cross sectional Area of the microplot (cm^2)	506.25	506.25	506.25
Soil water flux ($cm\ h^{-1}$)	0.094	0.071	0.082
Method of Nitrogen Application	Surface broadcasting	Surface broadcasting	Placement at 5 cm soil depth
Soil water Velocity ($cm\ h^{-1}$)	0.164	0.160	0.198
Form of Nitrogen	PU	NCPU	USG
Amount of N added (mg)	387.5	387.5	494.9
Flood water			
Maximum concentration of NH_2-N (mg/L)	60.10	51.11	16.62
Time required for NH_2-N attaining peak value (h)	9	18	18
Time taken for NH_2-N to attain negligible value	32	76	64
Percentage of NH_2-N	11.0	12.0	2.7
Maximum concentration of NH_4^+-N (mg/L)	127.52	98.72	26.67
Time required for NH_4^+-N attaining peak value (h)	9	28	24
Percentage of NH_4^+-N	3.81	4.87	2.5
At 10 cm Soil Depth			
Mean resident time, t_m, for NH_2-N (h)	47.3	52.8	28.1
Velocity of NH_2-N ($m\ h^{-1}$)	0.211	0.189	0.252
Percentage of NH_2-N leached	3.72	5.41	7.24
Urea hydrolysis rate constant, k (h^{-1})	0.0696	0.0552	0.0943
Mean resident time, t_m, for NH_4^+-N (h)	41.3	42.6	97.3
Velocity of NH_4^+-N ($cm\ h^{-1}$)	0.24	0.235	0.073
Percentage of NH_4^+-N leached	49.39	33.46	131.90

From the above investigation it has been observed that addition of nimin decreased urea hydrolysis resulting higher amount of NH_2-N in leachate fraction at 10cm depth in the rise field. On the other hand, application of USG resulted higher amount of NH_2-N in leachate fraction. Similarly, amount of NH_4^+-N leached at 10 cm soil depth was highest with T_3, intermediate with T_2 and lowest with T_1. So nimin to some extent can be used as an urease inhibitor and increased the nitrogen use efficiency than that of applied PU.

References

Dash, A.K., Mishra B.K., Jena, M.K. and Jena, D., 2003. Effect of dose and application methods of different urea based fertilizers on yield of lowland rice. *Annals of Agricultural Research*, New Series, 24(2): 308–331.

Kapre, M.M., Bisen, C.R. and Johar, M.S., 1996. Effect of slow released N fertilizers on rice under lowland condition. *Crop Research*, 11(1): 17–23.

Keeney, D.K. and Nelson, D.W., 1982. *Methods of Soil Analysis*, Part 2, 2nd Edition, Number 9, in the Agronomy Series, American Society of Agronomy, Soil Science Society of America, Medison, Wisconsin, USA, pp. 643–698.

Mishra, B.K., Mishra, S.R., Jena D. and Misra, C., 2000. Transport and transformation of urea super granule in an inceptisol under flooded condition. *Journal of Indian Society of Soil Science*, 48(2): 209–214.

Mulvaney, R.L. and Bremners, J.M., 1979. A modified diacetyl monoxime method for colarimetric determination of urea in soil extracts. *Communication of Soil Science and Plant Annals*, 10: 1163–1170.

Singh, K., Singh, A.N. and Singh, K.N., 1989. Effect of urea super granule depth of placement in irrigated transplanted rice. *International Rice Research Newsletter*, 14(3): 33.

Vimala, I. and Subramanium, S., 1994. Influence of nitiricatry inhibitors on nutrient availability, yield and uptake in rice soils. *Madras Agricultural Journal*, 81(10): 544–551.

Vyas, B.N., Godrej, N.B. and Mistry, K.B., 1991. Tracty the transformation of urea fertilizer during leaching. *Soil Science Society of American Journal*, 41: 896–902.

Chapter 9

Genotoxic Effects of 2,4-D and Modifying Effects of Vitamin C Tested in the Somatic Cells of High Bioactivation Strains of *Drosophila melanogaster*

Niraj K. Tripathy
Department of Zoology, Berhampur University, Berhampur – 760 007, Orissa, India

ABSTRACT

The genotoxicity of 2,4-dichlorophenoxyacetic acid, a phenoxy herbicide, was tested in the somatic cells (wing primordia) of *Drosophila melanogaster* through the wing mosaic assay. Larvae, transheterozygous for the recessive genetic markers *mwh* and *flr*^{3} obtained from the cross of high bioactivation strains, were used in the present experiments. The modifying effects of vitamin C on the genotoxicity of 2,4-D were assessed by way of pre-, concurrent and post-treatments with the vitamin. The results clearly demonstrate the protective effects of the vitamin against the genotoxicity of the herbicide.

Keywords: 2-4 D, Vitamin C, Genotoxicity, Drosophila melanogaster.

Introduction

2,4-dichlorophenoxyacetic acid (2,4-D) is a hormonic phenoxy herbicide, which has wide application for the control of a variety of plants. Although the evidence of carcinogenicity of this herbicide is inadequate in animals and humans (IARC, 1986). It is reported to be genotoxic in a number of short-term tests. This herbicide was mutagenic in *Salmonella typhimurium, Bacillus subtilis, Saccharomyces cerevisiae* and *Aspergillus nidulans* (Kappas, 1988; Klopman *et al.*, 1985; Siebert and Lamperle, 1974; Zetterberg *et al.*, 1977). In plants, *viz., Allium cepa, A. ascalonium, Vicia faba, Triticum aestivum, T. durum* and *Ambrosia milletolium,* 2,4-D was reported to be clastogenic and induce bridge formation, micronuclei and C-mitoses (Amar and Ali, 1974; Prasad and Das, 1977; Saalbach and Koblitz, 1977; Logvinenko and Morgun, 1978; Al-Najjar and Soliman, 1982; Tomkins and Grant, 1976; Suryakumari and Vaidyanath, 1989; Pavlica *et al.*, 1991). In the Drosophila sex-linked recessive lethal test, although it was reported to be nongenotoxic by Vogel and Chandlor (1974), Magnusson *et al.* (1977) and Tripathy *et al.* (1993) have obtained positive results. In the Drosphila wing mosaic assay also this compound was positive (Tripathy *et al.*, 1993). In *in vivo* tests, using mice and rats, 2,4-D induced chromosome aberrations and polyploidy (Turkula and Jalal, 1987; Adhikari and Grover, 1988). In *in vitro* assays using Chinese hamster V79 cells, Pavlica *et al.* (1991) have reported induction of mutation in the HGPRT locus by 2,4-D. In human VA-4 cells and lymphocytes in culture, this herbicide induced unscheduled DNA synthesis, chromosome aberrations and SCEs in large frequencies (Ahmed *et al.*, 1977; Korte and Jalal, 1982; Turkula and Jalal, 1985). However, there are a few reports wherein 2,4-D was reported to be nongenotoxic in different tests systems (Moriya *et al.*, 1983; Klopman *et al.*, 1985; Gollapudi *et al.*, 1999; Charles *et al.*, 1999).

Although oxygen is highly essential for sustenance of life systems, often under the action of various toxic chemical substances, it is compelled to produce oxygen-free radicals and reactive oxygen species which are highly toxic and damage biological systems by way of lipid peroxidation, protein oxidation and DNA damage (Chatterjee *et al.*, 1995). Mutations may arise directly by free radical-mediated oxidative deamination of DNA bases or indirectly through the reactions of free radicals with DNA polymerase, which may decrease the fidelity of replication or repair of the damaged DNA. Thus the effective method to prevent such degenerative changes would be the prevention of oxidative damages in a living cell. The primary cellular defence against such damages is certain antioxidants like vitamins, uric acid and certain enzymes (Yu, 1994). These cellular defences act as scavangers of free radicals or prevent the production of electrophilic metabolites by the toxic chemicals. Vitamin C (AA) is the most effective aquous phase antioxidant (Frei *et al.*, 1989) having strong antimutagenic potential (Bhattacharya *et al.*, 1987). At low concentrations AA acts as a free radical scavanger (Weitberg and Weitzman, 1985) but, under certain circumstances it may act as a pro-oxidant and promotes generation of free radicals and lipid peroxidation (Otero *et al.*, 1997). In the present work, the authors were interested to reevaluate the genotoxicity of 2,4-D in the wing mosaic assay following exposures of larvae obtained from high bioactive strains and to find out the modifying effects of AA, if any, on the genotoxicity of the herbicide by way of pre, concurrent and post-treatments with the vitamin.

Materials and Methods

The wing mosaic assay involves the exposure of a population of genetically marked wing primordial cells to a chemical mutagen so that genetic changes induced by the mutagen will be expressed as mutant mosaic spots (clones) on a normal wing of an eclosing adult. In the present experiments two high bioactive strains of *Drosophila melanogaster, viz.,* ORR; *mwh/mwh* and ORR; *flr*^{3}, *Ser* were used in contrast to the standard strains used in an earlier experiment (Tripathy *et al.*, 1993). The high bioactive strains carry chromosomes 1 and 2 from a DDT-resistant ORR-R (R) line and have high constitutive levels of cytochrome P-450 (Hallstrom and Blanck, 1985; Frolich and Wurgler, 1989). The allele *mwh* (multiple wing hairs) is a recessive marker mutation on the chromosome 3 (3-0.3), which is expressed as multiple trichomes in contrast to a single trichome per wing cell of the wild type fly. The allele *flr*^{3} (flare) is also a recessive marker on the same chromosome (3-38.8) and is expressed as short, thick and misshaped trichome on the adult wings. For details of the markers please refer Frolich and Wurgler (1989), Lindsley and Zimm (1992) and Graf and Singer (1992).

ORR; *flr*^{3}/TM3, *Ser* virgin females were mated to the ORR; *mwh*/*mwh* males (Frolich and Wurgler, 1991; Graf and Singer, 1992). Eggs were collected from optimally fertile parents of the above cross on standard Drosophila medium. Second and third instar larvae, corresponding to 48 hand 72 h of age (Graf *et al.*, 1984), were collected by floating them in 20 per cent NaCl solution. The *mwh* +/+ *flr*^{3} transheterozygous larvae were exposed to different concentrations of 2,4-D prepared with 3 per cent ethanol and 1 per cent tween in distilled water in batches with approximately 100 larvae per batch to determine the LD_{50}. The LD_{50}, where apprimately 50 per cent larvae hatched to adult stage was 5 mM. Thus the larvae were exposed to the LD_{50} and lower doses in instant food (Carolina Biological Supply Co., USA) to determine the genotoxicity of 2,4-D. A control set was run simultaneously where the larvae were exposed to the solvent alone.

For pretreatment of AA, 2nd instar larvae, transheterozygous for the above markers, were treated with 100 mM of AA for 24 h. Thereafter the AA-treated larvae were washed out of the food and exposed to LD_{50} or lower doses in different vials. Since AA was tested earlier for its mutagenicity in the Drosophila wing spot test (Tripathy *et al.*, 1990) and 100 mM of this vitamin was nongenotoxic, this concentration was chosen in the present experiments to study its modifying effects on the genotoxicity of 2,4-D. For concurrent treatment, 3rd instar larvae were exposed to mixtures of different concentrations of 2,4-D and 100 mM of AA. In another set of experiments 2nd instar larvae were treated with different doses of the herbicide for 48 h and then the treated larvae were floated out and exposed to 100 mM of AA for 24 h. All the experiments were repeated and conducted at a temperature of 25±10°C.

The wings of the closing adults were mounted in Faure's solution (Graf *et al.*, 1984). The mounted wings were screened under a compound microscope at 400 X magnifications. The size, *viz.*, small with 1–2 affected cells or large with more than 2 affected cells and file type, *viz.*, *mwh* or *flr*^{3} single or *mwh*/*flr*^{3} twin, were recorded separately. For statistical analyses of the data on the genotoxicity of 2,4-D, the frequencies of different types of spots in the treated and solvent control were compared. On the other hand, for the evaluation of antimutagenic effects of AA, data obtained after pre-, concurrent and post-AA treatment of the larvae were compared with the positive control (2,4-D treated). The data were subjected to conditional binomial test following Frei and Wurgler (1988).

Results

Each experiment was repeated and, since the data of the two experiments were not significantly different from one another, they were pooled. The data obtained on the genotoxicity of 2,4-D, following exposures of *mwh* +/+ *flr*^{3} transheterozygous larvae obtained from the cross of high bioactive strains are represented in the Table 9.1. In the present experiments the LD_{50} of 2,4-D for the larvae of both instars was 5 mM in contrast to 10 mM in the earlier studies (Tripathy *et al.*, 1993) suggesting that the larvae obtained from the cross of high bioactivation strains are more sensitive to the herbicide compared to the larvae of standard cross. Similar observations were made by Graf and Singer (1992) with respect to the sensitivity of larvae to N-nitrosopyrrolidine. It has been observed that the frequencies of both small and large single spots (either *mwh* or *flr*^{3}) on the wings of adults eclosing from larvae treated with different concentrations of 2,4-D were significantly higher than their corresponding frequencies in the control set. Only a very few *mwh*/ *flr*^{3} twin spots were induced following larval treatment with the herbicide. It was marked that the frequencies of single spots in the present experiments are at least five times higher than their frequencies when the larvae belonged to the standard cross (Tripathy *et al.*,

1993). This observation indicates an increased cytochrome P-450-dependent activation capacity in the larvae of the high bioactivation strain.

Table 9.1: Summary of Data on the Genotoxicity of 2,4-D.

Treatment Duration	*Concentration (mM)*	*No. of Tested Wings*	*Spots per Wing (No. of Spots) Stat. Diagn.**			
			Small Singles (s = 1–2) M = 2.0	*Large Singles (s > 2) m = 5.0*	*Twins (t) m = 5.0*	*Total (T) m = 2.0*
	Control	80	0.18 (14)	0.01 (1)	0.00 (0)	0.19 (15)
	0.5	80	0.46 (37)$^{+}$	0.25 (20)$^{+}$	0.00 (0)^{i}	0.71(57)^{i}
48 h	1.0	80	0.99 (79)$^{+}$	0.28 (22)$^{+}$	0.03 (2)^{i}	1.30 (103)$^{+}$
	2.5	80	1.21 (97)$^{+}$	0.66 (53)$^{+}$	0.0.5 (4)^{i}	1.93 (154)$^{+}$
	5.0	80	1.95 (156)$^{+}$	0.85 (68)$^{+}$	0.05 (4)^{i}	2.85 (228)$^{+}$

*: Statistical diagnoses following Frei and Wurgler (1988).

+: Positive; i: Inconclusive; m: Multiplication factor.

Probability levels: $\alpha = \beta = 0.05$. One-sided statistical tests.

Table 9.2: Summary of Data on Wing Spot Test After Larval Exposures to Different Concentrations of 2,4-D and 100 mM of AA

Treatment Schedule	*No. of Wings Tested*	*Spots per Wing (No. of Spots) Stat. Diagn.**			
		Small Singles (s = 1–2) m = 2.0	*Large Singles (s > 2) m = 5.0*	*Twins (t) m = 5.0*	*Total (T) m = 2.0*
2,4-D (5 mM)	80	1.95 (156)	0.85 (68)	0.05 (4)	2.85 (228)
PreAA + 2,4–D	80	0.54 (43)$^{-}$	0.04 (3)$^{-}$	0.00 (0)$^{-}$	0.58 (46)$^{-}$
AA + 2,4-D	80	0.30 (24)$^{-}$	0.20 (16)$^{-}$	0.08 (6)^{i}	0.58 (46)$^{-}$
Post AA + 2,4-D	80	0.51 (41)$^{-}$	0.28 (22)$^{-}$	0.00 (0)$^{-}$	0.79 (63)$^{-}$
2,4-D (2.5 mM)	80	1.21 (97)	0.66 (53)	0.05 (4)	1.93 (154)
Pre AA + 2,4-D	80	0.40 (32)$^{-}$	0.03 (2)$^{-}$	0.01 (1)$^{-}$	0.44 (35)$^{-}$
AA + 2,4-D	80	0.21 (17)$^{-}$	0.20 (16)$^{-}$	0.06 (5)$^{-}$	0.48 (38)$^{-}$
Post AA + 2,4-D	80	0.48 (38)$^{-}$	0.18 (14)$^{-}$	0.00 (0)$^{-}$	0.66 (52)$^{-}$
2,4-D (1 mM)	80	0.99 (79)	0.28 (22)	0.03 (2)	1.30 (103)
PreAA + 2,4-D	80	0.33 (26)$^{-}$	0.03 (2)$^{-}$	0.00 (0)^{i}	0.36 (28)$^{-}$
AA + 2,4-D	80	0.19 (15)$^{-}$	0.15 (12)$^{-}$	0.01 (1)^{i}	0.36 (28)$^{-}$
PostAA + 2,4-D	80	0.40 (32)$^{-}$	0.14 (11)$^{-}$	0.01 (1)^{i}	0.55 (44)$^{-}$
2.4-D (0.5 mM)	80	0.46 (37)	0.25 (20)	0.00 (0)	0.71 (57)
Pre AA + 2,4-D	80	0.26 (21)$^{-}$	0.01 (1)$^{-}$	0.00 (0)^{i}	0.27 (22)$^{-}$
AA + 2,4-D	80	0.19 (15)$^{-}$	0.07 (6)$^{-}$	0.01 (1)^{i}	0.27 (22)$^{-}$
PostAA + 2,4-D	80	0.31 (25)$^{-}$	0.13 (10)$^{-}$	0.00 (0)^{i}	0.44 (35)$^{-}$

*: Statistical diagnoses following Frei and Wurgler (1988).

i: Inconclusive, –: Negative, m: Multiplication factor. Probability levels: $\alpha = \beta = 0.05$. One-sided statistical tests.

Effect of AA on the Genotoxicity of 2,4-D

Data pertaining to the modifying effects ot AA on the genotoxicity of 2,4-D are summarised in the Table 9.2. It has been observed that AA has reduced the genotoxic potential of 2,4-D following pre-, concurrent as well as post-treatments. When the data obtained with AA treatments were compared with those of the positive control (2,4-D-treated) there was significant decrease in the spot frequencies and the statistical outcomes were negative or inconclusive.

Discussion

Single spots with *mwh* or flr^3 phenotype, following exposures of larvae transheterozygous for the markers *mwh* and flr^3, are observed on the wings of eclosing flies following induction of mutation or gene conversion in the corresponding wild type loci (Graf *et al.*, 1984). Again, *mwh* single spots may originate due to segmental aneuploidy caused by chromosome breakage (Haynie and Bryant, 1977). The origin of such spots due to nondis junction, as suggested by Graf *et al.* (1984), may not probably hold good in the present experiment as 2,4-D was reported not to induce nondis junction or X chromosome loss in *Drosophila melanogaster* (Magnusson *et al.*, 1977). Twin spots with *mwh* and flr^3 subclones originate due to the induction of mitotic crossing over in the region between the flr^3 locus and the centromere of chromosome 3 (Becker, 1976). In the present experiments, since twin spots were not observed in high frequencies, it is supposed that the test herbicide is non-recombinogenic or weakly so in the wing disc cells of Drosophila.

There has been a significant decrease m the frequency of wing spots following treatment with AA either before or after treatment with 2,4-D or along with the herbicide suggesting the modifying effects of AA. As suggested by Bhattacharjee and Pal (1982), AA might have shielded the genetic material from the toxic effects of the herbicide. The vitamin might have also interacted with 2,4-D preventing the generation of toxic metabolites responsible for induction of genetic changes. Since AA is nucleophilic in nature it might have intercepted electrophilic metabolites preventing them to attack the nucleophilic sites of the DNA (Bodo *et al.*, 1989; Liehr *et al.*, 1989) of the wing primordia. Further, since AA is an electron donor, it might have reacted with superoxides, peroxides and hydroxyl ions generated by the herbicide and nullified their actions on the DNA inside the cells. Alternately, since AA is an antioxidant it might have prevented oxidative metabolism of the test compound preventing the production of electrophilic metabolites as proposed by Goncharova and Kuzhir (1989). Further, AA since it is a part of the redox buffer system might have acted as a scavanger of electrophilic ions generated inside the cells due to the action of the test chemical as proposed by Sato *et al.* (1990), Ross (1990) and Sahu and Das (1994).

Thus although the exact mechanism(s) by which AA has minimised the genotoxic effects of 2,4-D following pre-, concurrent and post-treatments with the vitamin is difficult to comprehend, it may be concluded that this vitamin is protective against the genotoxicity of the test herbicide.

Acknowledgement

The authors are thankful to Dr. S.K. Udgata, Department of Computer Science, Berhampur University for statistical analysis of the wing spot data.

References

Adhikari, N. and Grover, I.S., 1988. Genotoxic effects of some systemic pesticides: *In vivo* chromosomal aberrations in bone marrow cells in rats. *Environ. Mot. Mutagen.*, 12: 235–242.

Ahmed, F.E., Hart, R.W. and Lewis, N.J., 1977. Pesticide-induced DNA damage and its repair in cultured human cells. *Mutat. Res.*, 42: 161–174.

Al-Najjar, N.R. and Soliman, A.S., 1982. Cytological effects of herbicides. I. Effect of 2,4-D and 2,4,5-T on meiotic cells of wheat and two related species. *Cytologia*, 47: 53– 61.

Amer, S.M. and Ali, E.M., 1914. Cytological effects of pesticides. V. Effects of some herbicides on *Vicia faba. Cytologia*, 39: 633–643.

Becker, H.J., 1976. Meiotic recombination. In: *Genetics and Biology of Drosophila*, (Eds.) Ashburner, A. and E. Novitski. Academic Press, New York, 1: 1020–1087.

Bhattarjee, S.B. and Pal, S., 1982. Tetracycline-induced mutation in cultured Chinese hamster cells. *Mutat. Res.*, 101: 329–338.

Bhattacharya, R.K., Francis, A.R. and Shetty, T.K., 1987. Modifying role of diatery factors on the mutagenicity of aflatoxin B_1: *In vitro* effect of vitamins. *Mutat. Res.*, 188: 121–128.

Bodo, M., Todisco, T., Pezzetti, P., Dottorini, M., Moggi, L. and Becchetti, E., 1989. Ability of retinoic and ascorbic acid to interfere with the binding of benzo(a) pyrine to DNA in explants from donors with bronchial cancer. *Onchology*, 46: 178–182.

Chatterjee, I.B., Mukhopadhyay, C.K. and Ghosh, M.K., 1995. Vitamin C: A potential saviour against free radical-induced oxidative damage. *Current Sc.*, 69: 747–751.

Charles, J.M., Cunny, H.C., Neilson, R.D., Ivett, J.L., Murli, H., Bus, J.S. and Gollapudi, B.B., 1999. *In vivo* micronucleus assay on 2,4-dichlorophenoxyacetic acid and its derivatives. *Mutat. Res.*, 444: 227–234.

Frei, B., England, L. and Ames, B.N., 1989. Ascorbate is an outstanding antioxidant in human blood plasma. *Proc. Natl. Acad. Sci.* (*USA*), 86: 6377–6381.

Frei, H. and Wurgler, F.E., 1988. Statistical methods to decide whether mutagenicity test data from Drosophila assays indicate a positive, negative or inconclusive result. *Mutat. Res.*, 203: 297–308.

Frolich, A. and Wurgler, F.E., 1989. New tester strains with improved bioactivation capacity for the Drosophila wing spot test. *Mutat. Res.*, 216: 179–187.

Frolich, A. and Wurgler, F.E., 1991. The high bioactivation cross for the SMART assay with the wing. *Drosophila Inf. Serv.*, 70: 246–247.

Graf, U. and Singer, D., 1992. Genotoxicity testing of promutagens in the wing somatic mutation and recombination test in *Drosophila melanogaster. Rev. Int. Contam. Ambient*, 8: 15–27.

Graf, U., Wurgler, F.E., Katz, A.J., Frei, H., Juon, H., Hall, C.B. and Kale, P.G., 1984. Somatic Mutation and Recombination Test in *Drosophila melanogaster. Environ. Mutagen.*, 6: 153–188.

Gollapudi, B.B., Charles, J.M., Ann Linscombe, V., Day, S.J. and Bus, J.S., 1999. Evaluation of the genotoxicity of 2,4-dichlorophenoxyacetic acid and its derivatives in mammalian cell cultures. *Mutat. Res.*, 444: 217–225.

Goncharova, R.I. and Kuzhir, T.D., 1989. A comparative study of the antimutagenic effects of antioxidants on chemical mutagenesis in *Drosophila melanogaster. Mutat. Res.*, 214: 257–265.

Halstrom, I. and Blanck, K., 1985. Genetic variation in cytochrome P-450 system in *Drosophila melanogaster.* I. Chromosomal determination of some cytochrome P-450-dependent reactions. *Chem. Biol. Interact.*, 56: 157–171.

Haynie, J.L. and Bryant, P.J., 1977. The effects of X-rays on the proliferation dynamics in the imaginal wing discs of *Drosophila melanogaster. Wilhelm Roux' Arch.*, 183: 85–100.

IARC, 1986. *IARC Monographs on the Evaluation of Carcinogenic Risk of Chemicals to Humans*, 41: 357–406.

Kappas, A., 1988. On the mutagenic and recombinogenic activities of certain herbicides in *Salmonella typhimurium* and *Aspergillus nidulans. Mutat. Res.*, 204: 615–621.

Klopman, G., Contreras, R., Rosenkranz, H.S. and Waters, M.D., 1985. Structure-genotoxic activity relationship of pesticides-comparison of the results from several short term assays. *Mutat. Res.*, 147: 343–356.

Korte, C. and Jalal, S.M., 1982. 2,4-D induced clastogenicity and elevated rates of sister chromatid exchanges in cultured human lymphocytes. *J. Hered.*, 73: 224–226.

Liehr, J.G., Roy, D. and Gladek, A., 1989. Mechanism of inhibition of estrogen-induced renal carcinogenesis in male Syrian hamster by vitamin C. *Carcinogenesis*, 10: 1983– 1988.

Lindsley, D.L. and Zimm, G.G., 1992. *The Genome of Drosophila melanogaster.* Academic Press, San Diego, pp. 1133.

Logvinenko, V.F. and Morgun, V.V., 1978. Study of the mutagenic effect of some pesticides on spring durum wheat. *Tsitologiya Genetica*, 12: 207–212.

Magnusson, J., Ramel, C. and Errikson, A., 1977. Mutagenic effects of chlorinated phenoxy acetic acids in *Drosophila melanogaster. Hereditas*, 87: 121–123.

Moriya, M., Ohta, T., Watanabe, K., Miyazawa, T., Kato, K. and Shirasu, Y., 1983. Further mutagenicity studies on pesticides in bacterial reversion assays. *Mutat. Res.*, 116: 185–216.

Otero, P., Viana, M., Herrera, L. and Bonet, B., 1997. Antioxidant and prooxidant effects of ascorbic acid and dehydroascorbic acid and flavonoids on LDL submitted to different degrees of oxidation. *Free Radic. Res.*, 27: 619–626.

Pavlica, M., Papes, D. and Nagy, B., 1991. 2,4-dichlorophenoxyacetic acid causes chromatin and chromosomal abnormalities in plant cells and mutation in cultured mammalian cells. *Mutat. Res.*, 263: 77–81.

Prasad, G. and Das, K., 1997. Effects of some growth substances on mitosis. *Cytologia*, 42: 323–329.

Ross, R.C., 1990. Ascorbic acid metabolism against free radicals: A radiation model. *Biochem. Biophys. Res. Communications*, 1691: 430–436.

Saalbach, G. and Koblitz, H., 1977. Karyological instabilities in callus cultures from haploid plants. *Biochemie and Physiologie Der Pflanzen*, 171: 469–473.

Sahu, K. and Das, R.K., 1994. Reduction of clastogenic effects of clofazimine, and antileprosy drug, by vitamin A and vitamin C in bone marrow cells of mice. *Fd. Chem. Toxicol.*, 32: 911–915.

Sato, K., Niki, E. and Shimasaki, H., 1990. Free radical-mediated chain oxidation of low-density lipoprotein and its synergistic inhibition by vitamin C and vitamin E. *Arch. Biochem. Biophys.*, 279: 402–405.

Sibert, D. and Lemperle, E., 1974. Genetic effect of herbicides: Induction of mitotic gene conversion in *Saccharomyces cerevisiae. Mutat. Res.*, 22: 111–120.

Suryakumari, T. and Vaidyanath, K., 1989. Testing of genotoxic effects of 2,4-D using multiple genetic assay systems in plants. *Mutat. Res.*, 226: 235–238.

Tomkins, D.J. and Grant, W.F., 1976. Monitoring natural vegetation for herbicide-induced chromosomal aberrations. *Mutat. Res.*, 36: 73–84.

Tripathy, N.K., Routray, P.K., Sahu, G.P. and Anandkumar, A., 1993. Genotoxicity of 2,4-dichlorophenoxyacetic acid tested in somatic and germ line cells of Drosophila. *Mutat. Res.*, 319: 237–242.

Tripathy, N.K., Wurgler, F.E. and Frei, H., 1990. Genotoxicity of six carcinogens and six noncarcinogens in the Drosophila wing spot test. *Mutat. Res.*, 242: 169–180.

Turkula, T.E. and Jalal, S.M., 1985. Increased rates of sister chromatid exchanges induced by the herbicide 2,4-D. *J. Hered.*, 76: 213–214.

Turkula, T.E. and Jalal, S.M., 1987. Induced clastogenicity in white rats by the herbicide 2,4–D. *Cytologia*, 52: 275–281.

Vogel, E.W. and Chandler, J.L.R., 1974. Mutagenicity testing of cyclamates and some pesticides in *Drosophila melanogaster. Experientia*, 30: 621–623.

Weitberg, A.B. and Wietzman, S.A., 1985. The effect of vitamin C on oxygen radical-induced sister-chromatid exchanges. *Mutat. Res.*, 114: 23–26.

Yu, B.P., 1994. Cellular defences against damage from reactive oxygen species. *Physiological Reviews*, 74: 139–162.

Zetterberg, G., Busk, L., Elovson, R., Starec-Nordenhamar, I. and Ryttman, H., 1977. The influence of pH on the effects of 2,4-D (2,4-dichlorophenoxy acetic acid) Na salt on *Saccharomyces cerevisiae* and *Salmonella typhimurium. Mutat. Res.*, 42: 3–18.

Chapter 10

Seasonal Analysis in the Water Quality of River Ganga

Afreen Zafr and Naheed Sultana
Department of Zoology, A.N.D. College, Kanpur – 208 012

ABSTRACT

The present communication deals on a seasonal variation in the water quality of river Ganga at Kanpur city (U.P.) during July 2002 to June 2004. The samples were collected from three different sites *viz.*, Bithoor, Bhairon Ghat and Jajmau. Various physico-chemical parameters such as temperature, turbidity, transparency, pH, dissolved oxygen, alkalinity, hardness, chloride, phosphate, biochemical oxygen demand, chemical oxygen demand were analyzed. The pollution level of the river water increased as from Site I to Site II and Site III. Significant seasonal variation in the physico-chemical parameters were observed during the study.

Keywords: Seasonal variation, Water quality, Pollution.

Introduction

Ganga is a largest and sacred river of India having an important place in Indian culture and tradition. As a result of the increasing industrialization and other anthropogenic activities in the Gangetic plain. Ganga water, quantity as well as quality has declined over the years. River is the main raw water source of the city. Several reports are available on water quality of river Ganga by many workers in Kanpur region (Saxena *et al.*, 1966; Pandey and Pandey, 1980 and Tare *et al.*, 2003) and other parts of the country (Pahwa and Mehrotra, 1966; Khan *et al.*, 1994; Mishra and Tripathi, 2003) in relation to some aspects of physico-chemical monitoring. The communication deals with the systematic study on seasonal variation in physico-chemical parameters in the selected stretch of the river Ganga at Kanpur.

Materials and Methods

The present investigation was carried out taking three sampling sites from Bithoor to Jajmau in the river Ganga. Site I: Bithoor is an outskirts of Kanpur and the main centre of pilgrimage and tourist activities. Site II: Bhairon Ghat is mainly used for cremation of human dead body. Site III Jajmau is the point of industrial and sewage discharge into river water.

The water samples were collected at the monthly intervals between 7:00 a.m. to 10:00 a.m. from July 2002 to June 2004 from each sampling sites. Water temperature was measured with the help of a mercury thermometer, transparency with secchi disc and pH with pH strips at the sites. While the rest of the physico-chemical parameters were analyzed in the laboratory following the standard methods APHA (1992).

The results were expressed on seasonal basis and the seasons were classified as monsoon (July and August), post-monsoon (September and October), winter (November to January), spring (February and March) and summer (April to June).

Results and Discussion

The results of various physico-chemical parameters of river Ganga are presented in Table 10.1. Water temperature ranged from a minimum of 17.13°C at Site I during winter to a maximum of 34.33°C at Site III during summer. In the present investigation water temperature is directly influenced by the ambient temperature. Similar seasonal trend was also reported by Khan *et al.* (1994), Kaur and Joshi (2003), Mishra and Tripathi (2003) in river Ganga. Turbidity values ranged between 1.85 JTU to 69.45 JTU. Highest values were observed during monsoon season at Site III and lowest during spring at Site I. High turbidity during monsoon was attributed due to surface runoff and suspended solids (Verma *et al.*, 1984). Transparency of water was recorded minimum of 23.1 cm during monsoon at Site III and maximum of 48.16 cm during winter at Site I. Similar seasonal light penetration was also observed by Srivastava and Kulshrestha (1990). Reduced transparency during monsoon may be due to soil erosion.

pH values ranged between 7.45 to 9.73. It was recorded lowest at Site III during monsoon and highest at the same site during winter. Lower values of pH during monsoon might be due to increased turbidity and higher during winter might be due to high algal population. Lowest pH during monsoon in river Ganga was also noted by Pahwa and Mehrotra (1966), Srivastava and Kulshreshta (1990) at Allahabad, Mishra and Tripathi (2003) at Varanasi. According to Venkateswarlu (1986) unpolluted sites of different rivers had a pH range between 7.7 to 8.7. In the present study the sewage polluted Site III had pH above 8.7 during winter and spring season, therefore, this site is considered as polluted based on pH.

A considerable seasonal variation in the concentration of dissolved oxygen (DO) was recorded in the present study ranging between 3.95 mg/l to 8.17 mg/l. DO was minimum during monsoon at Site III and maximum during winter at Site I. Pahwa and Mehrotra (1966), Saxena *et al.* (1966), Khan *et al.* (1994), Mishra and Tripathi (2003) have discussed the seasonal fluctuations of DO and uniformly reported winter peaks in oxygen levels in their studies in river Ganga. High DO in winters might be due to high amount of oxygen at low temperature and low DO in monsoon might be due to increased turbidity.

Alkalinity of the river fluctuated from 75 mg/l to 318 mg/l. It was observed lowest during monsoon at Site I and highest during spring at Site III. Similar seasonal observations have been reported by Khan *et al.* (1994), Mishra and Tripathi (2003), whereas Pahwa and Mehrotra (1966) repoiled high alkalinity during summer and low during monsoon in river Ganga. Hardness of the river water

Table 10.1: Seasonal Variation in Physico-chemical Parameters of River Ganga During July, 2002 to June, 2004 at the Sites Studied

Parameters	*Sites*	*Monsoon 2002*	*Post-Monsoon 2002*	*Winter 2002–03*	*Spring 2003*	*Summer 2003*	*Monsoon 2003*	*Post-Monsoon 2003*	*Winter 2003–04*	*Spring 2004*	*Summer 2004*
Temperature (°C)	Site I	26.250	22.800	17.800	20.450	34.130	25.600	23.350	17.130	21.000	34170
	Site II	25.800	22.100	17.560	20.650	33.870	26.100	23.450	17.500	20.750	34.030
	Site III	25.700	22.650	17.800	21.000	34.330	26.100	23.650	17.470	20.950	34.300
Turbidity (JTU)	Site I	59.250	35.950	4.500	1.850	5.830	52.150	34.550	6.900	2.050	5.000
	Site II	66.20	41.300	10.140	7.050	12.600	56.400	40.400	12.600	5.750	12.630
	Site III	69.45	42.700	19.930	9.950	19.860	63.750	43.200	13.600	8.350	18.260
Transparency (cm)	Site I	25.500	27.450	48.160	47.500	29.670	24.600	28.100	45.900	45.750	29.400
	Site II	25.50	26.600	47.340	44.500	30.300	23.600	26.500	46.400	44.400	29.300
	Site III	24.75	26.000	45.000	43.100	28.730	23.100	26.350	45.400	44.100	29.000
pH	Site I	7.550	7.650	8.100	7.800	7.700	7.550	7.800	8.230	7.900	7.700
	Site II	7.60	7.700	7.970	8.000	7.830	7.550	7.890	8.230	8.000	7.700
	Site III	7.45	7.700	9.730	9.450	8.630	7.600	7.900	9.330	9.150	8.630
Dissolved Oxygen (mg/l)	Site I	5.550	5.250	7.870	7.750	6.470	5.750	5.900	8.170	7.600	6.930
	Site II	4.85	5.250	7.700	6.660	5.700	5.200	5.500	7.500	6.650	6.530
	Site III	4.35	4.300	5.970	5.650	5.260	3.950	4.050	6.600	5.850	4.830
Alkalinity (mg/l)	Site I	104.000	113.500	218.670	289.000	183.330	75.000	99.000	210.670	221.000	221.330
	Site II	102.00	114.000	210.670	270.000	179.330	75.000	98.000	202.000	221.000	190.000
	Site III	165.00	172.000	280.000	318.000	214.670	86.000	115.000	214.000	275.000	206.000
Hardness (mg/l)	Site I	85.650	91.300	105.070	117.200	151.500	97.200	94.850	99.430	116.500	141.870
	Site II	81.70	90.350	105.270	120.750	157.900	97.450	97.150	101.930	119.750	143.500
	Site III	87.85	94.300	111.270	129.150	165.000	98.900	100.300	115.2300	132.850	161.600

Contd...

Table 10.1–Contd..

Parameters	*Sites*	*Monsoon 2002*	*Post-Monsoon 2002*	*Winter 2002–03*	*Spring 2003*	*Summer 2003*	*Monsoon 2003*	*Post-Monsoon 2003*	*Winter 2003–04*	*Spring 2004*	*Summer 2004*
Chloride (mg/l)	Site I	8.650	10.400	13.200	13.150	16.500	9.650	10.400	13.300	13.750	15.670
	Site II	9.75	11.350	14.970	14.850	15.500	9.900	11.000	14.970	15.350	16.530
	Site III	21.35	22.900	24.070	24.600	25.700	14.650	18.000	21.300	22.000	24.100
Phosphate (mg/l)	Site I	0.030	0.050	0.087	0.080	0.080	0.040	0.050	0.080	0.090	0.130
	Site II	0.045	0.050	0.116	0.120	0.150	0.055	0.600	0.116	0.115	0.140
	Site III	0.05	0.070	0.130	0.125	0.180	0.075	0.070	0.130	0.125	0.190
Biochemical Oxygen	Site I	2.150	1.750	2.170	2.400	2.830	2.250	1.100	1.670	2.800	3.070
Demand (mg/l)	Site II	2.30	1.950	2.330	2.400	3.170	2.350	1.450	1.970	2.800	3.230
	Site III	5.35	4.100	5.530	5.900	9.430	3.650	2.800	4.670	6.000	9.400
Chemical Oxygen	Site I	10.350	20.800	27.870	29.550	18.070	7.950	15.800	26.400	21.200	12.900
Demand (mg/l)	Site II	11.30	24.350	35.200	35.350	25.200	11.650	23.150	33.700	33.500	18.630
	Site III	45.350	55.250	54.700	64.900	49.430	50.100	50.300	61.000	65.400	55.700

ranged between 81.70 mg/l to 165.0 mg/l. It was minimum during monsoon at Site II and maximum during summer at Site III. Saxena *et al.* (1966), Mishra and Tripathi (2003) also reported similar seasonal trend of hardness in river Ganga.

Chloride content of river varied from 8.65 mg/l to 25.70 mg/l, the minimum at Site I during monsoon and maximum at Site III during summer season. Mishra and Tripathi (2003) reported low chloride during monsoon similar to present observation, but high during winter season. Chloride content was found to be high during summers might be due to decrease in amount of water and also high temperature which increases the decaying process as also reported by Cole, 1979. Low chloride during monsoon might be due to increased amount of water (Jana, 1973). Phosphate content varied from 0.030 ppm to 0.190 ppm, being minimum at Site I and Site II during monsoon and maximum at Site III during summer. Khan *et al.* (1994) reported maximum phosphate content during summer, similar to present observation but minimum during post-monsoon.

Biological Oxygen Demand (BOD) of the river during different seasons ranged between 1.10 mg/l to 9.43 mg/l. It was recorded lowest during post-monsoon at Site I and highest during summer at Site III. High amount of BOD at Site III might be due to large amounts of sewage and industrial effluents. According to Hynes (1960) water having BOD values of 3.0 mg/l or more is doubtful quality and that with more than 5 mg/l is bad quality. From this point of view, water in the river studied is rather unsatisfactory at all the sites during summer, but in worst condition at Site III. Chemical Oxygen Demand (COD) in the Ganga ranged between 7.95 mg/l to 65.40 mg/l being maximum at Site III during spring and minimum at Site I during monsoon. The upper permissible limit for COD for drinking water is 20 mg/l (ISI, 1982). In the present observation, COD values were more than 20 mg/l during winter and spring seasons at all the sites studied. BOD and COD have great importance in river water quality assessment since these are directly related organic matter content in the water and quantity of oxygen required for stabilization.

The analysis of river water at various sites revealed that the pollution level increased as from Site I to Site II and Site III. It may be concluded that water of river Ganga at Kanpur should not be used for drinking unless it is treated. pH values indicates the alkaline nature of the river. High values were recorded at the site where industrial effluents were discharged into the river.

References

APHA, 1995. *Standard Methods for Examination of Water and Wastewater*. American Public Health Association, New York.

Cole, G.R., 1979. *A Textbook of Limnology*, 2nd Ed. C.V. Mosley Company, pp. 283.

Hynes, H.B.N., 1960. *The Limnology of Polluted Waters*, Liv. Un. Press, pp. 1992.

ISI, 1982. *Tolerance Limits for Inland Surface Waters Subjected to Pollution*. Indian Standards Institute, New Delhi, pp. 14.

Jana, B.B., 1973. Seasonal periodicity of plankton in a fresh water pond, West Bengal, India. *Int. Rev. Gev. Hydrobiol.*, 58: 127-143.

Kaur, S. and Joshi, B.D., 2003. Seasonal variation in some physico-chemical parameters of river Ganga in and around Hardwar. *Him. J. Zool.*, 17(1): 45-55.

Khan, A.A., Haque, N., Instisar, A.S. and Narayanan, K., 1994. A comprehensive study on water quality parameters in the river Ganga between Narora and Kannauj, U.P. I. Physico-chemical characteristics. *J. Freshwater Biol.*, 6(4): 295-304.

Mishra, B.P. and Tripathi, B.S., 2003. Seasonal variation in physico-chemical characteristics of Ganga water as influenced by sewage discharge. *Indian J. Ecol.*, 30(1): 27-32.

Pahwa, D.V. and Mehotra, S.N., 1966. Observation on the fluctuation in abundance of plankton in relation to certain hydrobiological conditions of river Ganga. *Proc. Nat. Acad. Sci. India*, 368(2): 157-189.

Pandey, P.K. and Pandey, G.N., 1980. Physico-chemical characteristics of river Ganga at Kanpur. *J. Inst. Engr. India*, 60: 27-34.

Saxena, K.L., Chakrabarty, R.N., Khan, A.Q., Chattopadhya, S.N. and Chandra, H., 1966. Pollution studies of the river Gamga near Kanpur. *Environ. Health*, pp. 270-285.

Srivastava, U.S. and Kulshrestha, A.K., 1990. Seasonal variation in certain physico-chemical parameters in Ganga, Yamuna and Tons in Allahabad region (U.P., India). In: *Recent Trends in Limnology*, (Eds.) Agrawal, V.P. and Das P. Society of Biosciences, Muzaffarnagar, India, pp. 351-364.

Tare, V., Yadav, A.V.S. and Bose, P., 2003. Analysis of photosynthetic activity in the most polluted stretch of river Ganga. *Water Research*, 37: 67-77.

Venkateswarlu, V., 1986. Ecological studies on the rivers of Andhra Pradesh with special reference to water quality and pollution. *Proc. Ind. Sci. Acad.*, 96: 495-508.

Verma, S.R., Sharma, P., Tyagi, A., Rani S., Gupta, A.K. and Dalela, R.C., 1984. Pollution and saprobic status of eastern Kalinadi. *Limnologica*, Berlin, 15(1): 69-133.

Chapter 11

Effect of Acetazolamide on Body Weight, Organ Weight and Biochemical Parameters in the Kidney of Albino Rat

M. Maheshwar* and P.K. Singh

Department of Zoology, School of Life Sciences, Khandari Campus, Dr. B.R. Ambedkar University, Agra – 282 002

ABSTRACT

We have studied effect of acetazolamide on the body weight, kidney weight, cholesterol, total protein and alkaline phosphatase in the kidney of albino rats. The drug acetazolamide was given orally at dose of 250 $mgKg^{-1}$ body weight daily for 7, 14 and 21 days. The body weight recorded at start and after treatment days, while organ weight observed after decapitation of each rat. Total protein, cholesterol and alkaline phosphatase were assessed colorimetrically. The body weight, kidney weight, cholesterol, protein and alkaline phosphatase were increased significantly. In conclusion these results consistence with primary action of acetazolamide on the kidney to promote the cholesterol, protein and alkaline phosphatase due to metabolic side effects and ions disturbance in the proximal tubule. Therefore, the morphological parameters *i.e.* body, kidney weight and biochemical parameters such as cholesterol, protein and alkaline phosphatase in kidney after acetazolamide treatment would determine the significance and relationship to drug compound and clinical evidences of their metabolic side effects.

Keywords: Acetazolamide, Kidney, Protein, CHo, Alkaline phosphatase.

* Corresponding Author: C/o M/s Gupta Fertilizers Chharra, Aligarh – 202 130, India E-mail: maheshwari_m2004@yahoo.com

Introduction

Diuretics differ considerably in chemical derivation efficacy, site and mechanism of action, namely their pharmacology and pharmacodynamics. The diuretic in our therapeutic armament arum have predictable effects, which is based on their nephron site of action and may exert their side effects by interfering with cell metabolism (Breater, 2000). The various side effects of diuretic drugs depends upon their bioavailability, distribution, route and extent of elimination (Keller *et al.*, 1981). The most commonly countered clinical condition in which resistant to diuretic have many side effect on the target organs may be due to the mechanism of pharmacology of different diuretics (Rane *et al.*, 1978). Acetazolamide is prototype class of agent have limited usefulness as diuretics but have had a major role in the development of fundamental concepts of renal physiology and pharmacology (Kratzung, 1995). Inhibition of carbonic anhydrase activity, profoundly depress by carbonate reabsorption in the proximal tubule. At its maximal reabsorptive capacity of superficial proximal tubule is inhibited by acetazolamide. However, some bicarbonate can still be absorbed at other nephron site by carbonic anhydrase independent mechanism. The overall effect of maximal acetazolamide administration is about 45 per cent inhibition of whole kidney bicarbonate reabsorption. Nevertheless, carbonic anhydrase inhibition causes significant bicarbonate losses, resulting in hyperchloremic metabolic acidosis. Because of the toxicity of this acidosis and the fact that HCO_3^- depletion tube segments in the nephron, the diuretic effectiveness of acetazolamide decrease significantly with use over several days (Chapron *et al.*, 1985). The kidney is a main target organ for all diuretics drug because their interrelationship between drug and filtration the fluid through blood.

Although, it is originally assumed that the secretory mechanism with kidney should operates under same general principle that have been applied to hepatic clearance and dependence of its renal elimination and adverse side effects upon several physiological and biochemical features (Taft *et al.*, 1996). Keeping this idea in background of the present work taken to estimate the body weight, kidney weight, cholesterol, protein and alkaline phosphatase level in the kidney of albino rat.

Materials and Methods

Experimental Animals

Thirty male wistar albino rats of equal weight (175±10 gm) were obtained from small animal breeding center JALMA Research Institute, Agra. The animals were housed in Department animal house with maintained all good laboratory condition. The animals were kept in polypropylene cages. They were fed with Goldmohar brand rodents feed, manufactured by Lipton India Limited. Food and water were provided *ad libitum*.

Selection of Dose

In the present study, the drug acetazolamide (Trade name, diamox manufactured by Wyeth Lederle, Ltd, Valsad, Gujrat) has been used. Each dose of drug was given to the experimental group of rats at 250 mg/Kg body weight (Wallace *et al.*, 1997) daily for 7, 14 and 21 days.

Experimental Design

Body weight of animals were recorded initially before start the dose of drug and then they were divided into six groups of five rats each. From which three groups were treated as experimental groups and given acetazolamide orally through cathedral tube; while rest of three groups as control groups and were given only normal saline for 7, 14 and 21 days respectively. Animals were fasted overnight separately on 7th, 14th and 21st day. On the next day each experimental protocol, after recording the body weight, and animal were sacrificed by decapitation. The kidney were dissected out, blotted of blood and their weight was recorded, after that kidney were kept separately in pettrydishes and rinsed in phosphate buffered saline (pH 7.4).

Biochemical Estimations

The kidney cholesterol, was determined by a method of Zlatkis *et al.* (1955). The kidney protein was determined by method of Lowry *et al.* (1951). Alkaline phosphatase was estimated by a modified method of Kind and King (1954).

Statistical Analysis

All the data were statistically evaluated and the significance calculated using student 't' test to compare groups. All the results were expressed as mean and ±S.Em. (Fisher and Yates, 1963).

Results and Discussion

The body weight increase significantly ($p < 0.05$) after 7 days; while highly significant ($p < 0.01$) after 14 and 21 days of drug treatment. The kidney weight increase highly significant ($p < 0.01$) after 7 days and very highly significant ($p < 0.001$) after 14 and 21 days drug treatment. The kidney protein contents, cholesterol and alkaline phosphatase in kidney increase very highly significant ($p < 0.001$) after 7, 14 and 21 days of drug treatment (Table 11.1).

Table 11.1: Effect of Acetazolamide on the Body Weight, Kidney, Weight-biochemical Parameters in Kidney of Albino Rats

Sl.No.	*Parameters*	*Treatment Time (in Days)*	*No. of Rats*	*Control Group Mean±S.Em.*	*Treated Group Mean±S.Em.*
1.	Body weight (in gm)	7	5	178±3.5	192±2.9*
		14	5	176±3.6	192±2.5**
		21	5	183±4.6	203±2.5**
2.	Kidney weight (in gm)	7	5	1.85±0.01	1.97±0.02**
		14	5	1.87±0.02	2.21±0.01***
		21	5	1.82±0.04	2.30±0.01***
3.	Cholesterol (in mg/gm)	7	5	0.03±0.003	0.066±0.004***
		14	5	0.028±0.002	0.093±0.006***
		21	5	0.03±0.003	0.128±0.005***
4.	Total protein (in µg/ml)	7	5	483.33±4.9	844.44±47.79***
		14	5	474.44±8.7	822.21±47.78***
		21	5	517.77±10.3	999.99±47.65***
5.	Alkaline phosphatase (in K.A. units)	7	5	0.58±0.07	1.12±0.05***
		14	5	0.48±0.08	1.31±0.06***
		21	5	0.71±0.05	3.16±0.07***

*: Significant ($p < 0.05$); S.Em.: Standard error of mean; **: Highly significant ($p < 0.01$);
***: Very highly significant ($p < 0.001$).

The diuretics also cause many adverse side effects, and linked with sodium and potassium depletion (Ames, 1986). In the present study, the body weight and kidney weight increase may be due to the adverse side effects of acetazolamide and can also be correlated with an increase the cholesterol contents in kidney. As such Steven *et al.* (1988) have been reported an increase body weight and organ weight due to the side effect of acetazolamide; while Singh and Singh (2002) in rats due to spironolactone and Fanestile *et al.* (1997) and Costanzo (1998) in the kidney of rats after thiazide administration.

Alteration in lipid, play an important role in the development and progression of renal disease and induced specific changes in the metabolism of cholesterol. Therefore, in albino rats, total cholesterol in kidney increase due to adverse side effect of acetazolamide. Similar findings have been supported by Arthur (2000), Monhiyan *et al.* (1990) and Miner *et al.* (1983) in human and rats due to adverse side effects of diuretic drug compounds which is stimulated the lipolysis due to the intracellular cyclic AMP either by inhibition or by phosphodiester esterase and which is might catacholamine level. In the present study, the total protein contents in the kidney increase, due to adverse side effect of acetazolamide. Like that, Taft and Sweeney (1995) have been reported an increase protein in serum and tissue due to cellular hypertrophy. Tucker (1981), Wilkinsson and Schand (1975) in rats due to intrinsic organ clearance which is linked with ion disturbance. An increase the alkaline phosphatase may indicates the serious effect of toxic compound and diseases. In the present investigation alkaline phosphatase in kidney increase due to adverse side effect of acetazolamide. Similar observations have been reported by Breater (2000) and Arthur (2000) due to the disturbance in metabolic activity and acid-base disorders by the activity of 17-α-hydroxylase in tissue and carbonic anhydrase inhibitor character of acetazolamide and can also be produced metabolic alkalosis.

In conclusion the data were indicated that the acetazolamide given in large dose for short and long periods of time that have profound and specific interesting morphological and biochemical parameters of kidney. Some of the action described the significance in relation to the effect of drug seen clinically. However, studies now needed which smaller doses of drug are given for longer period of time in an attempt to approximately more closely to the clinical use of acetazolamide. Additional study are now needed to establish the identify of tissue metabolites *in vivo* and to determine the sites of production of various known metabolites of acetazolamide. In addition the distribution of acetazolamide should be examined. Such studies should contribute particularly with respects to the role of metabolites in therapeutic action and side effects of the drug.

References

Ames, R.P., 1986. The effect of antihypertensive drug on serum, lipid and lipoproteins. *Drug*, 32: 260–278.

Arthur, G., 2000. Diuretic complications. *Am. J. Med. Sci.*, 319(1): 10–24.

Breater, D.C., 2000. Pharmacology of diuretics. *Am. J. Med. Sci.*, 319(1): 38–59.

Chapron, D.J., Sweeny, K.R., Feig, P.V. and Kramer P.A., 1985. Influence of advanced age on the disposition of acetazolamide. *Br. J. Clin. Patho.*, 19: 367–371.

Costanzo, L.S., 1998. Mechanism of action of thiazide diuretics. *Somin Nephrol*, 18: 234–241.

Fenestill, D.D., Vaughn, D.A., Ma, L., Hyde, R.H. and Printz, M.P., 1997. Thiazide diuretics normalize urinary calcium spontaneous hypertensive male rats. *Kidney Int.*, 51(4): 1018–1020.

Fisher, R.A. and Yates, F., 1963. *Statistical Tables for Biological Agriculture and Medical Research*, 6th Edn. Hing Yip Printing Co., Hong Kong, pp. 1–146.

Keller, E., Hoppe, G. and Schollever, P., 1981. Influence of liver, cirrhosis and anelstage renal disease on pharmacolokinetics and pharmacodynamics of furasemide. *Eor. J. Lin. Pharmacol.*, 20: 27–30.

Kind, P.R.N. and King, E.J., 1954. Colorimetric method for determination of alkaline and acid phosphatase in tissue and serum. *J. Clin. Patho.*, 7: 322–329.

Kratzung, B.G., 1995. *Principles of Pharmacology: Basic Concepts and Clinical Applications*. Chapman and Hall Medical Publishing Company, New York, pp. 657–672.

Lowry, D.H., Rosebrough, N.S., Farr, A.L. and Randal, R.J., 1951. Protein measurement with folin phenol reagent. *J. Biol. Chem.*, 193: 265–275.

Miner, P.B., Sneller, M. and Crawford, S.S., 1983. Spironolactone and conrenon induced change in hepatic Na^{+}, K^{+}, ATPase activity. Surface membrane cholesterol and phospholipid and flurosonea polarization in rats. *Hepatol.*, 3(4): 481–488.

Moynihan, J.B. and Ennis, S., 1990. Acetazolamide insensitive carbonic anhydrase activities in liver and tonic skeletal muscles of adult male rats with streptozotocin induced diabetes mellitus. *Biochem. J.*, 4(2): 616–620.

Rane, A.J.P., Willeneuve, W.J., Stone, A.S., Nies, A.S. and Branch, R.A., 1978. Plasma binding and disposition of furasemide in the nephritic syndrom and in uremia. *J. Pharmacol. Exp. Ther.*, 19: 77–81.

Singh, D.P. and Singh, P.K., 2002. Increase in body weight and testes weight in the albino rat after exposure to spironolactone. *Bionotes*, 4(1): 19.

Steven, C., Zell, M.P. and Philip, H., 1988. Acetazolamide and dexamethason in prevention of acute mountain sickness. *West J. Med.*, 148: 541–545.

Taft, D.C. and Sweeny, K.R., 1995. Influence of protein binding on the elimination of acetazolamide by the isolated perfused rat kidney: Evidence of albumin mediated 'tubular secretion'. *J. Pharmacol. Exp. Ther.*, 26(5): 752–760.

Taft, D.C., Chapron, D.S., Sweeney, K.R. and Fourvier, D.J.F., 1996. Concentration dependent tubular secretion of acetazolamide and its inhibition by salcyclic acid in isolated perfused rat kidney. *Drug Metab. Dispo.*, 27(4): 752–760.

Tucker, G.I., 1981. Measurement of the renal clearance of drug. *Br. J. Clin. Pharmacol.*, 12: 761–770.

Wallace, S.M., Shah, V. and Reielman, S., 1997. GLC analysis of acetazolamide in blood. Plasma and saliva following oral administration to normal subjects. *J. Pharm. Sci.*, 66: 527–530.

Wilkinsson, G.R. and Schand, D.G.A., 1975. A physiological approach to hepatic drug clearance. *Clin. Pharmacol. Ther.*, 18: 377–390.

Zlatkis, A., Zac, B.B. and Bong, G.J.B., 1955. A colorimetric method for determination of cholesterol. *J. Lab. Clin. Med.*, 41: 486– 496.

Chapter 12

Study of Flowers of Various Varieties of *Hibiscus rosa-sinesis* L. as an Anti-Solar

S.B. Patil, M.S. Kondawar, N.S. Naikwade and C.S. Magdum

Appasaheb Birnale College of Pharmacy, South Shivajinagar, Sangli – 416 416

ABSTRACT

This article evaluates UV absorption ability of flowers from various varieties of *Hibiscus rosa-sinesis* L. (Malvaceae) as an anti-solar agent. Extracts were prepared by maceration with a mixture of distilled water and ethanol (1 : 1). The method is performed by UV spectrophotometry in the range of 200 to 340 nm and results of all the extracts shows effective UV absorption in the given range.

Keywords: *UV protective, Concord, Kingkalakaua, Lahina, China rose.*

Introduction

Exposed sun ultraviolet light classify in to 3 types, by its wavelengths: UVA, UVB and UVC. The dimensions of their wavelengths are roughly 400 to 320 nm for UVA, 320 to 290 nm for UVB and 290–200 nm for UVC. Although it may seen backwards, the shorter the wavelength and the lower the number, the greater the energy level of the light and the more damage it can do (Saraf, 2005).

Direct exposure to UVC for a length of time would destroy the skin. Fortunately, UVC is completely absorbed by gases in the atmosphere before it reaches the ground (Saraf, 2005). In any time the longer wavelength of UVB and UVA pass right through the atmosphere. The molecules in sunscreen absorb most UVB and prevent it from reaching the skin just as the molecules of the atmosphere absorb UVC and prevent it from reaching the ground.

Various herbal formulations and chemicals are available to block UV rays and always prevent all types of skin from various types of damages. Our objective is to find out such flowers that are widely used as sunscreen from ancient time. *Hibiscus rosa-sinesis* is a small plant belonging to family Malvaceae, flowers are red coloured, large, showy flowers, cyclic hypogynous, Ebracteates, pedicilate, complete, hermaphrodite, actinomorphic, peutamerous (Sharma, 1985). Flowers are refrigerant, emollient, demulcent and aphrodisiac; also emmenagogue (Kulkarni, 1991).

Materials and Methods

Hibiscus rosa-sinesis L. flowers were freshly obtained from nursery at Sangli, botanical identification was performed at the department of Botany in Padambhushan Dr. Vasantdada Patil Mahavidyalay, Tasgaon. Later the petals were separated and dried by circulating cool air. 100 g powdered petals (4 different species) were extracted with distilled water : ethanol (1 : 1) by maceration. The extracts were evaporated to dryness on steam bath. The general flavonoid identification test was performed on all the four extracts (Khandelwal, 2004).

Test 1 (Shinoda test)

To dry extract, add 5 ml 95 per cent ethanol, few drops concentrated hydrochloric acid and 0.5 g magnesium turnings. Pink colour observed.

Test 2

To small quantity of extract, add lead acetate solution, yellow coloured precipitate is formed.

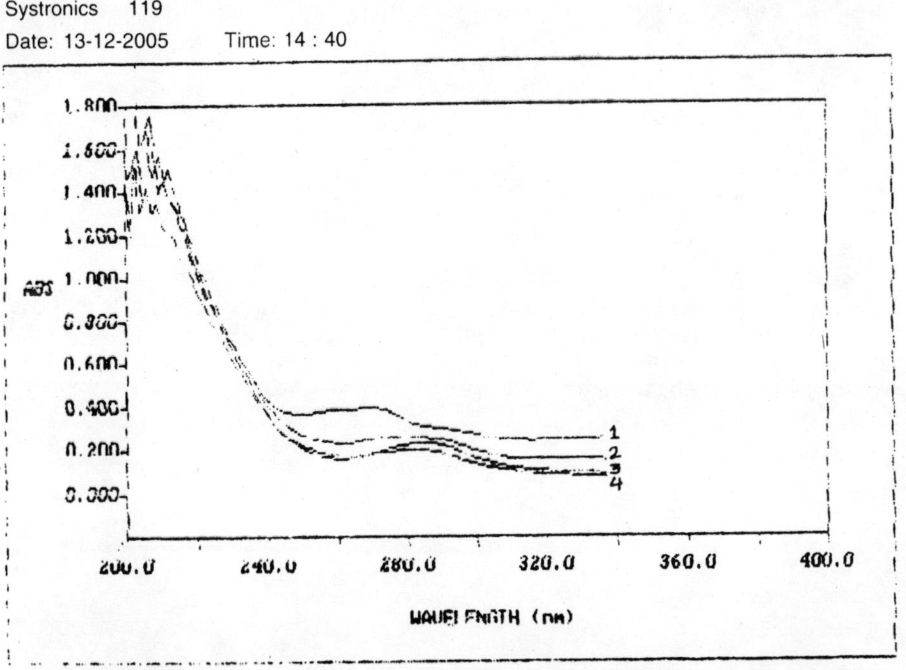

Figure 12.1: Scanning Spectra of Extracts on UV-spectrophotometer
1, 2, 3 and 4 indicated the spectrum obtained from Lahina, Concord, Kingkalakaua and China rose respectively.

Preparation of Sample

Samples were prepared in 8 mg per cent w/v concentration by using distilled water (8 mg/100 ml).

UV absorption spectra for all the four extracts were obtained in the range of 200–340 nm using Systronic UV spectrophotometer 119. Figure 12.1 indicates absorption spectra for all the 4 extracts in the given range.

Results and Discussion

The results obtained showed the ability of extracts to absorb UV radiation and hence proved its UV protection ability. All extracts showed a prominent absorbance at 200–240 nm while good absorbance at a range of 240–310 nm. The little absorbance was noted at the range 310–340 nm. Qualitative investigation indicated the presence of flavonoids in all the extracts and this protection ability may observe due to the presence of flavonoids.

References

Khandelwal, K.R., 2004. *Practical Pharmacognosy*, 12th edn. Nirali Prakashan, Pune, pp. 153.

Kulkarni, K.M., 1991. *Indian Materia Medica*. Popular Prakashan, Mumbai, pp. 630: 1225.

Saraf, S., Saraf Swarnlata, S. and Ashawat, M., 2005. Anti solar activity of flowers extracts of *R. damascena* and *R. erecta* Linn. *Planta indica*, 1(3): 26–27.

Sharma, O.P., 1985. *A Manual of Practical Botany*, 3rd edn. Pragati Praashan, Meerut. 2: 323.

Chapter 13

Studies on the Impact of Lignite Flyash on Greengram in Pond Ash

N. Senthilkumar[1], S. Srinivasan[1] and S. Krishnaprabu[2]
[1]Department of Soil Science and Agricultural Chemistry, [2]Department of Agronomy, Faculty of Agriculture, Annamalai University, Annamalai Nagar – 608 002, Tamil Nadu, India

ABSTRACT

In India alone nearly 80 million tonnes of ash is produced per year. In the thermal power station of the Neyveli Lignite Corporation Ltd., Neyveli ash pond spreads over an area of more than 45 ha and causes serious environmental hazards besides occupying fertile cultivable land. In future, the ash pond area is likely to be increased. Hence there is an immediate need to convert this ash pond area into a cultivable land. In the study, 50 surface pond ash samples and a profile was excavated in the ash pond area and the depth wise samples were collected and analysis physico-chemical properties. Preliminary experiment I (Mini pot culture study) was conducted in the glass house of the Department of Soil Science and Agricultural Chemistry, Annamalai University to study the influence of lignite flyash on yield and yield attributes of greengram (CO_3). The experiment was laid out in split plot design with three replications. The main treatments are (Three levels of LFA) M_1–control, M_2–150 t ha^{-1} and M_3–300 t ha^{-1} and the sub-treatments (three sources of phosphorus) include S_1–100 per cent P as rock phosphate, S_2–100 per cent P as super phosphate and S_3–50 per cent P as rock phosphate + 50 per cent P as super phosphate. Among the levels of LFA application, the treatment receiving 150 t ha^{-1} (M_2) was found to be beneficial in increasing the plant height (48.52 cm), LAI (2.89), number of pods plant^{-1} (12.50), number of seeds pod^{-1} (8.55), 100 grain weight (4.29 g) and seed yield plant^{-1} (4.49 g) of green gram in pond ash.

Keywords: *Pond ash, Lignite flyash, Rock phosphate, Super phosphate, Pressmud, Greengram.*

Introduction

The natural resources constitute majority of the environment. The plentiful nature of the resources till now made us to take it for granted in all respects. But the growing population, and the growing industrialization slowly but steadily has increased the pressures. This has resulted in a dramatic change of natural resources into scarcity from the once plenty situation (Ramachandran, 2003).

In India alone nearly 80 million tonnes of ash is produced per year. In the thermal power station of the Neyveli Lignite Corporation Ltd., Neyveli ash pod spreads over an area of more than 45 ha and causes serious environmental hazards besides occupying fertile cultivable land in future the ash pod area is likely to be increased. Hence there is an immediate need to convert this ash pod area into a cultivable land.

Materials and Methods

In the study, fifty surface pond ash samples and a profile was excavated in the ash pod area and the depth wise samples were collected for studying the physico-chemical properties. Mini pot culture experiment was conducted with a view to optimize the LFA and pressmud added and to select the suitable crop.

Main Treatments	*Subtreatments*
M_1–100 per cent PA + 300 g LFA	S_1–20 g PM/pot
M_2–80 per cent PA + 30 per cent RS + 240 g LFA	S_2–25 g PM/pot
M_3–60 per cent PA + 40 per cent RS + 180 g LFA	S_3–30 g PM/pot
M_4–40 per cent PA + 60 per cent RS + 120 g LFA	S_4–35 g PM/pot
M_5–20 per cent PA + 80 per cent RS + 60 g LFA	S_5–40 g PM/pot

The pots were filled with 6 kg of pond ash collected from ash pond area, Neyveli. The pressmud 90 g pot^{-1} (@ 30 t ha^{-1}), Lignite flyash (LFA) at calculated amounts 0, 450 and 900 g pot^{-1} (0, 150 and 300 t ha^{-1}) and Rock phosphate, superphosphate and Rock phosphate + superphosphate to supply P @ 100 kg P_2O_5 ha^{-1} were incorporated, watered, maintaining moisture at field capacity for 15 days. Then the pot were mixed well. At the time of sowing nitrogen and potassium fertilizers were applied @ 200 and 100 kg N and K_2O ha^{-1} respectively, then 6 bold seeds of green gram cv. CO-$_3$ were sown and watered. The pots were watered at times of need. The germination percentage was recorded on 10 DAS. Leaving two healthy plants the others were removed and incorporated in the respective pots.

Treatment details

Main treatments (Three levels of LFA)	*Sub Treatment (Three sources of phosphorus)*
M_1–Control	S_1–100 per cent P as Rock phosphate
M_2–150 t ha^{-1}	S_2–100 per cent P as super phosphate
M_3–300 t ha^{-1}	S_3–50 per cent P as Rock phosphate + 50 per cent P as super phosphate

The experiment was conducted in a split plot design replicated thrice. The plants in each pot were used for recording the following observations *viz.*, number of seeds pot^{-1} and seed yield plant^{-1}.

Results and Discussion

The pond ash surface samples (0–30 cm) were collected from 50 different places in the ash pod area of Thermal Power Station 1, Neyveli. These samples were analysed for pH and EC. Mini pot culture experiment was conducted with a view to optimize the LF A and pressmud added and to select the suitable crop. Rice and greengram were selected as test crops.

The crops were grown in plastic pots and maintained upto maturity. The growth parameters were recorded periodically. The post harvest pond ash sample was analysed for bulk density, particle density, pore space and water holding capacity. The methods of analysis adopted for determination of the physico-chemical properties, germination percentage and plant height are given in the Tables 13.1 and 13.2.

Table 13.1: Physico-chemical Properties of Lignite Flyash

Colour	–	Grey (10 YR5/1) (Munsell colour notation)
Bulk density (Mg m^{-3})	–	1.22
Particle density (Mg m^{-3})	–	2.14
Particle size (per cent)		
2.0 to 0.5 mm	–	0.50
0.5 to 0.25 mm	–	3.50
< 0.25 mm	–	96.00
pH (1 : 2.5)	–	9.80
EC (dS m^{-1})	–	1.10
Organic carbon (per cent)	–	0.12
Nitrogen (per cent)	–	Nil
Phosphorus (per cent)	–	Nil
CaO (per cent)	–	7.77
MgO (per cent)	–	5.25
SO_4 (per cent)	–	3.00
Na_2O (per cent)	–	3.05
K_2O (per cent)	–	2.99
Fe_2O_3 (per cent)	–	5.44
Al_2O_3 (per cent)	–	15.10
Zn (mg kg^{-1})	–	110.00
Cu (mg kg^{-1})	–	120.00
Mn (mg kg^{-1})	–	200.00
B (mg kg^{-1})	–	85.00
SiO_2 (per cent)	–	41.20

Preliminary study with greengram and paddy in pond ash pot mixtures (2000 g PA + 300 g LFA + 25 g pressmud), the result revealed that bulk density (1.39 Mg m^{-3}), particle density (2.06 Mg m^{-3}), pore space 32.5 per cent, water holding capacity 40 per cent, greengram germination percentage (100

per cent), plant height (25 cm), and paddy crop germination percentage (30 per cent) and plant height (41 cm).

Table 13.2: Physio-chemical Properties of Pond Ash

Colour	–	Light brownish grey (10 YR 6/2) (Munsell colour notation)
Particle size (per cent)		
> 2.0 mm	–	0.53
2.0 to 0.2 mm	–	38.91
0.2 to 0.02 mm	–	60.56
pH (1 : 2.5)	–	3.36
EC (dS m^{-1})	–	0.41
Bulk density (Mg m^{-3})	–	1.39
Particle density (Mg m^{-3})	–	2.00
Pore space (per cent)	–	30.50
Moisture retentive capacity (per cent)	–	42.00
CEC [cmol (P^+) Kg^{-1}]	–	7.00
Exchangeable cations [cmol (p^+) Kg^{-1}]		
C^{2+}	–	4.50
Mg^{2+}	–	2.02
Na^+	–	0.22
K^+	–	0.16
Organic carbon (per cent)	–	1.65
CaO (per cent)	–	1.50
MgO (per cent)	–	0.55
Fe_2O_3 (per cent)	–	0.81
Al_2O_3 (per cent)	–	2.19
SiO_2 (per cent)	–	91.63
Available nitrogen (kg ha^{-1})	–	126.00
Available phosphorus	–	Nil
Available potassium (kg ha^{-1})	–	60.00
Available sulphur (kg ha^{-1})	–	250.00
Available zinc (mg kg^{-1})	–	14.74
Available copper (mg kg^{-1})	–	1.18
Available manganese (mg kg^{-1})	–	0.18
Available iron (mg kg^{-1})	–	772.02
Available boron	–	Traces.

Among the levels of LFA, M_2 (150 t ha^{-1}) recorded the beneficial in increasing the plant height (48.52 cm), LAI (40 DAS) 2.89, number of pods $plant^{-1}$ (12.50), number of seeds pod^{-1} (8.55), 100 grain weight (4.29 g) and seed yield $plant^{-1}$ 4.49 g (27 per cent increases over control) followed by M_3 (300 t ha^{-1}) 4.05 g, while the control (M_1) recorded the least (3.52 g).

Table 13.3(a): Effect of LFA and Sources of Phosphorus on Plant Height, LAI (40 DAS) and Number of Pods Plant^{-1} of Greengram at Harvest in Pond Ash

Main Treatments	*Plant Height (cm)*			*Mean*	*Leaf Area Index*			*Mean*	*Number of Pods Plant^{-1}*			*Mean*
	Sub Treatments				*Sub Treatments*				*Sub Treatments*			
	S_1	S_2	S_3		S_1	S_2	S_3		S_1	S_2	S_3	
M_1	38.66	39.33	42.13	40.04	2.55	2.58	2.66	2.59	8.83	9.66	9.83	9.44
M_2	47.73	48.53	49.30	48.52	2.86	2.90	2.93	2.89	12.00	12.16	13.33	12.50
M_3	43.66	46.66	44.80	45.05	2.67	2.78	2.72	2.72	10.66	11.50	11.16	11.11
Mean	43.35	44.84	45.41	–	2.69	2.75	2.79	–	10.50	11.11	11.44	–
	$S.E_D$		CD (p = 0.05)		$S.E_D$		CD (p = 0.05)		$S.E_D$		CD (p = 0.05)	
M	0.20		0.41		0.005		0.01		0.18		0.36	
S	0.17		0.35		0.015		0.03		0.15		0.30	
M × S	0.28		0.57		0.021		0.04		0.34		0.68	
S × M	0.21		0.43		0.026		0.05		0.35		0.70	

Table 13.3(b): Effect of LFA and Sources of Phosphorus on 100 Grain Weight, Number of Seeds Pot^{-1} and Yield Plant^{-1} of Greengram at Harvest in Pond Ash

Main Treatments	*100 Grain Weight (g)*			*Mean*	*Number of Seeds Pot^{-1}*			*Mean*	*Seed Yield Plant^{-1} (g)*			*Mean*
	Sub Treatments				*Sub Treatments*				*Sub Treatments*			
	S_1	S_2	S_3		S_1	S_2	S_3		S_1	S_2	S_3	
M_1	3.92	3.94	3.96	3.94	5.50	5,83	6.50	5.94	3.14	3.62	3.80	3.52
M_2	4.27	4.29	4.31	4.29	8.16	8.66	8.83	8.55	4.34	4.42	4.72	4.49
M_3	4.16	4.21	4.18	4.18	6.83	7.50	7.16	7.16	3.95	4.18	4.02	4.05
Mean	4.12	4.14	4.16	–	6.83	7.33	7.50	–	3.81	4.07	4.18	–
	$S.E_D$		CD (p = 0.05)		$S.E_D$		CD (p = 0.05)		$S.E_D$		CD (p = 0.05)	
M	0.004		0.008		0.07		0.14		0.03		0.06	
S	0.008		0.010		0.07		0.15		0.03		0.07	
M × S	0.013		0.020		0.09		0.18		0.06		0.12	
S × M	0.015		0.030		0.08		0.16		0.06		0.13	

The combined application of P sources S_3 (50 per cent P as rock phosphate + 50 per cent P as superphosphate) recorded the highest plant height (45.41 cm), LAI (40 DAS) 2, 79, number of pods plant^{-1} (11.44), number of seeds pod^{-1} (7.50), 100 grain weight (4.16.g) and seed yield plant^{-1} 4.18 (9 per cent increase over control) followed by S_2 (100 per cent P as superphosphate) followed by S_1 which (100 per cent P as rock phosphate) recorded the least (3.81 g). The interaction effects between main at same/different sub treatments were also significant in influencing the yield and yield attributes of greengram. The main treatment M_2 performed best at same/different sub treatments followed by M_3 and M_1. The beneficial effect of LFA in improving the yield of crops was well established by many workers Arivazhagan (1989) in sorghum, Raghupathy (1990) in rice and maize, Naveen Saviour (1997) in ragi and Christy Nirmala Mary (1999) in rice.

Conclusion

From the results of the pot culture experiment, it was evident that the combined application of LFA @ 150 t ha^{-1} along with P through 50 per cent P as rock phosphate + 50 per cent P as super phosphate recorded the highest plant height, LAI, number of pods $plant^{-1}$, number of seeds pod^{-1}, 100 grain weight and yield $plant^{-1}$. The present study established the potentiality of using lignite flyash in reclaiming phosphorus deficient acidic pond ash from the ash pond area of Thermal Power Station 1, Neyveli Lignite Corporation Ltd., Neyveli.

References

Arivazhagan, K., 1989. Field evaluation of lignite fly ash on yield and nutrient content of sorghum cv. CO_{26} in latertic soil. *M.Sc. (Ag.) Thesis*, Annamalai University.

Christy Nirmala Mary, P., 1999. Effect of integrated plant nutrient supply system and the nutrition, growth and yield of rice cv. ADT-38 in open cast mine spoil. *Ph.D. Thesis*, Annamalai University, Annamalainagar.

Naveen Saviour, M., 1997. Studies on the forms of sulphur and the relative effect of gypsum and lignite fly ash as sources of sulphur on ragi, cv. CO_2 in Annamalainagar farm soil. *M.Sc.(Ag.) Thesis*, Annamalai University, Annamalainagar.

Raghupathy, B., 1990. Effect of lignite fly ash (as source of silica and other cationic nutrients) and phosphorus on rice, maize and sugarcane in a lateritic soil. *Ph.D. Thesis*, Annamalai University, Annamalainagar.

Ramachandran, S., 2003. Environmental challenges: Problems and prospects. *Kisan World*, 30(5): 17–18.

Chapter 14

Study on Mycotoxigenic Fungi and Mycotoxins on Seeds of Mustard (*Brassica nigra* Koch)

S.K. Ghosh[1] and C. Sengupta[2]

[1]Myco-pathology Lab, Department of Botany, Ramakrishna Mission Vivekananda Centenary College, Rahara, Kolkata – 700 118, West Bengal, India

[2]Professor, Department of Botany, University of Kalyani, Kalyani, Nadia, W.B., India

ABSTRACT

Seven districts of southern part of West Bengal, India were surveyed for mycotoxic fungi infested seeds of Mustard (*Brassica nigra* Koch). These districts were Birbhum, Burdwan, Nadia, Hoogly, Howrah, 24-Parganas (N) and 24-Parganas (S). Maximum percentage (50 per cent) of infested samples of seeds was obtained from Hoogly and Howrah. Minimum percentage (32.0 per cent) of fungal infested seeds was found in Birbhum. From all samples of seeds collected from different districts, ten (10) fungi were isolated. Out of them, four (4) fungi were mycotoxin producing. These four fungi were *Aspergillus niger*, *A. flavus*, *A. nidulans* and *Penicillium citrinum*. Various strains of these four fungi were also isolated and screened for mycotoxins. Aflatoxin-B_1 was detected from three (3) strains of *A. niger* and four strains of *A. flavus*. Aflatoxin-B_2 was screened from six strains of *A. flavus* and seven strains of *A. nidulans*. But citrinin was only detected from *Penicillium citrinum* (3 strains).

Keywords: *Mycotoxin, Aflatoxin, Citrinin, Mustard seed.*

Introduction

Mycotoxins are secondary metabolites of some fungi. Various mycotoxin producing fungi were reported from several dry fruits and spices (Giridhar and Reddy, 2001 and Bilgrami, 1985) in India. Mycotoxins are hazardous to higher animals including human being causing various ailments when taken with contaminated food (Casteel and Braun, 1992; and Betina, 1989). But there is little or no report on the mycoloxic mycoflora on seeds of mustard, although mustard oil is widely used in cooking purpose.

The aim of the present investigation is to detect mycotoxin producing fungi and mycotoxins on the seeds of mustard.

Materials and Methods

Seed samples of mustard (*Brassica nigra* Koch) were collected from farmers house after harvesting of this crop in sterilized polythene bags and brought to the laboratory. For isolation of seed mycoflora two standard methods *viz.*, Standard Blotter Method (SBM) and Agar Plate Method (APM) (Muskett, 1948; De temp, 1953 and Anonymous, 1985) were used. To isolate internal seed borne fungi seeds were surface sterilized by 0.1 per cent mercuric chloride solution for three minutes followed by three times washing with sterilized distilled water. Sterilized and unsterilized seeds were separately placed on separate sterilized Petri-plates containing sterilized agar medium (APM method) or containing sterilized moist blotting paper (SBM). These Petri-plates were incubated for 5 days at 20±2°C.

The fungi grown on the seeds of mustard were isolated and examined under microsope and were identified with the help of standard fungal monographs (Domesch, 1980 and Burnett and Hunter, 1972).

Percentage of frequency or incidence of individual fungus was calculated by the following formula:

$$\text{Percentage of Frequency} = \frac{\text{No. of Seeds Containing Particular Fungus}}{\text{Total Number of Seeds}} \times 100$$

The isolated fungal strains were screened for their mycotoxin producing potential by the method as suggested by Rao *et al.* (1985) and Scott *et al.* (1970).

Results and Discussion

Table 14.1 indicated that out of seven districts of West Bengal percentage of fungal infested seed samples (50 per cent) was highest in both Hoogly and Howrah and lowest percentage of fungal infested seed sample (32.0 per cent) was in Birbhum. It may be due to favourable environmental condition of both Hoogly and Howrah.

It was evident from the Table 14.2 that ten fungi belonging to seven genera were isolated from unsterilized seeds of mustard. It also showed that APM is better than SBM for isolating seed mycoflora. Similarly Ghosh and Das (1999) reported that APM method was most efficient. It was interesting to note that *Alternaria alternata, A. brassicicola, Aspergillus niger* and *A. flavus* were both externally and internally seed borne. The percentage of frequency of *A. brassicicola* was highest (18 per cent and) followed by *A. alternata* (15 per cent), *Aspergillus niger* (12 per cent) and *A. flavus* (10.1 per cent) in USS/ APM. Both *A. alternata* and *A. brassicicola* causes leaf spot and leaf blight of mustard and also causes serious loss of yield of mustard (Anonymous 1987, and Rangaswami, 1984). Percentage of frequency of *Penicillium citrinum* and *A. nidulans* were five and three respectively in USS/APM. It was revealed

(Table 14.3) that out of ten fungi, four fungi secreted mycotoxins. Aflatoxin-B_1 was detected from three (3) strains of *Aspergillus niger* and four (4) strains of *A. flavus* on seeds of mustard while Aflatoxin-B_2 was screened from six (6) strains of *A. flavus* and seven (7) strains of *A. nidulans.* Citrinin mycotoxin was detected from *Penicillium citrinum.* These observations were at par with the investigation of Giridhar and Reddy (2001) on raison and Giridhar and Reddy (1997) on mustard.

Table 14.1: Area Surveyed and Percentage of Fungal Infested Samples

Area	*Number of Samples Collected*	*Number of Infested Sample*	*Percentage of Infested Sample*
Birbhum	25.0	08.0	32.0
Burdwan	20.0	08.0	40.0
Hoogly	20.0	10.0	50.0
Nadia	15.0	05.0	33.3
24-Parganas (N)	25.0	12.0	48.0
24-Parganas (S)	15.0	07.0	46.6
Howrah	10.0	05.0	50.0

*: Single sample = 200 seeds.

Table 14.2: Fungi Isolated from Mustard Seeds and their Percentage of Frequency

Name of the Fungus	*Percentage Frequency*			
	USS		*SSS*	
	APM	*SBM*	*APM*	*SBM*
Alternaria atternata	15.0	8.0	6.5	4.7
Alternaria brassicicola	18.0	5.0	7.1	1.0
Aspergillus niger	12.0	5.0	8.0	1.0
Aspergillus flavus	10.1	2.1	4.0	0.0
Aspergillus nidulans	03.0	2.0	0.0	0.0
Curvularia sp.	10.0	3.1	0.0	0.0
Mucor sp.	07.5	4.0	0.0	0.0
Penicillium citrinum	05.0	2.0	0.0	0.0
Rhizopus sp.	02.0	1.0	0.0	0.0
Helminthosporium sp.	04.0	1.0	0.0	0.0

USS: Unsterilized seeds; SSS: Surface sterilized seeds.

Earlier several authors (Bilgrami, 1985 and Wyllie *et al.*, 1978) natural contamination and storage spoilage of starchy, oily and proteinaceous seeds by mycotoxins. On the other hand, it has been established that the region of the world with highest incidence of liver cancer correspond to areas where climatic factors favour aflatoxin contamination of food commodities (Casteel and Braun 1992). Therefore, it may be advocated that unscreened mustard oils or seeds have to be avoided.

Table 14.3: Mycotoxic Fungi and Mycotoxins on Mustard

Name of Mycotoxic Fungi	Number of Strains Screened	Number of Toxigenic Strains	Mycotoxins
Aspergillus niger	10.0	3.0	Aflatoxin-B_1
Aspergillus flavus	15.0	4.0	Aflatoxin-B_1
		6.0	Aflatoxin-B_2
Aspergillus nidulans	12.0	7.0	Aflatoxin-B_2
Penicillium citrinum	06.0	3.0	Citrinin

Acknowledgement

First author is thankful to UGC, Delhi for providing financial assistance.

References

Anonymous, 1985. International rules for seed testing Annexes. *Seed Sci. and Technol.*, 13: 356–513.

Anonymous, 1887. Studies on seed pathology of certain vegetables. Final Report ICAR Research Scheme. Dept. of Botany, AMU, Aligarh.

Betina, V., 1989. *Mycotoxins Chemical, Biological and Environmental Aspects.* Elsevier, Amsterdam, Oxford.

Bilgrammi, K.S., 1985. Mycoflora associated with dry fruits and spices. In: *Mycotoxins in Dry Fruits and Spices,* (Ed.) K.S. Bilgrami. Allied Press, Bhagalpur, pp. 16–20.

Burnett, H.L. and Hunter, B.B., 1972. *Illustrated Genera of Imperfect Fungi.* Burgess Publishing Co., USA.

Casteel, S.W. and Braun, W.F., 1992. Mycotoxicoses. In: *Encyclopedia of Microbiology,* (Ed.) J. Lederberg. Academic Press, 3: 217–227.

De Temp, J., 1953. The Blotter method of seed health testing. In: *Proc. Int. Seed Test Ass.*, 28: 133–135.

Domesh, K., Gams, W. and Adderson, T.H., 1980. *Compendium of Soil Fungi.* Academic Press, New York.

Ghosh, S.K. and Das, N., 1999. Effect of fungicides on seed mycoflora and seed germination of mustard and cauliflower *J. Mycopathol. Res.*, 37(1): 37–39.

Giridhar, P. and Reddy, S.M., 1997. Combined activity of temperature and water activity (aw) on biodeteriotion and mycotoxin production by two storage mounds in mustard seeds. *Adv. Plant Sci.*, 10: 125–128.

Giridhar, P. and Reddy, S.M., 2001. Incidence of mycotoxigenic fungi on raisons. *Adv. Plant Sci.*, 14(1): 291–294.

Muskett, A.E., 1948. Technique for the examination of seed for the presence of seed borne fungi. *Trans. Br. Mycol Soc.*, 30: 76–82.

Rangaswami, G., 1984. *Diseases of Crop Plants in India.* Prentice-Hall of India Pvt. Ltd., New Delhi, India.

Rao, G.V., Rao, P.S. Girisham, S and Reddy, S.M., 1985. A novel spray reagent for chromatographic detection of trichothecene toxin. *Curr. Sci.*, 54: 507–509.

Scott, P.M., Lawrence, J.W. and Van Walbecfk, W., 1970. Detection of mycotoxin by thin layer chromatography, application to screening of fungal extracts. *Appl. Environ. Microbiol.*, 19: 839–842.

Wyllie, T.D. and Morehouse, L.G., 1978. (Eds.) Mycotoxic fungi. *Mycotoxins, Mycotoxicoses: An Encyclopedic Handbook. Mycotoxicoses of Man and Plants: Mycotoxin Control and Regulatory Aspect.* Marcel Dekker, Inc., New York and Bassel, 202 p.

Chapter 15

Growth Performance and Mass Multiplication of Antidiabetic Gymnema Species of Sirumalai Hills

M. Poonkothai[1] and R. Kumuthakalavalli[2]

[1]Department of Life Sciences, Avinashilingam Deemed University, Coimbatore – 641 043

[2]Department of Biology, Gandhigram Rural University, Gandhigram, Dindigul – 624 302, Tamil Nadu

ABSTRACT

Sirumalai hills are bestowed with number of antidiabetic plants. The knowledge on the medicinal property of these plants is known by the tribals, villagers, and these plants are used as folklore medicine. Continuous exploitation of the particular plant species namely *Gymnema elegans* and *Gymnema sylvestre* resulted in a decreased population size. Regeneration studies revealed that these two species could be conserved through micropropagation and vegetative propagation. The growth performance of these propagules could be accelerated in plains also, by growing them in red soil enriched with cowdung. *In vitro* conservation could be done by supplementing MS media with auxin and cytokinin in different concentration.

Keywords: *Gymnema species, Antidiabetic, Conservation, Mass multiplication.*

Introduction

Plants with medicinal properties, the gift of Mother Nature to mankind, are in use for centuries, in the traditional system of medicine. The rural folk in Tamil Nadu have immense faith in traditional herbal medicines (Saraswathy, 1998). The indigenous systems of medicine are largely dependent on plant drugs, for treatment of various human ailments, which were gathered and marketed from wild sources (Sharma, 1999). Many plant species have been exploited indiscriminately by human beings and there over exploitation has resulted in the depletion of natural population, which was now facing an increased risk of extinction. So conservation status of each species is the need of hour (Mamgain *et al.*, 1998). Conservation can be done by constructing ethanoherbal gardens or by means of plant tissue culture and vegetative propagation, which has largely been adopted for mass multiplication of selected elite varieties of endangered and threatened species (Manju Sharma *et al.*, 1999).

Materials and Methods

The population of *Gymnema sylvestre* and *Gymnema elegans* at Sirumalai Hills, Dindigul district were studied through periodic field visits and administering by random quadrat (10 m × 10 m) sampling method. Micropropagation and vegetative propagation methods were attempted for mass multiplication.

The uninodal segments of 1–1.5 cm in diameter were collected and washed under tap water and then washed twice with distilled water. They were sterilized with 50 per cent (v/v) ethanol for two minutes followed by 0.1 per cent (w/v) freshly prepared mercuric chloride for two minutes and then washed five times with sterile distilled water. Murashige and Skoog (1962) medium with 3 per cent sucrose autoclaved at 121°C, 15 lbs for 15 minutes after adjusting the pH to 5.8. The uninodal segments were placed on MS medium supplemented with different concentrations of 2, 4-Dichlorophenoxy acetic acid (0.5, 1.0, 1.5, 2.0 mg/l) and 6-Benzyl aminopurine (2 mg/l) for callus induction. The tubes were maintained under 16 h photoperiod at 25±2°C. The well-developed calli were transferred to the MS medium fortified with BAP (1, 2, 3 mg/l) and kinetin (1 mg/l) for multiple shoot bud differentiation. The elongated shoots were transferred to MS medium containing IBA (0.5, 1.0, 1.5, 2.0 mg/l) for rooting. Regenerating plants having well developed root system were initially maintained in plastic cups containing in sterile soil and sand (2 : 1) for hardening. Each treatment consisted of 15 replicates and the experiments were repeated in triplicates.

Using stem cuttage of *Gymnema sylvestre* and *Gymnema elegans* vegetative propagation technique was carried out. These cuttings were planted in poly bags containing red soil as base supplemented with cow dung, forest humus and coir pith in two different ratios (2 : 1, 1 : 1). The cuttings were dipped in 0.1 M IAA and IBA for an hour before planting. This study was carried out for 45 days and the performance was analyzed in terms of number of buds, number of leaves, size and length of leaf, and number of roots. Phytochemistry of the leaves of wild and cultivated plants was carried out through thin layer chromatography.

Results and Discussion

Simpsons diversity index revealed that the species of Gymnema namely *Gymnema sylvestre* and *Gymnema elegans* were about 0.01 in a quadrat and this showed that these species are decreasing in a rapid manner and hence calls for immediate conservation assessment and strategies (Bajpai, 1997).

As per earlier reports, it was observed that seed viability of *Gymnema sylvestre* and *Gymnema elegans* was found to be very low and hence mass multiplication through micropropagation (tissue culture method) would augment the population of these species. While analyzing the performance of callus culture callus initiation was observed on uninodal explants, within 15–23 days of culture. The calli obtained were semi translucent, with patches of creamish white color among the different

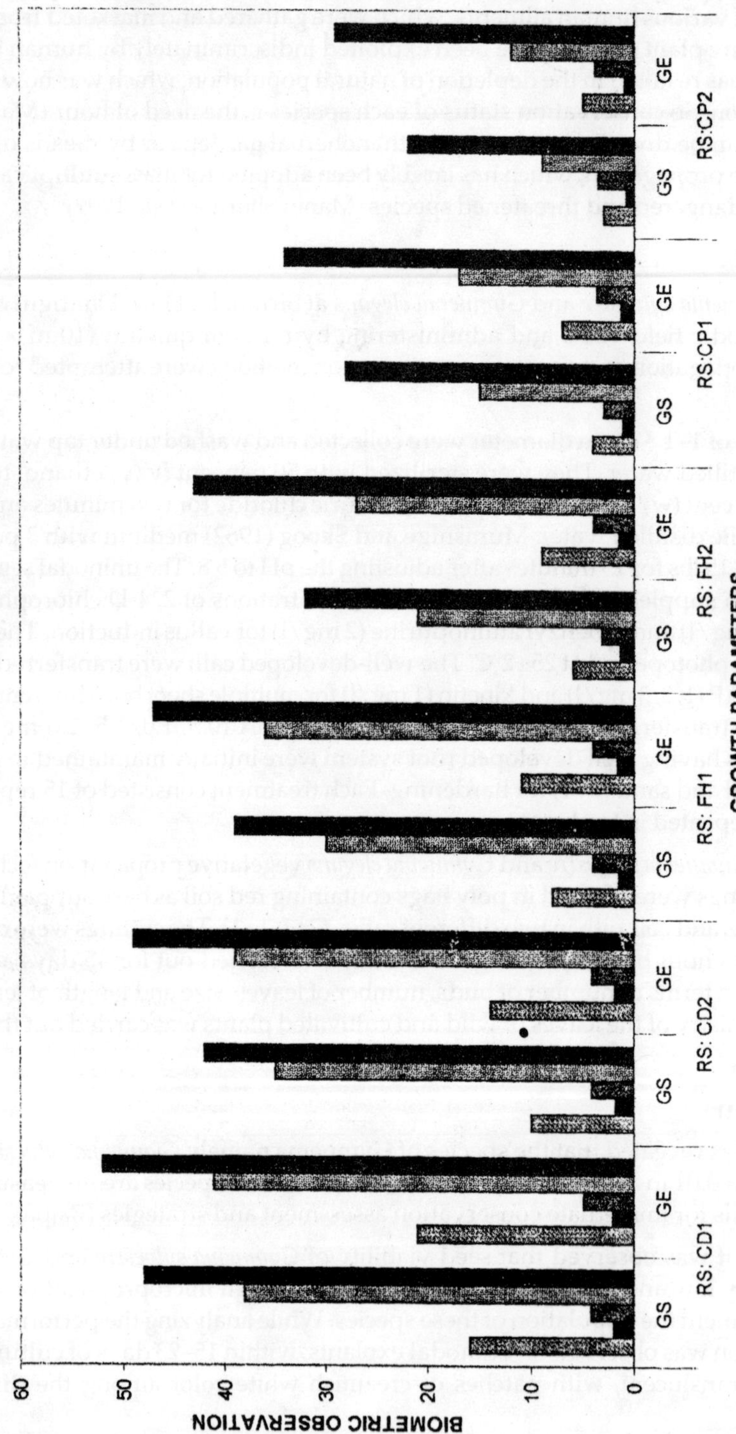

Figure 15.1: Growth Performance of *Gymnema sylvestre* and *Gymnema elegans* in Different Treatments after 45 Days

combinations of 2, 4 D and BAP. Maximum proliferation of callus was observed on the uninodal segments cultured on MS medium containing 2, 4 D 2 mg/l + BAP 1 mg/l. The well-developed callus remained healthy and fresh for 25–30 days after initial culture. When the calli was transferred to shoot induction medium containing BAP (3 mg/l) and Kinetin (0.5 mg/l), there was higher percentage of shoot bud differentiation. The elongated shoots were transferred to rooting medium containing IBA (2 mg/l). The rooted plants were successfully acclimatized in the soil.

While analyzing the performance of vegetative propagation for regeneration, the performance of *Gymnema elegans* was found to be quite appreciable and encouraging than *Gymnema sylvestre,* which may be due to the texture of the plant. Various parameters such as number of buds, number of leaves, size and length of leaf, and number of roots were analyzed which could perform the regeneration in a good manner. Of different treatments tested RS: CD1 and RS: CD2 performed on par with FH1, FH2 and CP1, CP2 (Figure 15.1). The good results were due to the treatment of the cuttages with auxin (IAA and IBA), which supported the best growth, and development of the plant and also helps in the induction of adventitious roots on cuttings. Auxin is preferred to a wide range because it is non-toxic to plants and effective in promoting roots (Hudson *et al.*, 1998). It also supported the development of lateral branches on most of the newly formed shoots (Philip *et al.*, 1992). Cowdung mixed with red soil showed a good result, which was due to the richness of NPK in the cowdung, which enrich the crop production effectively compared to commercial fertilizer. Cowdung has high moisture retention capacity, improves the soil structure with corresponding increase in infiltration rate and decrease in soil bulk density. Due to the presence of high amount of organic matter it also improves the soil fertility. These results give scope for mass multiplication of forest inhabiting *Gymnema sylvestre* and *Gymnema elegans* by using red soil and cowdung. Studies through thin layer chromatography showed that the relative rate of movement of triterpenoids was found to be 0.98 and 1.7 in wild and cultivated variety of *Gymnema sylvestre* and *Gymnema elegans* that were found to be similar.

Acknowledgement

The authors wish to place their record of thanks to Dr. Thilagavathi Daniel, Head, Department of Biology, Gandhigram Rural Institute, Gandhigram, Dindigul for the encouragement given to conduct this study.

References

Bajpai, H.R., 1997. Indigenous medicinal practices of hill Korwas of MP. *Journal of Human Ecology*, 9(3): 295.

Bakhru, H.K., 1996. Handling diabetes without medicines. *Health Action*, 6(3): 15.

Hartmann, Hudson, T., Dale E. Kestre, Fred T. Davies and Robert L. Geneve, 1997. *Plant Propagation: Principles and Practices*, 6th Edition. Jay Print Pack Private Ltd., New Delhi, pp. 288.

Mamgain, S.K., Goel, A.K. and Sharma, S.C., 1998. Conservation assessment of some important threatened medicinal plants of India. *Journal of Non-timber Forest Products*, 5(1&2): 1–9.

Murashige, T. and Skoog, F., 1962. A revised medium for rapid growth and bioassays with tissue culture. *Physiol. Plant*, 15: 473–497.

Philip, V.J., Dominic Joseph, Triggs, G.S., and Dickinson, N.M., 1992. Micropropagation of black pepper through shoot tip culture. *Plant Cell Reports.*, 12(10): 41–44.

Saraswathy, A., 1998. Some folk remedies of Tamil Nadu in the management of common ailments. Golden Jubilee National Symposium on Spices, Calicut, India, 64(1): 34.

Sharma, Manju, Nandhi, S.K., Kumar, A. and Palani, L.M.S., 1999. *Current Science*, 77(10): 1230.

Sharma, P.C., 1999. Conservation of medicinal plants: A Challenging task. In: *South-east Asian Seminar on Herbs and Herbal Medicine*, Patna, India, p. 118.

Chapter 16

Role of Proton Driven Antiporters in Na^+ Resistance and pH Homeostasis in the Lithium-resistant (Li^+-R) Mutant of the Cyanobacterium *Nostoc muscorum*

Santosh Bhargava[1]* and Ritu Thakur Bais[2]

[1]Department of Botany, Swami Vivekanand Government College, Raisen – 464 551, M.P., India

[2]Department of Botany, Government Sarojani Naidu Autonomous Girl's College, Shivaji Nagar, Bhopal – 462 016, M.P., India

ABSTRACT

Nostoc muscorum and its spontaneously occurring lithium-resistant (Li^+–R) mutant strains were characterized with regard to NaCl and alkaline pH stresses. These stress conditions resulted in severe impairment in per cent survival, nitrogenase activity, glutamine synthetase (OS) activity and oxygenic photosynthesis of the wild type. The lithium-resistant (Li^+-R) mutant was found to occur spontaneously with a frequency characteristic of single mutational event and was found to show resistance to NaCl and alkaline pH stresses by Na^+ efflux system driven by the H^+ electrochemical gradient across the membrane. The present findings are quite encouraging for understanding the antiporters at molecular level for the generation of enhanced salt tolerance in agriculturally important microbes and plants.

***Keywords**: Alkalophilly, Diazotrophic, Heterocystous, Mutant, Per cent survival, Protonophore, Salinity/osmotic stress.*

* Corresponding Author: E-mail: santoshbhargava@)hotmail.com.

Introduction

Cyanobacteria are photosynthetic prokaryotes, many of which are diazotrophic. They grow and multiply in water-logged rice agro-ecosystem of tropical countries. It is believed that they contribute immensely to the nitrogen-budget of such agricultural ecosystem. Modern agricultural practices are rendering the agricultural ecosystem increasingly saline. Therefore it is desirable to understand the biochemical mechanism conferring resistance against ionic stress in cyanobacteria. Microbial adaptations to salinity stress are known to include biochemical mechanism such as ion exclusion, ion export, cell wall modification, osmotic adjustment.

The primary biochemical mechanism of cyanobacterial cells adaptating them to NaCl stress is prevention of intracellular Na^+ accumulation by curtailment of Na^+ influx along with efficient efflux of the cation (Apte *et al.*, 1987; Apte and Thomas, 1985). There is also a role of intracellular K^+ and organic solutes in mediating cyanobacterial adaptation to salinity/osmotic stresses (Blumwald *et al.*, 1983; Mackey *et al.*, 1983; Reed *et al*, 1984). The degree of accumulation of compatible solutes in response to salinity/osmotic stress is a measure of degree of cellular adaptation to the stresses imposed (Higgins *et al.*, 1987; Warr *et al.*, 1988; Csonka, 1989). Further studies have revealed intracellular accumulation of K^+ as a primary intracellular salinity/osmotic signal activating genes conferring resistance to salinity/osmotic stress through their activity on intracellular level of compatible solutes (Csonka, 1989). Another adaptive mechanism to high salinity is the exclusion of Na^+ ion from cells which has been proposed as a function of Na^+/H^+ antiport and Na^+ ATPase (Blumwald *et al.*, 1984; Fry *et al.*, 1986; Serrano, 1996).

The present study on mutational aspects of salinity tolerance includes physiological parameters like growth, per cent survival, time for heterocyst differentiation, heterocyst frequency, photosynthetic O_2 evolution and nitrogenase activity, for characterizing the parent and its spontaneously occurring LiCl resistant mutant strains. In addition studies have been made to characterize further this mutant strain with regards to the Na^+ efflux capacity and pH tolerance capacity.

Materials and Methods

Organism and Maintenance of the Culture

In the present study, the experimental organism was *Nostoc muscorum* (a freshwater cyanobacterium) grown axenically in Chu No. 10 medium (Gerloff *et al.*, 1950) without any combined nitrogen source. Cultures were incubated in a culture room at 28°C±2°C and illuminated with daylight fluorescent tubes having a photon fluence rate of 50 µmol $m^{-2}S^{-1}$. The culture medium was buffered to pH 7.5 with 10 mol m^{-3} HEPES-NaOH.

Estimation of per cent survival, isolation of LiCl-resistant mutant, growth, chlorophyll *a* contents, protein contents, heterocyst frequency, nitrogenase activity and photosynthetic O_2 evolution were measured as described previously by Bhargava *et al.* (2003).

Determination of $^{22}Na^+$ Extrusion

The method as described by Bhargava *et al.* (2003) was used for measurement of $^{22}Na^+$ extrusion. The cyanobacterial cells were centrifuged, washed and resuspended in 10 mol m^{-3} HEPES-NaOH buffer (pH 7.5) and equilibrated for 30 min at 28°C±2°C at a photon fluence rate of 50 µmol m^{-2} S^{-1}. ^{22}NaCl (carrier free, specific activity 7.4 MBq mol^{-1}) was added to the incubation mixture at a concentration of 0.5 mol m^{-3}. Samples were withdrawn at regular time intervals and reaction was terminated by the addition of 0.1 ml of 2N HCl. 2.5 ml of scintillation cocktail containing four part 0.8

per cent PPO (2,5-diphenyloxazole) plus 0.01 per cent POPOP [1,4-bis (4-methyl-5phenyl-2-oxazole)-benzene] in toluene and three parts of ethanol was added to each sample vial. ^{22}Na extrusion was counted in a liquid scintillation counter (Wallac 1409, Finland). Non-specific binding of the radiolabel was determined by measuring its incorporation in toluene-treated cells. This value was always subtracted from the value obtained from the untreated samples.

Assay of Glutamine Synthetase (Biosynthetic) Activity

Cultures were harvested by centrifugation at 2000 × g and washed twice with 50 mol m^{-3} Tris-HCl buffer, pH 7.5 (Buffer-A) followed by washing and resuspension in Buffer B (Buffer-A + 5 mol m^{-3} $MgCl_2$, 10 mol m^{-3} sodium glutamate, 5 mol m^{-3} β-mercaptoethanol, 1 mol m^{-3} EDTA), pH 7.5. Cells were broken by Soniprep in the suspension at 4°C and the cell free extract thus obtained was centrifuged at 36,000 × g for 30 min at 4°C. The supernatant was used for Mg^{2+}-dependent assay of GS activity according to the method of Sampaio *et al.* (1979).

NaCl and pH stresses were given to the ammonium grown cultures for 12 h before transferring them to fresh diazotrophic medium under photoautotrophic growth conditions. All given parameters except time for heterocyst differentiation were measured in 6-day-old diazotrophically grown culture. The time for heterocyst differentiation was regularly monitored microscopically between 12–84 h old cultures. Such treated cultures along with their respective controls were incubated under diazotrophic growth conditions and then examined for heterocyst formation, NH_4^+-assimilating glutamine synthetase activity, H^+ driven Na^+ efflux system, photosynthetic O_2 evolution and assayed for nitrogenase activity.

The pH of the diazotrophic growth medium was maintained at 7.0 and 8.0 by addition of HEPES (N-2-hydroxyethylpiperzine-N-2-ethansulfonic) buffer. Growth medium of pH 9.0 was achieved by addition of AMPSO [3-(1,1-Dimethyl-2-hydroxy-thylamino)-2-hydroxypropanesulfonic acid buffer, while pH 10.0 was achieved by addition of CAPSO [3-(cyclohexy amino)-2-hydroxy-1-propanesulfonic acid], and pH 11.0 by addition of CAPS (Cyclohexy laminopropane sulfonic acid) buffer.

Results

In order to determine the lethal dose of LiCl for cyanobacterial growth, it was necessary to determine first the dose effect curve, which showed that 10 mol m^{-3} LiCl was lethal to diazotrophic growth of the cyanobacterium. Therefore, it was thought proper to make use of a stronger dose of LiCl *i.e.* 15 mol m^{-3} to select out spontaneously occurring cyanobacterial mutants exhibiting resistance to growth inhibitory action of LiCl. Such mutant clones were repeatedly tested for Li^+-resistance phenotype on LiCl containing nutrient agar plates. Finally a stable Li^+-R mutant clone was isolated and stored as stock on nutrient slants containing 15 mol m^{-3} LiCl. The mutational frequency of the cyanobacterium to Li^+-R phenotype ($2.5 \sim 2.9 \times 10^{-6}$) to Li^+-R phenotype suggested it was a product of single mutational event. Subsequent experiments were performed with the parent and mutant strains in order to know the physiological difference between the two strains.

In the parent strain, growth and per cent survival decreased with increasing NaCl stress. In addition GS activity was also adversely affected with increasing concentration of NaCl. The differentiation of N_2-fixing heterocysts was delayed with increase in salinity stress and a dose of 100 mol m^{-3} NaCl for 12 h resulted in complete prevention of the formation of N_2-fixing heterocysts. Parallel studies were conducted to examine the effect of increasing concentration of NaCl on photosynthetic O_2 evolution in the parent strain (Table 16.1). The findings suggested that a dose of 100 mol m^{-3} NaCl completely inhibited the nitrogenase activity in the parent strain, while oxygenic photosynthesis was still significant at this concentration. It was therefore concluded that photosynthetic O_2 evolution is also sensitive to NaCl stress but its sensitivity is significantly less than that of the nitrogenase activity. It is therefore concluded that cyanobacterial sensitivity to salinity stress under diazotrophic growth condition is primarily because of greater sensitivity of its N_2-fixing process to the stress parameters.

Table 16.1: Table Showing NaCl Tolerance Characteristic of the Parent *N. muscorum* in Terms of Growth (OD change at 663 nm), Per cent Survival, Heterocyst Frequency (HF per cent), Time for Heterocyst Differentiation in h (^{t}Het), Nitrogenase Activity (nmol C_2H_4 formed mg^{-1} Chl *a* h^{-1}), Photosynthetic O_2 Evolution (μmol O_2 evolved mg^{-1} Chl *a*), GS (biosynthetic) Activity (μmol NADH oxidized g^{-1} protein min^{-1})

Parameter	*Parent Control*	*Parent + 30 mol m^{-3} NaCl*	*Parent + 60 mol m^{-3} NaCl*	*Parent + 90 mol m^{-3} NaCl*
Growth	0.82	0.68	0.46	0.12
Per cent survival	100	84	70	10
HF per cent	7–8	7–8	5–6	0–0
^{t}Het	24–30	36–58	72–78	α
Nitrogenase activity	12.28	8.82	7.80	0.0
O_2 evolution	540	512	320	120
GS activity	88	69	42	6

Non-heterocystous NH^{+}_{4}-grown cultures were stressed with NaCl at different concentration for 12 h, later washed and used as inoculum for incubation on diazotrophic growth medium and then periodically examined for their respective characteristic.

Each reading is an average of three independent experimental determinations. The variation from the mean value was < 5 per cent (same for the Tables 16.2 and 16.3)

Li^+-R mutant of the wild type *N. muscorum* grows diazotrophically as well as on nitrite, nitrate or ammonium as nitrogen source in fact mutation to Li^+-R did not seems to influence much of its inorganic N_2 nutrition ability (data not shown). The Li^+-R mutant was therefore examined for its physiological characteristics under normal as well as under NaCl stress conditions. The results of such studies are given in Table 16.2, showing that growth, time for heterocyst differentiation, heterocyst frequency, nitrogenase activity, GS activity, photosynthetic O_2 evolution and per cent survival were all less affected by a given NaCl stresses in the mutant strain.

The next series of experiment was carried out to compare the efflux rate of $^{22}Na^+$ from pre equilibrated samples (with different concentration of NaCl) of the parent and its Li^+-R mutant strains. The results as shown in the Figure 16.1 clearly indicate that $^{22}Na^+$ efflux rate was manifold higher in the Li^+-R mutant strain than in the parent strain. The protonophore CCCP (Carbonyl cyanide m-chlorophenyl hydrazone) is an inhibitor of H^+-gradient formation in cyanobacteria as well as in all biological membranes. It was used to ascertain the role of H^+-gradient in driving the Na^+/H^+ antiporter activity in the *N. muscorum*. The results (data not shown in the Figure 16.1) of the parent and its Li^+-R mutant strain clearly indicates that Na^+ efflux system in the cyanobacterium is sensitive to CCCP and H^+ driven.

Table 16.3 compares the pH survival characteristics of the parent and its Li^+-R mutant strain under different alkaline pH of the diazotrophic growth medium. The pH tolerance characteristics of the parent and its Li^+-R mutant strain remained indistinguishable up to pH 9.0 of the growth medium above which the survival of the parent strain started declining with increase in pH and was about 40 per cent at pH 11.0. However, the pH survival characteristic of Li^+-R mutant strains at growth medium of pH 11.0 remained almost 100 per cent. Further the diazotrophic characteristics in terms of nitrogenase activity and photosynthetic O_2 evolution of the parent and its Li^+-R mutant strain at different alkaline pH was compared. As shown in the table the parent strain showed a significant decline in nitrogenase activity as well as in photosynthetic O_2 evolution. The mutant strain was more tolerant in maintaining its activities under various alkaline pH stresses. These findings further indicate the operation of a common physiological mechanism regulating NaCl tolerance and alkaline pH tolerance in Li+-R mutant of the cyanobacterium *N. muscorum*.

Table 16.2: Table Showing NaCl Tolerance Characteristic of Li^+-R Mutant Strains of *N. muscorum* in Terms of Growth (OD change at 663 nm), Per cent Survival, Heterocyst frequency (HF per cent), Time for Heterocyst Differentiation in h (^{t}Het), Nitrogenase Activity (nmol C_2H_4 formed mg^{-1} Chl; *a* h^{-1}), Photosynthetic O_2 Evolution (μmol O_2 evolved mg^{-1} Chl *a*), GS (biosynthetic) Activity (μmol NADH oxidized g^{-1} protein min^{-1})

Parameter	*Li^+-R Control*	*Li^+-R + 30 mol m^{-3} NaCl*	*Li^+-R + 60 mol m^{-3} NaCl*	*Li^+-R + 90 mol m^{-3} NaCl*
Growth	0.72	0.68	0.71	0.71
Per cent survival	100	100	90	88
HF per cent	7–8	7–8	7–8	7–8
^{t}Het	26–32	26–32	26–32	26–34
Nitrogenase activity	10.58	9.82	9.60	8.8
O_2 evolution	520	520	500	480
GS activity	82	82	78	78

Non-heterocystous NH_4^+-grown cultures were stressed with NaCl at different concentration for 12 h, later washed and used as inoculum for incubation on diazotrophic growth medium and then periodically examined for their respective characteristic.

Table 16.3: Nitrogenase Activity (nmol C_2H_4 formed mg^{-1} Chl *a* h^{-1}), Photosynthetic O_2 Evolution (μmol O_2 evolved mg^{-1} Chl *a*) and Per cent Survival of the Parent *N. muscorum* and its Li^+-R Mutant Strain Under Varying pH Stress Conditions

pH	*Nitrogenase Activity*		*O_2 Evolution*		*Per cent Survival*	
	Parent	*Li^+-R*	*Parent*	*Li^+-R*	*Parent*	*Li^+-R*
7.0	12.28	11.68	540	524	100	100
8.0	12.10	11.66	530	524	96	100
9.0	10.20	11.68	488	51.8	86	100
10.0	7.80	11.20	222	512	71	100
11.0	6.28	10.98	108	512	40	100

Six-day-old diazotrophic cultures growing at pH 7.0 were the source of inocula for the experiments.

N. muscorum showed about 40 per cent survival during its growth at pH 11.0 in diazotrophic growth medium. Cyanobacterial systems growing at alkaline pH normally exhibit a pH homeostasis

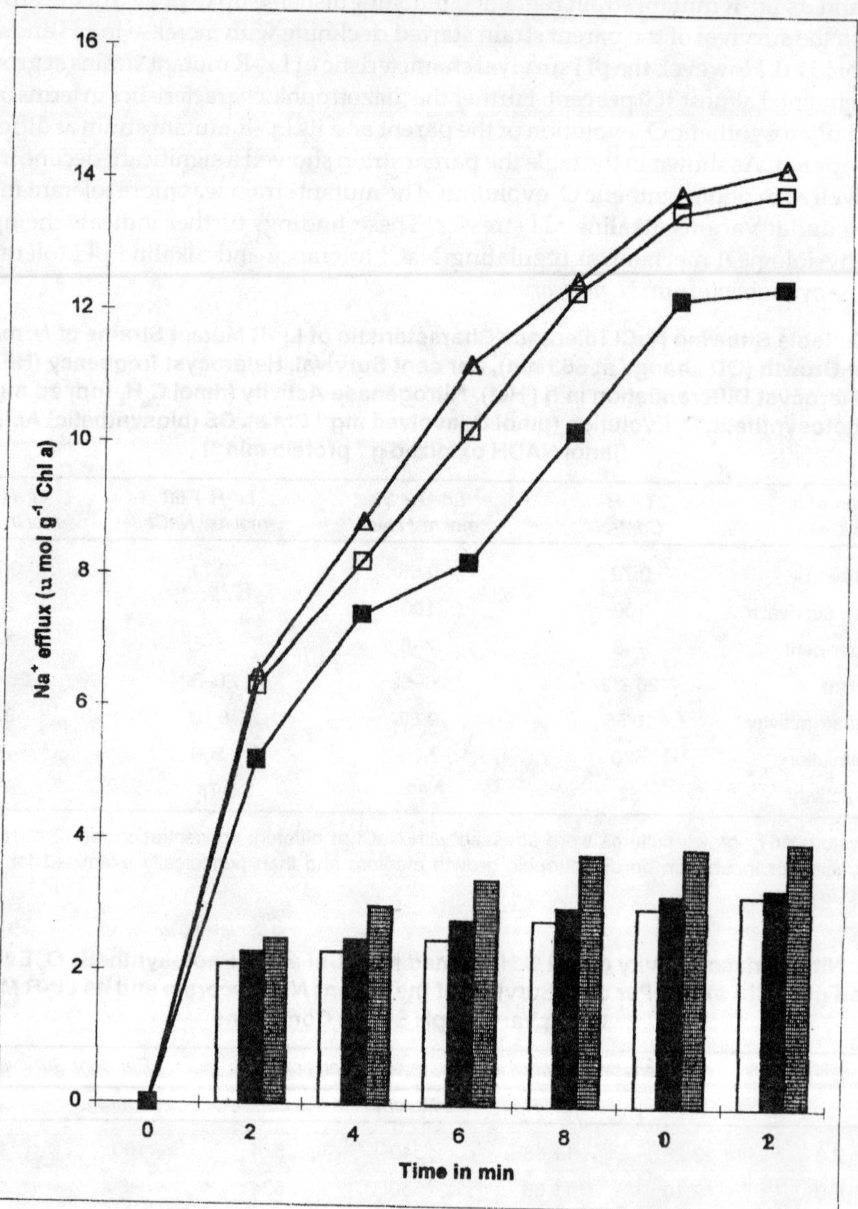

Figure 16.1: Na⁺ Efflux Pattern in the Parent *N. muscorum* [parent control (blank bars), parent + 30 mol m^{-3} NaCl (dark bars)] and its Li^+-R Mutant [Li^+-R control (checks bars), Li^+-R + 30 mol m^{-3} NaCl (■), Li^+-R + 60 mol m^{-3} NaCl (□), Li^+-R + 90 mol m^{-3} NaCl (Δ)] Strain Under Varying Salinity Stress Conditions. The data on 60 and 90 mol m^{-3} stress conditions in the parent strain are not mentioned in the Figure because the efflux pattern was similar to similarly grown culture under 30 mol m^{-3} stress condition.

Each reading is an average of three independent experimental determinations. The variation from the mean value was < 5 per cent (same for the Figure 16.2 and 16.3).

mechanism maintaining their cytoplasmic pH in a physiologically functional range driven by Na^+/H^+ antiporter. The experiments were therefore planned to examine the role of the Na^+/H^+ antiporter in *N. muscorum* in maintaining its cytoplasmic pH and providing at the same time reasons for cyanobacterial allow per cent survival at pH 11.0. The results of such study are shown in Figure 16.2. The figure represents a Na^+ efflux trend in the parent cyanobacterium growing diazotrophically at alkaline pH range of 7.0–11.0. The results clearly indicate that cyanobacterial Na^+ efflux system is rendered inefficient at pH 11.0 but not in the pH range of 7.0–9.0.

Parallel studies were conducted with Li^+-R mutant of the parent *N. muscorum*. The results are given in Figure 16.3. The mutant cyanobacterium exhibits more Na^+ efflux in the pH range 9.0–11.0 than in the pH range 7.0–8.0, thus implying a role of Li^+-R gene in enabling the cyanobacterium to grow at pH 11.0 as efficiently as in growth at pH 7.0–8.0. CCCP pretreatment resulted in complete abolition of the $^{22}Na^+$ extrusion as shown in the Figures 16.2 and 16.3.

Discussion

Alkali cation stress in plants and in microbial cells mainly consists of two components, the ionic stress component resulting from cellular accumulation of alkali cations to toxic levels and the osmotic stress component resulting from cellular loss of water (Csonka, 1989). The ionic stress component in plants and microbes can overcome by an efllux system (Blumwald *et al.*, 1984), on the other hand the non-ionic (osmotic) component can overcome by actively accumulating compatible solutes (Bartels and Nelson, 1994). Na^+/H^+ antiporter activity in regulating Na^+ stress or alkaline pH stress has already been documented for cyanobacteria (Blumwald *et al.*, 1984) and enterobacteria (Padan and Schuldiner, 1994). Li^+ toxicity in cyanobacteria results ftom its inhibitory action on Na^+- bicarbonate symport system (Espie *et al.*, 1988). In *E. coli* Li^+ toxicity is a function of its inhibitory action on melibiose transporter and on pyruvate kinase (Schuldiner and Padan, 1992). Li^+-R mutant of *E. coli* are known in which Li^+ toxicity disappear as a result of enhanced Na^+/H^+ antiporter activity (Schuldiner and Padan, 1992).

The Li^+-R mutant reported here occurred spontaneously with a frequency characteristic of a single mutational event. The mutant strains resistant to 15 mol m^{-3} LiCl would be a mutant reducing intracellular Li^+ level thus rendering the cyanobacterium resistant to growth inhibitory action of Li^+. Since such Li^+-R mutant also exhibit resistance to 100 mol m^{-3} NaCl, the mechanism reducing the intracellular Li^+ level would be more effective in reducing intracellular Na^+ level, thereby exhibiting resistant even to 100 mol m^{-3} NaCl. The osmotic stress component in 100 mol m^{-3} NaCl may not be as deleterious as its ionic component. This hypothesis explains the origin of 100 mol m^{-3} NaCl resistance in the L^+-R strains.

When the parent strain was examined for relative sensitivity of oxygenic photosynthesis and nitrogenase activity to NaCl stress, the latter was found to be more sensitive than the former. This is in agreement with a previous report on osmoregulation of oxygenic photosynthesis and N_2-fixation (Tel-Or, 1980). The cyanobacterial strains grew diazotrophically at the expense of oxygenic photosynthesis and the ammonium assimilatory activity of GS. Like oxygenic photosynthesis, the GS (biosynthetic) activity of the parent strain was also adversely affected under salinity stress conditions. The inhibition of these activities under salinity stress is the primary reason for cyanobacterial death.

Since Li^+ inhibition in cyanobacteria could result from multiple inhibitory action, a Li^+-R mutant could involve a mechanism that effluxes out Li^+ much faster than the system causing Li^+ influx. Since Li^+ is an analog of Na^+ it can enter the cyanobacterial cells by all transport system transporting Na^+.

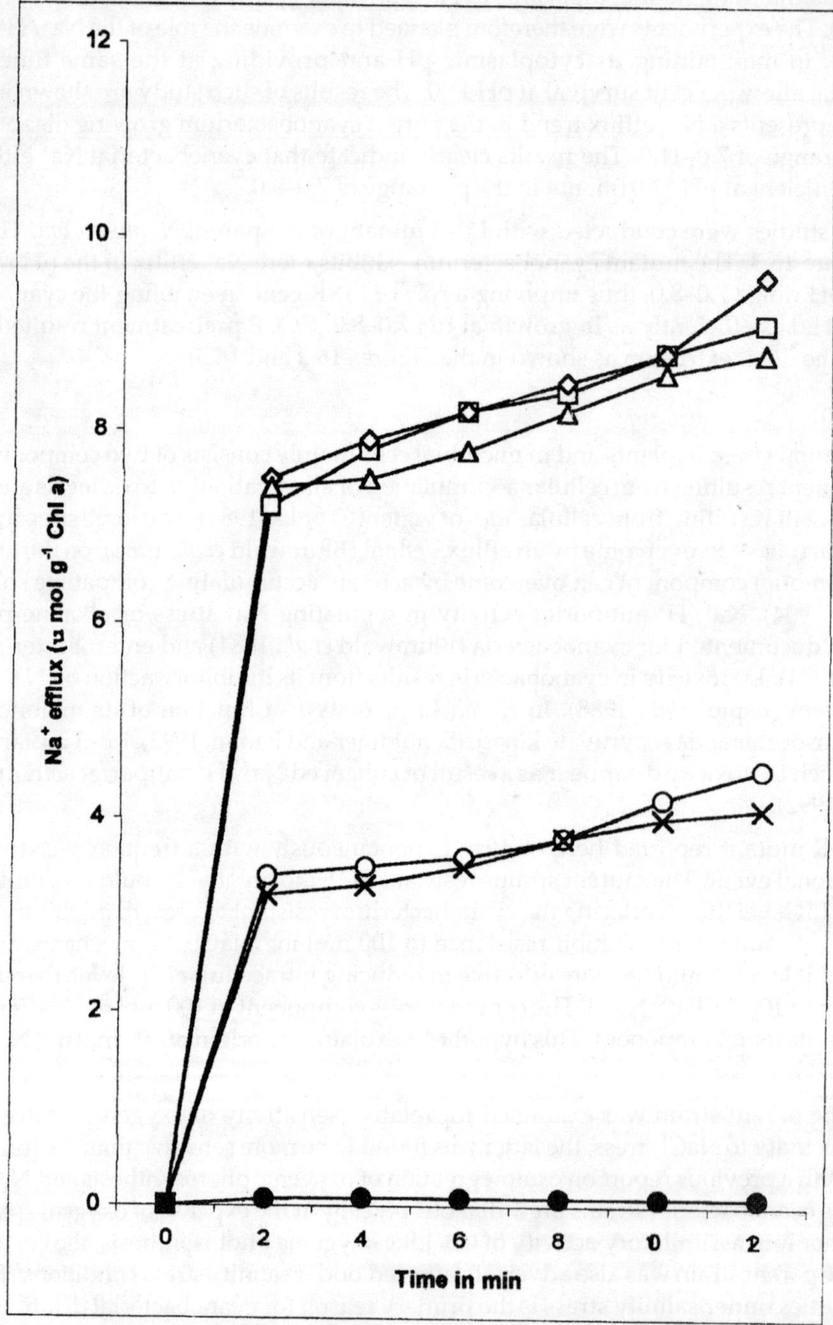

Figure 16.2: Na⁺ Efflux Pattern in *N. muscorum* Under Varying pH Stress [pH 7 (□), pH 8 (◊), pH 9 (Δ), pH 10 (O), pH 11 (x)] Conditions. Effect of CCCP on Na⁺ efflux was examined by pretreating the NaCl equilibrated samples with 0.01 mol m⁻³ CCCP for 20 min. The data on pH 7, 8, 9, 10 are not mentioned in the Figure because they are similar to pH 11 (●) data.

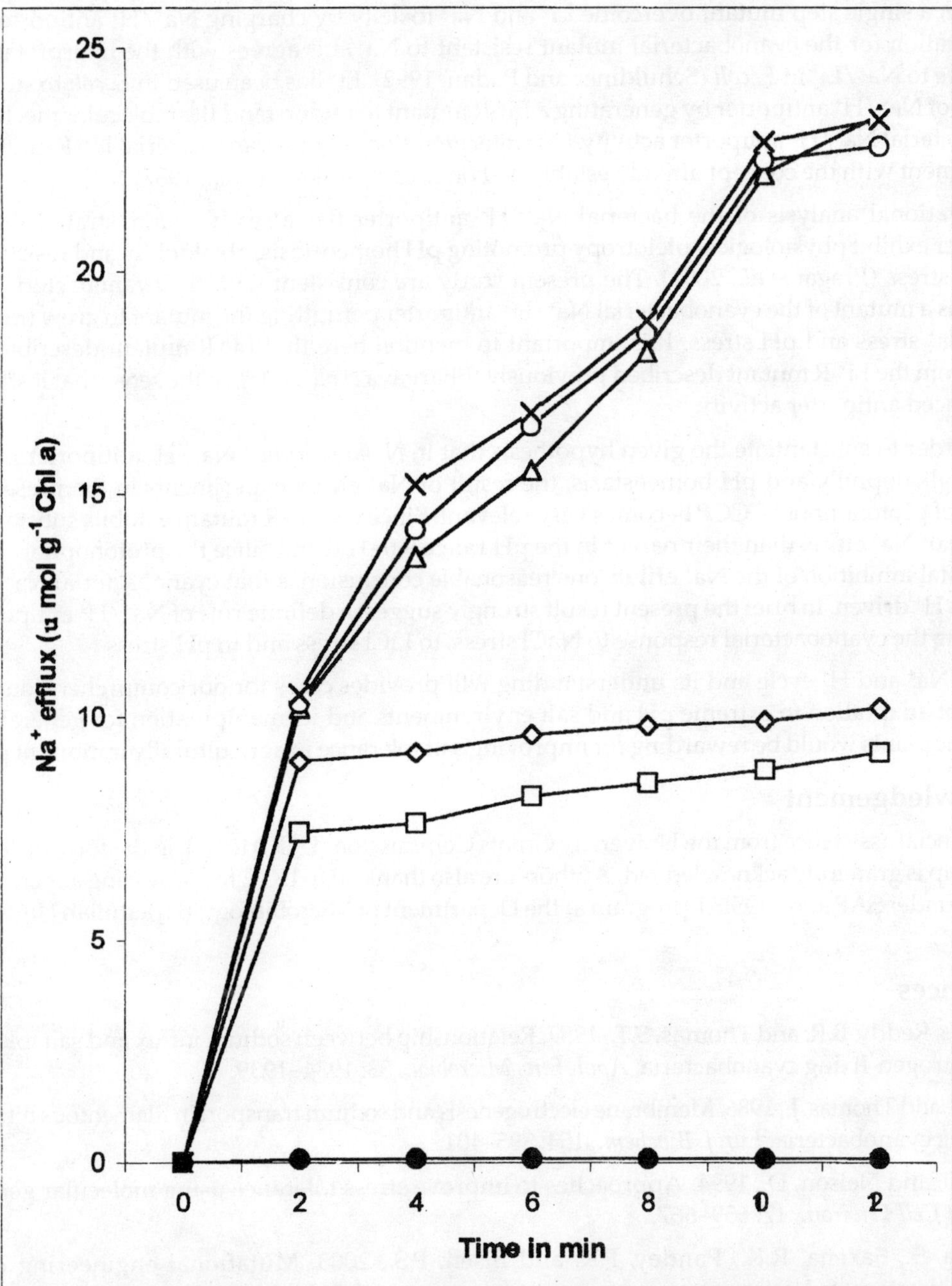

Figure 16.3: Na⁺ Efflux Pattern in Li⁺-R Under Varying pH Stress [pH 7 (□), pH 8 (◊), pH 9 (Δ), pH 10 (O), pH 11 (×)] Conditions. Effect of CCCP on Na⁺ efflux was examined by pretreating the NaCl equilibrated samples with 0.01 mol m⁻³ CCCP for 20 min. The data on pH 7, 8, 9, 10 are not mentioned in the Figure because they are similar to pH 11 (●) data.

Cyanobacteria like bacteria have Na^+/H^+ antiporter which effluxes Na^+ as well as Li^+. In the present organism a single step mutant overcome Li^+ and Na^+ toxicity by charging Na^+/H^+ antiporter. This interpretation for the cyanobacterial mutant resistant to Na^+/Li^+ agrees with the interpretation of resistance to Na^+/Li^+ in *E. coli* (Schuldiner and Padan, 1992). Li^+ has been used in *E. coli* to study the genetics of Na^+/H^+ antiporter by generating a Li^+-R mutant to understand the molecular mechanism of the bacterial Na^+/H^+ antiporter activity. Our interpretations of the cyanobacterial Li^+-R mutant are in agreement with the concept already established on *E. coli* (Ishikawa *et al.*, 1987).

Mutational analysis of the bacterial Na^+/H^+ antiporter has already demonstrated that this antiporter exhibit physiological pleiotropy promoting pH homeostasis, alkalophily and resistance to Li^+/Na^+ stress (Pragai *et al.*, 2001). The present study are consistent with the cyanobacterial Li^+-R mutant as a mutant of the cyanobacterial Na^+/H^+ antiporter permitting the mutant to grow under Li^+ stress, Na^+ stress and pH stress. It is important to mention here that Li^+-R mutant described here differs from the Li^+-R mutant described previously (Bhargava *et al.*, 2003), in the sense that it showing an enhanced antiporter activity.

In order to substantiate the given hypothesis that in *N. muscoruma*, Na^+/H^+ antiporter is at the route of alkalophilly and pH homeostasis, the result of Na^+ efflux experiments in the presence or absence of protonophore CCCP becomes very relevant. Since the Li^+-R mutant exhibits substantially high rate of Na^+ efflux than their parent in the pH range 9.0–11.0 and since the protonophore CCCP causes total inhibition of the Na^+ efflux, one reasonable conclusion is that cyanobacterial Na^+ efflux system is H^+ driven. In brief the present result strongly suggest a definite role of Na^+/H^+ antiporter in regulating the cyanobacterial response to NaCl stress, to LiCl stress and to pH stress.

The Na^+ and H^+ cycle and its understanding will provides clues for our comprehension of the process of adaptation to extreme pH and salt environments and its manipulation for generation of transgenic plants would be rewarding for improving salt tolerance in agriculturally important plants.

Acknowledgement

Financial assistance from the University Grants Commission (UGC) to S.B. in the form of Teacher Fellowship is gratefully acknowledged. Authors are also thankful to UGC for providing experimental facilities under SAP and COSIST program at the Department of Microbiology, Barkatullah University, Bhopal.

References

Apte, S.K., Reddy, B.R. and Thomas, S.T., 1987. Relationship between sodium influx and salt tolerance of nitrogen-fixing cyanobacteria. *Appl. Env. Microbiol.*, 53: 1934–1939.

Apte, S.K. and Thomas, J., 1986. Membrane electrogenesis and sodium transport in filamentous nitrogen-fixing cyanobacteria. *Eur. J. Biochem.*, 154: 395–401.

Bartels, D. and Nelson, D., 1994. Approaches to improve stress tolerance using molecular genetics. *Plant Cell Environ.*, 17: 659–667.

Bhargava, S., Saxena, R.K., Pandey, P.K. and Bisen, P.S., 2003. Mutational engineering of the cyanobacterium *Nostoc muscorum* for resistance to growth-inhibitory action of LiCl and NaCl. *Curr. Microbiol.*, 47: 5–11.

Blumwald,.E., Mehlhorn, R.J. and Packer, L., 1983. Studies of osmoregulation in salt adaptation of cyanobacteria with ESR spin probe techniques. *Pro. Natl. Sci. USA*, 80: 2599–2602.

Blumwald, E., Wolosin, J.M. and Packer, L., 1984. Na^+/H^+ exchange in the cyanbacterium *Synechococcus* 6311. *Biochem Biophys. Res. Comm.*, 122: 452–459.

Csonka, L.N., 1989. Physiological and genetic responses of bacteria to osmotic stress. *Microbiol. Rev.*, 53: 121–147.

Espie, G.S., Miller, A.G. and Canvin, D.T., 1988. Characterization of Na^+-requirement in cyanobacterial photosynthesis. *Plant Physiol.*, 88: 757–763.

Fry, I.V., Hufleit, M., Erber, W.W.A., Peschek, G.A. and Packer, L., 1986. The role of respiration during adaptation of the freshwater cyanobacterium *Synechococcus* 6311 to salinity. *Biochem. Biophys. Acta*, 244: 686–691.

Gerloff, G.C., Fitzerald, G.P. and Skoog, F., 1950. The isolation, purification and culture of blue-green algae. *Am. J. Bot.*, 37: 216–218.

Higgins, C.F.J., Cairney, D.A., Stirling, L., Sutherland, L. and Booth, I.R., 1987. Osmotic regulation of gene expression: Ionic strength as an intracellular signal? *Tren. Biochem. Sci.*, 12: 339–344.

Ishikawa, T., Hama, H., Tsuda, M. and Tsuchiya, T., 1987. Isolation and properties of a mutant of *E coli* possessing defective Na^+/H^+ antiporter. *J. Biol. Chem.*, 262: 7443.

Padan, E. and Schuldiner, S., 1994. Molecular physiology of the Na^+/H^+ antiporter in *Echerichia coli. J. Exp. Bot.*, 196: 443–456.

Pragai, Z., Eschevins, C., Bron, S. and Harwood, C.R., 2001. *Bacillus subtilis*, an Na^+/H^+ antiporter, influences expression of the *phoPR* operon and production of alkaline phosphatases. *J. Bacteriol.*, 183: 2505–2515.

Reed, R.H., Richardson, D.L., Warr, S.R.C. and Stewart, W.D.P., 1984. Carbohydrate accumulation and osmotic stress in cyanobacteria. *J. Gen. Microbiol.*, 130: 1–4.

Sampaio, M.J.A.M., Rowell, P. and Stewart, W.D.P., 1979. Purification and some properties of glutamine synthetase: from the nitrogen-fixing cyanobacteria *Anabaena cylindrica* and *Nostoc* sp. *J. Gen. Microbiol.*, 111: 181–191.

Schuldiner, S. and Padan, E., 1992. Na^+/H^+ antiporters in *E. coli*. In: *Alkali Cations Transport Systems in Prokaryotes,* (Ed.) Bakker, E.P. CRC Press, Boca Raton, FL, Chap. 1B.

Serrano, R., 1996. Salt tolerance in plants and microorganisms: Toxicity targets and defense responses. *Int. Rev. Cytol.*, 165: 1–51.

Tel-Or, E., 1980. Adaptation of salt of the photosynthetic apparatus in cyanobacteria. *FEBS Lett.*, 110: 253–256.

Warr, S.R.C., Reed, R.H. and Stewart, W.D.P., 1988. The compatibility of osmotica in cyanobacteria. *Plant Cell Environ.*, 11: 137–142.

Chapter 17

Effects of Neem Leaf Extract on Growth Characteristics and Organic Constituents of *Trigonella foenum-graceum*, L.

S.L. Khapke, R.K. Aher, D.A. Bandal, G.V. Vyavahare, Y.B. Agre, D.B. Shelake and Miss R.K. Shaikh
Department of Botany, New Arts, Commerce and Science College, Parner, Dist. Ahmednagar, M.S., India

ABSTRACT

Trigonella foenum-graceum, L. plant has been valued as a medicinal in many parts of the world. The ayurvedic pharmacopoeia has ascribed many therapeutic uses to it. It has wide spread use as a therapeutic in chronic diseases. The leaves of the plant are commonly used as a vegetable in India. Morphological characteristics and biochemical investigation was conducted to study effects of neem leaf extract on *Trigonella.* Four per cent of leaf extract of neem was prepared as stock solution. Growth characteristics such as root length, shoot length and root nodules increased at 0.8 per cent concentration of leaf extract. There was significant increase in the organic constituents such as chlorophyll and proteins over control at 0.6 and 0.8 per cent concentrations.

Keywords: *Trigonella, Neem leaf extract, Morphological and biochemical changes.*

Introduction

Methi (*Trigonella foenum-graceum* L.) is an important food crop as well as a quick cash crop all over India especially in Maharashtra. The duration of crop varies from 25 to 40 days. It is grown for it's green leaves and young pods. The leaves are quite rich in protein (4.4 g), minerals (1.5 g), vitamin A (6450 IU) and vitamin C (54 mg) per 100 g of edible portion. Seeds are used in condiments. The whole plant is medicinally too much important.

The use of neem leaves (*Azadirachta indica* Juss) in pest control has been known since ancient time. But recently it has been studied scientifically and reported that it contains different chemicals which have insect repellant, insecticide antifeedant, nematicide, and antimicrobial properties. The chemical ingredients reported are azadirachtin, meliantrol and salanin.

Fujii *et al.* (2002) while taking the review of recent work in allelopathy explained that allelopathy now refers to any process involving secondary metabolites produced by plants, viruses, microorganisms and fungi that influences the growth and development of agricultural and biological systems. According to Bhat and Chauhan (2000) and Singh and Narsing Rao (2003) the chemicals present in the extract (aqueous or alcoholic) have stimulatory or inhibitory influence on seed germination, seedling growth, yield and physiology of subsequent crops.

Materials and Methods

Trigonella seeds were obtained from market. The earthen pots (20 × 30 × 40 cm) were used for growing the *Trigonella* seeds. The pots were filled with mixture containing garden soil and well decomposed compost (3 : 1). About 500 seeds of uniform size from cultivar were selected for the experiments. The seeds were surface sterilized with 0.1 per cent $HgCl_2$, washed thoroughly 3–4 times in sterilized distilled water and then soaked in distilled water for 12 hours. 25 well imbibed seeds were sown in each earthen pot. The seedlings were grown under natural and uniform conditions.

For the preparation of extract freshly mature green leaves of neem are used. About 4 g leaflets of neem were uniformly homogenized in 100 ml of 80 per cent ethanol and condensed on rotary evaporator to 10 ml at temperature below 60°C to avoid evaporation of volatile compound and denaturation of bioactive chemicals. This was filtered and the filtrate was made to 100 ml with sterile distilled water. This extract was diluted with distilled water to require concentrations and used for foliar spraying.

The experimental plants of *Trigonella* grown in pots were treated with freshly prepared alcoholic extract at the concentrations of 0.2 per cent, 0.4 per cent, 0.6 per cent, 0.8 per cent, 1.0 per cent. Spraying was usually done in the morning.

The controlled and treated seedlings were then used for further biochemical analysis. Chlorophylls were extracted and estimated by Arnon's (1949) method. The leaf proteins were estimated by using Lowry *et al.* (1951) method.

Results and Discussion

Results recorded in Table 17.1, indicate that the concentration from 0.6 per cent to 1.0 per cent of neem leaf extract has caused significant increase in root as well as shoot length in *Trigonella* seedlings over control. The maximum root length and shoot length were recorded in 0.8 per cent concentration (16.2±1, 25.1±1) over control.

Similar trend was observed in case of total number of leaves per plant. There is increase in total leaves found in 0.8 per cent and 1.0 per cent concentration. Highest number of leaves are found in 1.0 per cent concentrations pot (21 leaves).

Also the fresh weight and dry weight was goes on increasing simultaneously in increasing concentration of extract over control. Maximum fresh weight and dry weight will be noted in 1.0 per cent concentrations seedlings (2.3 g, 0.4 mg).

As the nodulation is the part of productivity in legumes, the effect of neem leaf extract on nodulation is also studied. There is also a promotion in nodulation in treated plants over control. The maximum number of nodules are found in 0.8 per cent concentration plants.

Table 17.1: Effect of Neem Leaf Extract on Seedling Root Length, Shoot Length, Total Number of Leaves, Fresh Weight and Dry Weight and Root Nodules

Growth Parameters Per Plant ± SD	*Days from Spraying*	*Concentration of Neem Leaf Extract*					
		C	*0.2%*	*0.4%*	*0.6%*	*0.8%*	*1.0%*
Root length, cm	9	7.5±1	7.5±1	7.4±1	7.5±1	8±1	7.8±1
	16	10.5±1	14±1	12.2±1	10.9±1	13.9±1	14.1±1
	23	12.7±1	12.7±1	14.3±1	14.1±1	16.2±1	15.7±1
Shoot length, cm	9	10.5±1	10.4±1	10.4±1	10.7±1	10.8±1	10.7±1
	16	15.3±1	18.1±1	16.7±1	17.8±1	19.5±1	18.1±1
	23	20.1±1	21.3±1	21.5±1	21.6±1	25.1±1	22.2±1
Total no. of leaves	9	6	6	6	6	6	6
	16	12	12	12	15	15	15
	23	15	18	18	18	21	21
Fresh weight mg	9	0.7	0.7	0.8	0.7	0.9	0.8
	16	1.2	1.5	1.5	1.6	1.7	1.8
	23	1.6	1.8	2.1	2.3	2.3	2.4
Dry weight mg	9	0.08	0.08	0.09	0.08	0.1	0.1
	16	0.1	0.1	0.1	0.1	0.2	0.2
	23	0.2	0.3	0.3	0.3	0.4	0.4
Root nodule	9	8	9	10	9	11	10
	16	10	11	12	12	13	12
	23	11	13	13	14	15	14

Average of three determinations.

Sukul and Chaudhuri (1995 and 2001) reported increase in number of leaves per plant in rice due to treatment of leaf extracts of *Lantana camara*. Yamada *et al.* (2002) reported improved number of leaves per plant in various plant species due to alleochemicals present in *Lepidium sativum*.

Thakur *et al.* (1999) reported significant increase in plant height in bell pepper, when treated with bioextracts. Reddy and Rao (1995–96) also reported increased plant height in *Cajanus cajan* at lower concentrations, when treated with aqueous extracts of different plants. Sukul and Chaudhuri (1995 and 2001) noted increase plant height in rice when treated with the leaf extracts of *Lantana*.

The results of photosynthetic pigment indicated that the cultivar showed variation in their total chlorophyll contents as a effect of foliar spray. In general the increase in concentration showed the increase in total chlorophyll contents upto the 0.8 per cent concentration than control. But at 1.0 per cent concentration there is decrease in total chlorophyll content than 0.8 per cent concentration (Table 17.2). Bansal *et al.* (1976) reported enhanced chlorophyll contents in different plants like *Boerravia, Cenchrus, Convolvulus* and *Sesamum* due to allelopathic treatments. Rizvi *et al.* (1989) also recorded increased chlorophyll contents in maize due to treatment of nicotine.

Tripathi *et al.* (1999) reported that root extracts of *Tectona grandis* applied through soil stimulated the synthesis of chlorophyll *a* and chlorophyll *b* in soybean. Venkata Raju (1999) reported that low concentrations of aqueous extracts of root and shoot of *Plumbago zeylanica* stimulated the chlorophyll contents in *Amaranthus.* Bhalerao *et al.* (2003) observed same trend in strawberry with foliar application of *Tectaria* frond extract.

Table 17.2: Effect of Neem Leaf Extract on Increase in Total Chlorophyll Content and Total Protein Content in *Trigonella* in mg/g

Treatment	*Control*	*0.2%*	*0.4%*	*0.6%*	*0.8%*	*1.0%*
Chlorophyll	0.83	0.86	0.88	0.90	0.96	0.93
Protein	0.037	0.045	0.055	0.061	0.072	0.075

Average of three determinations.

The results regarding protein contents in *Trigonella* leaves as influenced by different concentrations of neem leaf extract recorded in Table 17.2. There is increase in the protein content over the control in different concentration and protein content goes on increasing simultaneously over control. The highest protein content should be found in 1.0 per cent concentration (0.075 mg/g).

Tripathi *et al.* (2000) studied the effects of leaf extracts of *Dalbergia sissoo* and *Eucalyptus* leaf litter or *Vigna radiata.* He noted decrease in protein content. However, Padhy *et al.* (2000) reported increase in protein content in 12 days old seedlings of finger millet, when treated with *Eucalyptus* leaf leachates. Similarly Bhalerao *et al.* (2001) noted increase in protein content in Mentha leaves when treated with *Pteridium* and *Aspidium* leaf leachates. Bhalerao *et al.* (2003) also reported increase protein content in strawberry treated with *Tectaria* frond extract.

Sonali Bhargava and Ashok K. Bhargava (2004) observed increase in root nodules in *Cicer arietinum* treated with neem oil.

Acknowledgement

Authors are thankful to Principal of New Arts, Commerce and Science College Parner. Also authors are thankful to Department of Botany, New Arts, Commerce and Science College, Parner.

References

Bendre and Kumar. 1994–95. *Economic Botany.* p. 52, 53.

Bhalerao, E.B., S.L. Laware, R.R. Vaidya and K.N. Dhumal. 2001. Influence of leaf leachates of *Pteridium* and *Aspidium* in physiology of Mehtha. *J. Med. and Arom. Plant Sci.*, 22(4A) & 23(IA): 502–504.

Bhargava, Sonali and Ashok K. Bhargava. 2004. Effects of Neem oil on Growth characteristics of *Cicer arietinum* and control of pod borer. *Ad. Plant Sci.*, 17(11): 713–715.

Bhatt, B.P. and D.S. Chauhan. 2000. Allelopathic effects of *Quercus* spp. on crops of Garhwal Himalaya. *Allelopathy J.*, 7: 265–273.

Fujii, Y., S. Hirdate and H. Hrya. 2002. In: *Abstr. 3rd World. Cong on Allelopathy: Challenge for the New Millenium*, (Ed.) Tsukuba, Japan.

Kokate, C.K., A.P. Purohit and S.B. Ghokhale. 1999. *Pharmacognocy*. p. 250, 498.

Lowry, O.H., N.J. Rosebrough, A.L. Farr and R.J. Randall. 1951. Protein measurement with Folin phenol reagent. *Biochemistry*, 15: 529–536.

Nandal, D.P.S., S.S. Bisla and S.S. Narwal. 1992. Allelopathic influence of Eucalyptus and *poplar* leaf extract on the germination and seedlings growth of winter vegetables. In: *Proc. First National Symposium on Allelopathy in Agroecosystems: Agriculture and Forestry*, (Eds.) P. Tauro and S.S. Narwal. Indian Society of Allelopathy, Hisar, India, p. 98–100.

Narswal, S.S., R. Pillaniraj, S.C. Sati, H.S. Kadian and D.S. Dahiya. 2003. Allelopathic plants: 8. *Parthenium hysterophours* L. *Allelopathy J.*, 11(2): 151–170.

Padhy, B., P.K. Patnaik and A.K. Tripathy. 2000. Allelopathic potential of Eucalyptus leaf litter leachates on germination and seedling growth of Finger millet. *Allelopathy J.*, 7(1): 69–78.

Panda, H. *Medicinal Plants: Cultivation and their Uses*. p. 201–213, 401–402.

Pandey, B.P. 1995. *Economic Botany*. p. 63, 306,307.

Singh, Deena and Y.B. Narsing Rao. 2003. Allelopathic evaluation of *Andrographis paniculata* aqueous leachates on rice (*Oryza sativa* L.). *Allelopathy J.*, 11(1): 71–76.

Sukul, S. and S. Chudhuri. 2001. Effect of foliar application of leaf extract containing plant growth promoting activity from *Lantana camara* on growth and yield of rice. *Indian J. of Plant Physiol.*, 6(2): 127–130.

Thakur, Anju, P.S. Thakur, Surekha Attri and S.P. Kanaujia. 1999. Effect of bioextracts on growth and yield of bell pepper under water stressed condition. *Indian J. Hort.*, 56(1): 46–51.

Tripathi, S.A., D.C. Tripathi, Kori and S. Paroha. 2000. Effect of *Dalbergia sissoo* extracts, rhizobium and nitrogen on germination, growth and yield of *Vigna radiata*. *Allelopathy J.*, 7(2): 255–264.

Venkata Raju, R.R. 1999. Allelopathic effect of *Plumbago-zeylanica* L. on it's associated plants. In: *Abstracts, 2nd World Congress on Allelopathy: Critical Analysis and Future Prospects*. (Ed.) A.U. Malik, pp. 181.

Yamada, K., K. Miyamoto, N. Goto, H. Kato-Naguchi, S. Kosemura, S. Yamamura and K. Hesegawa. 2002. Chemical and biological analysis of a hovel allelopathic substance, lepidimoide. In: *Abstr. 3rd World Congress on Allelopathy*, p. 237.

Chapter 18

Toxic Effect of High Chromium Intake on the Human Body and Chromium Removal from Water with Low Cost Adsorbents

Jyotsna Lal* and Kanchan Sachan
Chemistry Department, Christ Church Post Graduate College, Kanpur – 208 001, U.P., India

ABSTRACT

The heavy metal chromium is emitted from various sources for example electroplating industry, paint and pigment waste, cotton textiles dyeing, glassm chrome leather tanning industries. Hexavalent form of Chromium is more toxic than trivalent form, Cr (VI) causes bronchial cancer in man. The maximum permissible limit of Cr in drinking water as recommended by WHO is 0.05 mg\l. The industrial town of Kanpur has a many cotton textiles, chrome leather tanning, chemical manufacturing industries. A survey was conducted and found that in comparison to other industrial areas of India the groundwater of this area contains a very high concentration Cr (VI). Most water treatment technologies suitable for the developed countries are expensive for towns and small industries of the developing countries because of limitation of funds and lack of skilled man power. There is a increasing search for low cost adsorbent material made from locally available waste material and plant substrata. Some low cost adsorbents selective for Cr (VI) have been tested in the laboratory by batch and column technique pH, concentration and flow rate on break-through capacity have been investigated.

Keywords: *Virons or carcinogenic metals, Chromium, Sawdust, Tealeaves carbon, Red mud rice husk carbon; Rice straw, Brick kiln ash, Flyash, Groundnut husk carbon, Baggase.*

* E-mail: jyotsna_lal@yahoo.com.

Introduction

Carcinogen chromium is emitted from waste effluents of electroplating, paint, pigment waste, cotton textiles dyeing, glass, chrome leather tanning industries (Table 18.1) (Poonam Tyagi 1999). A survey was conducted and found that in comparison to other industrial areas of India the groundwater of industrial town of Kanpur in Uttar Pradesh contains a very high concentration Cr(VI) (Figure 18.1) (GWQS/8/1996–97). The max permissible limit of Cr(VI) in drinking water, recommended by WHO is 0.05 mg\l Hepato and nephrato toxic effects have been reported in rats (Meenakshi, 1989). Cr is highly toxic to fish and other aquatic life, causes changes in oxygen uptake and blood parameters in the fish *Sarotherodan mossambicus* (Peters) (Pamila D., 1991).

Table 18.1: Chromium in the Groundwater of Industrial Areas of India

Area	*mg/L*	*Industry*
Rourkela, Bihar	0–0.01	Steel plant
Cochin, Kerala	0.008–0.019	Tanneries
Jharia, Bihar	0.01–0.09	Ore-mining
Himachal Pradesh	0.001–0.019	Ore-mining
Haryana		Bicycles
Punjab		Woollen garments

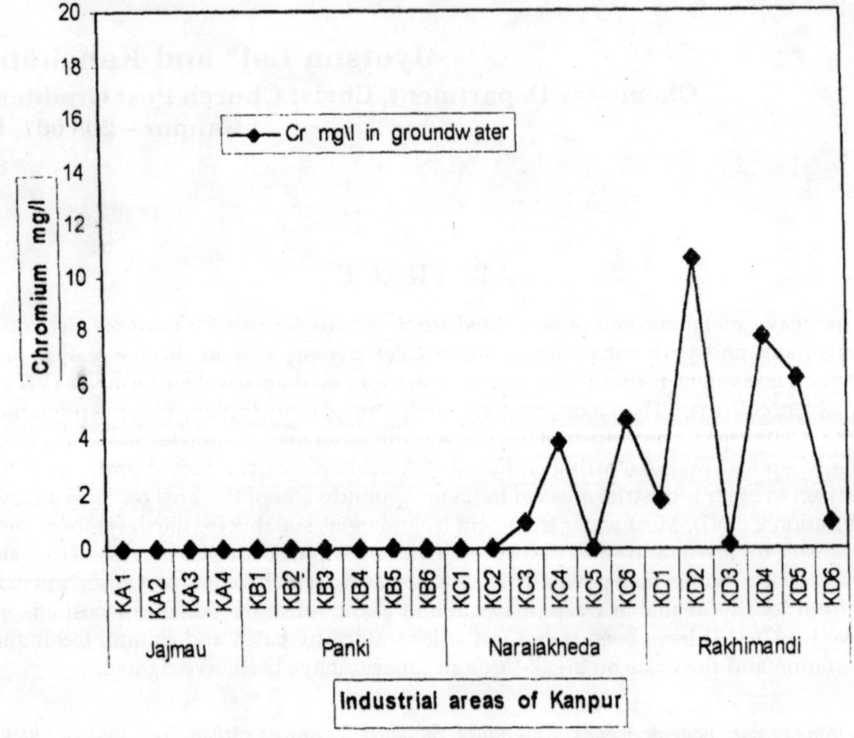

Figure 18.1: Chromium in Groundwater of Industrial Areas of Kanpur City, Uttar Pradesh (India)

The virons or carcinogenic metals with their tendency for chelation are associated with vitamins; protein and nucleic acid. These metals ions may replace the (ZnII, MgII) from the normal enzyme system or nucleic acid and may alter the structure and function of genetic material resulting in the formation of cancer. It has been shown that specific metals are required for genetic regulation and wrong metal ions, even the essential metal ions in wrong concentrations can lead to errors resulting in metal carcinogenesis. The fashioned DNA molecule require only deoxy nucleotides and not ribonucleotide be incorporated in to the nucleic acid. The ability of DNA polymerase 1 to distinguish between these two monomers types has been shown to depend on the nature or the metal used to activate the enzyme thus the presence of Mg(II) insure incorporation of deoxynucleotides, but Mn(II) and Cr(VI) permits the concurrent incorporation or ribonucleotides. Thus the presence of Mn(II) and Cr(VI) lead to incorporation of the wrong sugar (Ribonucleotidees) in place or deoxynucleotides in to the nucleic acid with both type of enzyme (RNA and DNA polymerase) carcinogenic metals are suppose to cause error (Chakrawarti, 2003).

Precipitation, ion-exchange, adsorption are technologies are available for removing Cr(VI) from water. Reduction and precipitation of chromate with barium or lead, the major problem with this type of treatment is the disposal of precipitated chromium hydroxide. Ion-exchange treatment does not have a sludge disposal problem with the advantage of reclamation of Cr(VI) but has disadvantages of being expensive, poorly selective for chromate over foreign anions and a critical flow rate while the process of adsorption has an edge over other methods due to its simplicity and sludge free clean operation. There is an increasing search for low cost adsorbent material made from locally available waste material and plant substrata. Some adsorbents selective for Cr(VI) have been tested in the laboratory by batch and column technique, pH, concentration and flow rate on break-through capacity have been investigated.

Materials and Methods

Reagents and Methods

Solutions of Cr(VI) were prepared in tapwater using $K_2Cr_2O_7$ (BDH, India) pH was adjusted with HCl. Presence of Cr (VI) in the effluent was determined by Diphenyl Carbazide. All the reagents used were analytical grade.

Locally available sawdust was sieved (60–100 mesh) was treated with water to remove the impurities, then the water was filtered off using cloth bag.

Waste tea leaves (TL) were treated hot demineralised water, dried at 100°C. A pseudo activated carbon was prepared by reacting 10 gm TL/10 ml H_2SO_4. The carbonised material was washed with DMW and dried at 100°C sieved (60–100 mesh) gave the adsorbent waste tea leaves carbon. Rice straw was procured from the fodder of the local market cut 1 cm long size, soaked in water overnight, washed free of colored impurities and dried sieved (60–100 mesh). Adsorption capacity by column and batch technique, Breakthrough capacity of Cr(VI) was determined by passing Cr(VI) solution through a glass column (i.d. 1.25 cm) packed with 5 gm adsorbent on a glasswool support. The flow rate was maintained at 125 ml/h. Batch studies were carried out in glass stoppered conical flasks containing 100 ml solution with different dosage and contact time in a mechanical shaker.

Results and Discussion

Some low cost adsorbents selective for Cr(VI) have been tested in the laboratory by batch and column technique pH, concentration and flow rate on break-through capacity have been investigated. The author compared some adsorbents which are available through out the year. Human Hair was a poor adsorbent at pH range 1–5 removed 1.1 mg Cr/gm (Tan, 1985). Similarly Brick kiln ash showed only 50 per cent removal as compared to Fly ash at pH 1.3 from chromeplating wastewater as shown in Figure 18.2 (Rai, 1999). Bituminous coal at pH 2 removed 2.5mg/gm (Narayan, 1989) Adsorbent made from a mixture Fly ash–Chinaclay showed 100 per cent removal at pH 2 from solution 4.3mg Cr(VI)/L (Pandey, 1984).

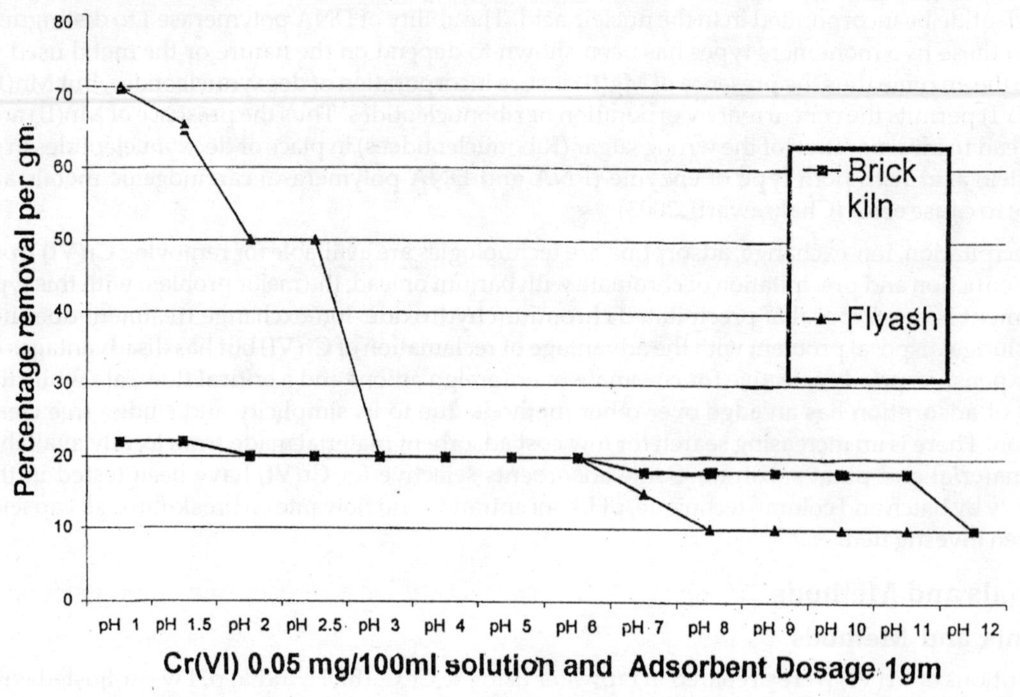

Figure 18.2: Cr(VI) Removal by Adsorption on Brick Kiln Ash and Fly Ash

Columns packed with Sawdust a natural inexpensive material, available everywhere can be utilized as low cost adsorbent for the removal of Cr(VI) from aqueous solution, the adsorption capacity was determined as 97.5 mg/5 gm or 19.5 mg/gm at pH~2. Break through capacity as a function of pH is plotted in Figure 18.3. Rice straw an inexpensive material can be utilized as low cost adsorbent for the removal of Cr(VI) from aqueous solution (Figure 18.4), the adsorption capacity was determined 25mg Cr/gm Baggase is reported to remove 80.80 per cent at pH 6–7 from solution 3.5gm Cr(VI)/L (Rao, 1999). Tea leaves carbon (TLC) can play an important role in treatment processes or metal bearing industrial effluents, a cheap carbonaceous adsorbent made from waste tea leaves Break through capacity 39.3 mg/m. In Figure 18.5, Cr(VI) removal as a function of pH and carbon dosage can be seen. The amount of TLC required for 100 per cent Cr(VI) removal 433 mg Cr(VI)/L carbon dosage 1200 mg/L at pH 1.5–2.0 Author compares TLC to Groundnut husk carbon GHC. It can be seen that, Activated Groundnut husk carbon at pH range 0.5–1.7 showed 100 per cent removal from 10 mg Cr(VI)/L solution, carbon dosage 2400 mg/L in Figure 18.6 (Periswamy, 1991).

Presence of holocellulose and lignin on sawdust and rice straw can be explained for removal of Cr(VI) from aqueous solution. Sigworth and Smith 1972 have proposed the possibility of heavy metal precipitation on the carbon surface by nucleation as one of the pathways for heavy metal removal by

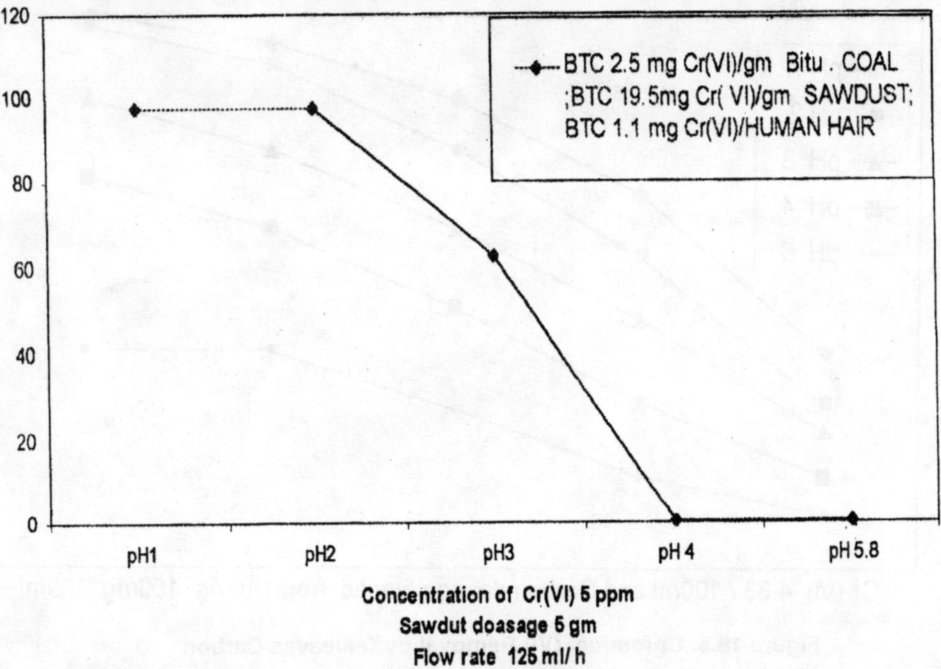

Figure 18.3: Chromium (VI) Removal by Sawdust

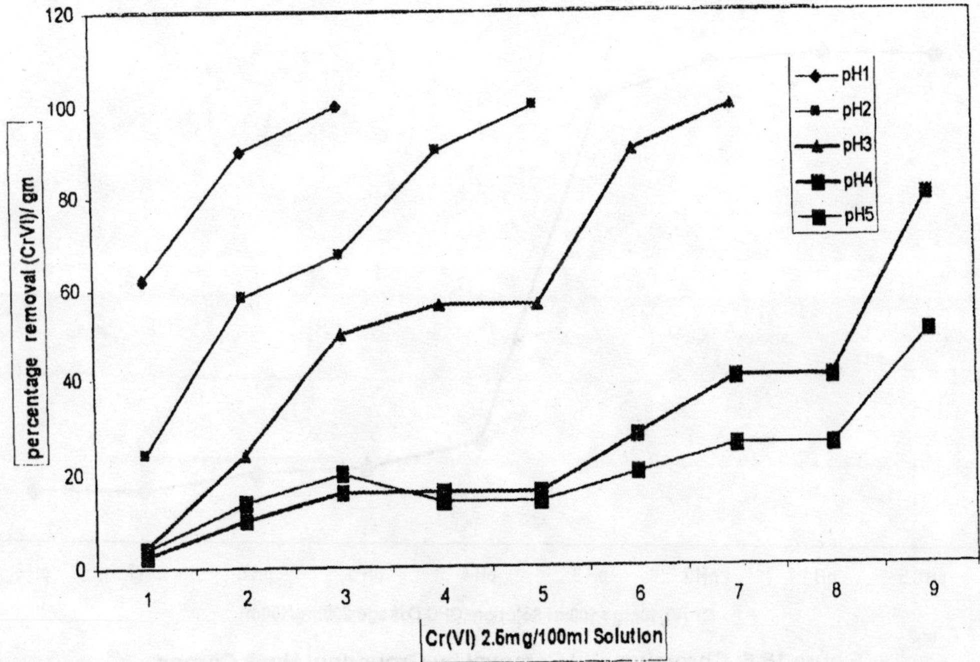

Figure 18.4: Chromium (VI) Removal by Rice Straw

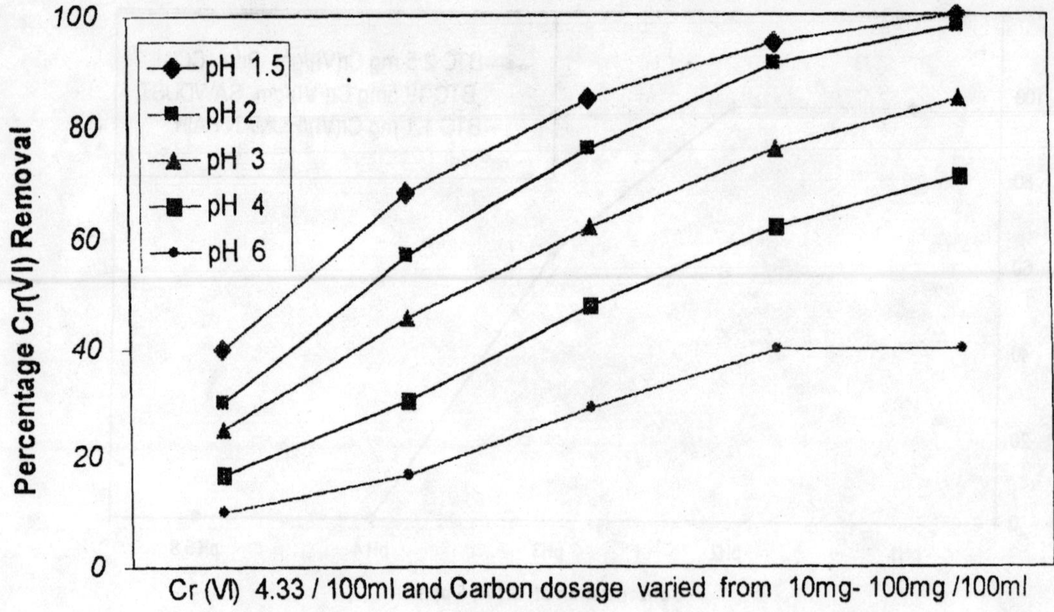

Figure 18.5: Chromium (VI) Removal by Tealeaves Carbon

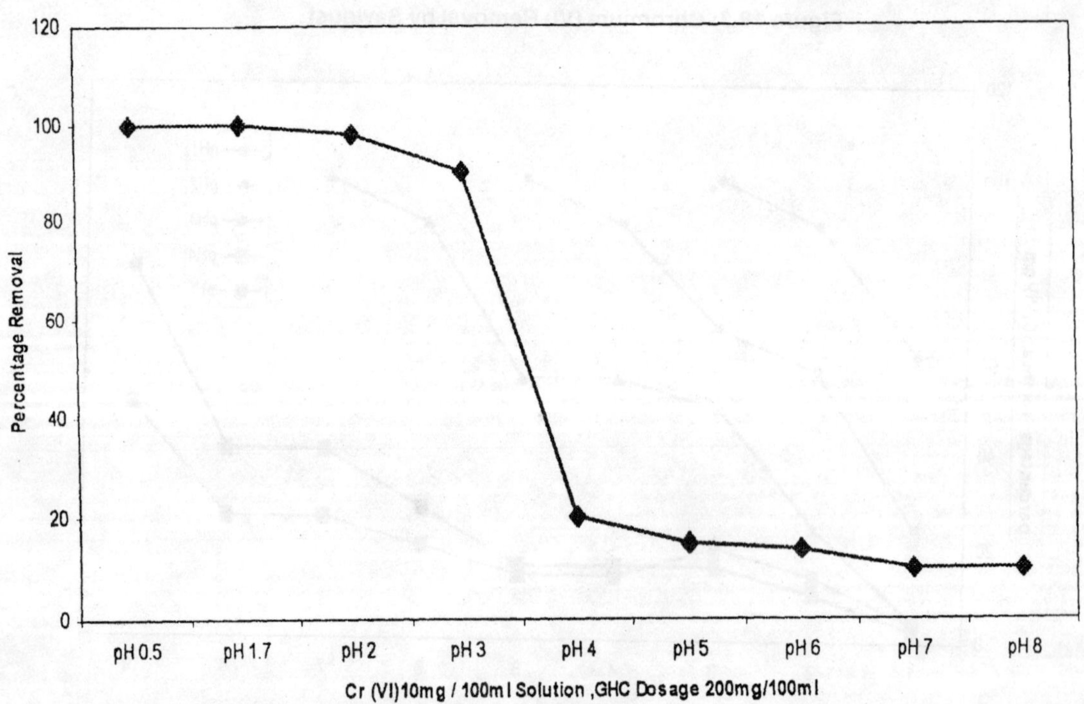

Figure 18.6: Chromium (VI) Removal by Groundnut Husk Carbon

activated carbon. The adsorption of Cr(III) on red mud is highly dependent on pH = 5–6 Cr(III) adsorption on red mud also depends the upon contact time, sorbent dose and initial concentration or Cr(III) Red mud showed 99.6 per cent Cr(III) removal from a solution containing 150 mg Cr (III)/100 ml (Figure 18.7) (Kumar Raj, 1999).

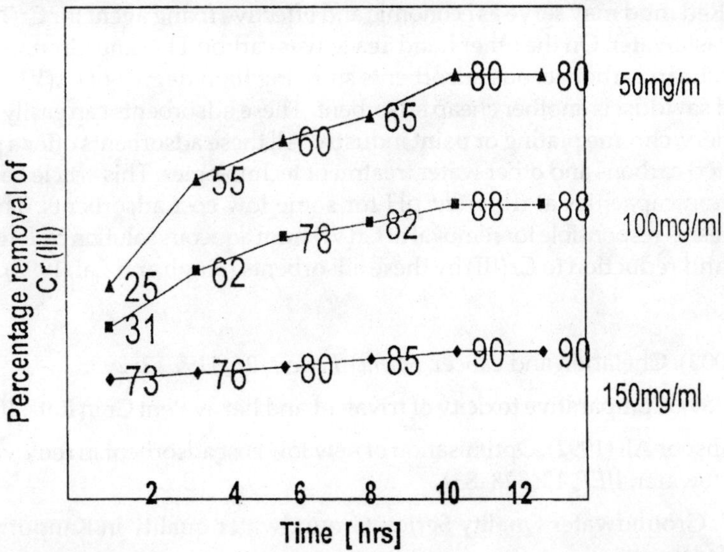

Figure 18.7: Effect of Agitation Time and Initial Concentration of Cr(III) on its Adsorption on Red Mud

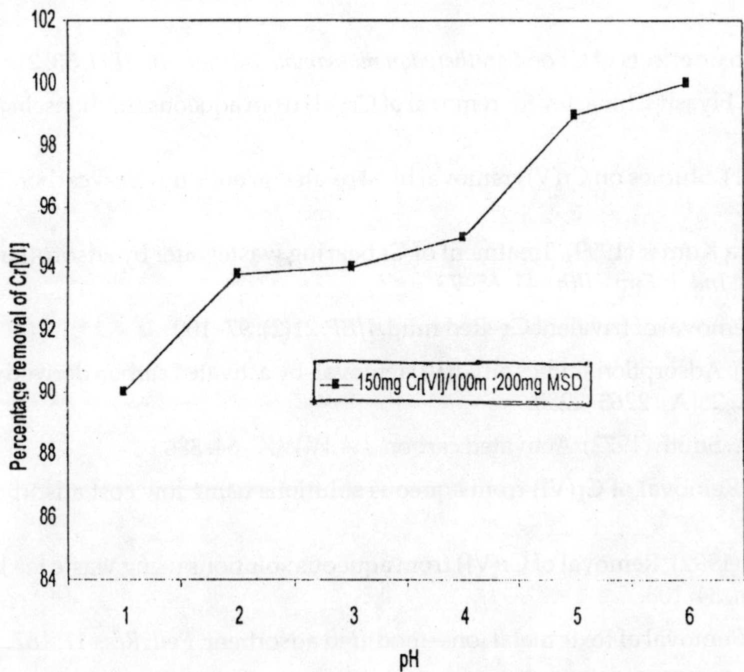

Figure 18.8: Cr(VI) Removal by Iron Hexamine Modified Sawdust

Iron hexamine modified sawdust removed 19.53 mg Cr(VI)/gm in presence of other ions in pH range 2–6 (Figure 18.8) (Singh, 1992) Rice straw, Baggase and Sawdust are natural inexpensive materials, which are easily available and do not require chemical pre-treatment, exhibit exceptionally high degree of Cr(VI) adsorption, thus can be utilized for treatment of industrial waste effluents containing Cr(VI). Red mud may serve as economic and effective fixing agent for Cr(III) removal from chromium laden wastewater. On the other hand Tea leaves carbon TLC and Groundnut husk carbon GHC are excellent cheap carbonaceous adsorbents showing high degree of Cr(VI) adsorption. Iron hexamine modified sawdust is another cheap adsorbent. These adsorbents can easily prepared in any laboratory of a tannery, chrome plating or paint industry. All these adsorbents offer a good alternative to expensive activated carbons and other water treatment technologies. This article has been limited to a study of adsorption capacities at different pH for some low cost adsorbents. There is a need to describe the mechanisms responsible for removal of Cr(VI) from aqueous solution by low cost adsorbents, Cr(VI) adsorption and reduction to Cr(III) by these adsorbents (Singh and Lal, 1992).

References

Chakravati, P.B. (2003). Chelation and cancer. *Science Letters*, 26, 11 & 12: 295.

Meenakshi, C.E. (1989). Comparative toxicity of trivalent and hexavalent Cr in Rats. *IJEH*, 31: 250–256.

Deo, Nandita & Mansoor Ali (1992). Optimisation of new low cost adsorbent in removal of Cr(VI) from industrial wastewater. *IJEP*, 12: 828–834.

GWQS/8/1996–97. Groundwater Quality Series, Groundwater quality in Kanpur: Status, Sources and Control Measures.

Narayan, N. & K.A. Abburi (1989). Study of removal of Cr(VI) by adsorption on bituminous coal. *IJEH*, 31: 304.

Pamila, D. (1991). Toxic effects of Cr on *Sarotherodan mossambicus* (Peters). *IJEH*, 33: 218–224.

Pandey, K.K. (1984). Flyash-Chinaclay for removal of Cr(VI) from aqueous solutions. *Ind. J. Chem.*, 234: 514.

Periswamy, K. (1991). Studies on Cr(VI) removal by activated groundnut huskcarbon. *IJEH*, 33: 433–439.

Rai, A.K. & Surendra Kumar (1999). Treatment of Cr bearing wastewater by adsorption on brick kiln ash and fly ash. *Ind. J. Env. Hlth.*, 41: 65–73.

Rajkumar (1999). Removal of trivalent Cr–Red mud. *IJEP*, 21(2): 97–100.

Shantha, G.M. (1993). Adsorption studies on Cr(VI) removal–by activated carbon derived from Baggase. *J. Env. Sci. Hlth.*, 28(A): 2263–2280.

Sigworth, E.A. & J.A. Smith (1972). Activated carbon. *J.A.W.W.A.*, 64: 386.

Singh, D.K. (1990). Removal of Cr(VI) from aqueous solutions using low cost adsorbent. *JIE(I)–CH Eng.*, 70: 90.

Singh, D.K. & J. Lal (1992). Removal of Cr(VI) from aqueous solutions using waste tea leaves carbon. *Ind. J. Env. Hlth.*, 34: 108.

Singh, D.K. (1992). Removal of toxic metal ions–modified adsorbent. *Poll. Res.*, 11: 187.

Tan, T.C., Chia, C.K. & C.K. Teo (1985). Uptake of metal ions by chemically treated human hair. *Water Res.*, 19: 157.

Chapter 19

Influence of pH, Temperature, Carbon and Nitrogen Sources on the Growth of Fungal Species Isolated from Textile Dye Effluent Contaminated Soils

S. Devarajan and S.M. Hameed Sulaiman
Department of Environmental Sciences,
Tamil Nadu Agricultural University, Coimbatore – 641 003

ABSTRACT

The fungal cultures *Trametes* sp. and *Phanerochaete* sp. isolated from the textile dye effluent contaminated soil were selected and standardized for their growth and biomass production in different carbon and nitrogen sources to find the influence of operational conditions such as temperature, pH and agitation. The *Trametes* sp. showed higher growth and biomass production in fructose as carbon and ammonium chloride as nitrogen source at the optimum temperature and pH of 35ºC and 5.0 and 6.0 respectively at static condition. Similarly the *Phanerocheate* sp. showed higher growth and biomass production when fructose was given as carbon source and peptone as nitrogen source at the optimum temperature and pH of 35ºC and 5.0 and 6.0 respectively at static condition.

Keywords: Trametes sp., Phaenerocheate sp., Carbon, Nitrogen, Temperature, pH, Agitation.

Introduction

Textile effluent discharged into the river adversely affected the water quality rendering the down stream water unsuitable for drinking, bathing, irrigation and fish culture (Mohapatra *et al.*, 1990).

Even though the dye concentration may be well below 1 ppm *i.e.*, lower than many other chemicals found in wastewater the colour is predominant and visible. Biological treatment methods are cheap and often the best alternative with proper analysis and environmental control. Establishing suitable condition for microbial degradation may mean that adding nutrients to promote the growth of particular organism, adjusting moisture condition, temperature and pH etc. The basic concept is to provide critical environmental requirements which may be adverse in particular site from where it is isolated.

Materials and Methods

Factors Influencing the Growth of Fungi

Two cultures of fungi *viz., Trametes* sp and *Phanerochaete* sp were selected for further studies. Factors which influenced the growth of fungi like carbon and nitrogen sources, temperature, pH of the medium and agitation were studied.

Effect of Different Carbon Sources on Biomass Production

The effect of different carbon sources on the growth and biomass production of the fungi was studied. In the Czapek-Dox mineral medium, various sources of carbon were added at one per cent level.

The carbon sources include glucose, fructose, xylose, sucrose, maltose, mannitol, starch and galactose. One hundred ml of the medium was dispensed in to 250 ml Erlenmeyer flask and steam sterilized. The pH was adjusted to 5.5 before sterilization. Mycelial disc of 0.6 cm diameter was inoculated aseptically.

The flasks were incubated at 35°C for seven days under static conditions. After seven days of incubation the biomass production was estimated. The mycelial growth was separated by filtration, dried at 80°C for 3 h and estimated as fungal biomass and expressed in gL^{-1}.

Effect of Different Nitrogen Sources on the Biomass Production

The effect of different nitrogen sources on the growth and biomass production of the fungal isolates were studied. The nitrogen source in Czapek-Dox medium was replaced with various organic and inorganic nitrogen sources at 0.3 per cent level. The inorganic nitrogen sources were ammonium chloride (NH_4Cl), sodium nitrate ($NaNO_3$), potassium nitrate (KNO_3), ammonium sulphate [$(NH_4)_2 SO_4$] and diammonium phosphate. Organic nitrogen sources were yeast extract, glycine and peptone. One hundred ml of culture medium amended with different nitrogen sources was taken in 250 ml Erlenmeyer flask and steam sterilized. The pH was adjusted to 5.5 before sterilization. Mycelial disc of 0.6 cm diameter was inoculated aseptically.

The flasks were incubated at 35°C for seven days under static conditions. The growth and biomass production was observed. The mycelial growth was l separated by filteration, dried at 80°C for 3 hours and estimated as fungal biomass and expressed in gL^{-1}.

Effect of Temperature

Czapek-Dox broth with one per cent glucose as carbon source was prepared. The pH of the medium was adjusted to 5.5. One hundred ml of the broth was dispensed in 250 ml Erlenmeyer flask and steam sterilized. The flasks were inoculated with 0.6 cm diameter of mycelial disc of the isolated fungi and incubated at different temperatures of 15°C, 35°C and 45°C for seven days under static condition. The growth and biomass production was observed. The mycelial growth was separated by filteration, dried at 80°C for 3 hours and estimated as fungal biomass and expressed in gL^{-1}.

Effect of pH

Czapek-Dox broth with one per cent glucose as carbon source was prepared. The broth was maintained at different pH levels adjusted to 5.0, 6.0, 7.0, 8.0 and 9.0 by adding calculated quantities of 0.5 N HCl or 0.5 N KOH. One hundred ml of the broth was dispensed in 250 ml Erlenmeyer flask and steam sterilized. The flasks were inoculated with 0.6 cm diameter of mycelial disc of the isolated fungi and incubated at 35°C for seven days under static condition. The growth and biomass production was observed. The mycelial growth was separated by filteration, dried at 80°C for 3 hours and estimated as fungal biomass and expressed in gL^{-1}.

Effect of Agitation

Czapek-Dox with one per cent glucose as carbon source was prepared. The pH of the broth was adjusted to 5.5. One hundred ml of the broth was dispensed in 250 ml Erlenmeyer flask and steam sterilized. The flasks were inoculated with 0.6 cm diameter of mycelial disc of the isolated fungi and kept at different frequency of shaking *viz.*, 0, 50, 100, 150 strokes per min at 35°C. The growth and biomass production was observed. The mycelial growth was separated by filteration, dried at 80°C for 3 hours and estimated as fungal biomass and expressed in gL^{-1}.

Results and Discussion

Factors Influencing the Optimum Growth of Fungal Isolates

Effect of Different Carbon Sources on the Biomass Production of Isolated Fungi

The influence of different carbon sources on the biomass production of the isolated fungi are presented in the Table 19.1.

Table 19.1: Effect of Different Carbon Sources on Growth of *Trametes* sp. and *Phanerochaete* sp.

Sources of Carbon	*Biomass Production (g l^{-1} dry weight)*	
	Trametes sp.	*Phanerochaete sp.*
Glucose	5.54	3.15
Fructose	5.71	4.71
Xylose	3.73	1.46
Sucrose	3.39	2.57
Lactose	1.47	1.58
Mannitol	5.55	3.11
Starch	2.05	2.61
Cellulose	3.01	2.41
SEd.	0.06	0.04
CD (0.05)	0.13	0.09

In general, all the carbon sources used in the study encouraged fungal growth. Fructose (5.71 g L^{-1}), glucose (5.54 g L^{-1}) and mannitol (5.55 g L^{-1}) recorded increased biomass production as compared to other carbon sources in *Trametes* sp. Lactose recorded poorest biomass production

(1.47 g L^{-1}). Similar in *Phanerochaete* sp fructose, glucose and mannitol recorded increased biomass production of 4.71, 3.15 and 3.11 g L^{-1} respectively when compared to other carbon sources.

Effect of Different Nitrogen Sources on the Biomass Production of Isolated Fungi

The effect of different sources of nitrogen on the biomass production of the isolated fungi is presented in the Table 19.2. The nitrogen sources included nitrate nitrogen, ammonical nitrogen and organic source of nitrogen.

Table 19.2: Effect of Different Nitrogen Sources on Growth *Trametes* sp. and *Phanerochaete* sp.

Sources of Nitrogen	*Biomass Production (g l^{-1} dry weight)*	
	Trametes sp.	*Phanerochaete sp.*
Ammonium chloride	2.48	1.85
Sodium nitrate	3.54	1.79
Diammonium phosphate	3.52	2.28
Ammonium sulphate	4.16	3.87
Potassium nitrate	3.06	2.86
Yeast extract	3.45	2.78
Peptone	3.03	4.43
Glycine	3.48	3.11
SEd.	0.05	0.05
CD (0.05)	0.11	0.10

Ammonium sulphate recorded highest biomass production of 4.16 g L^{-1}. Sodium nitrate (3.54 g L^{-1}), diammonium phosphate (3.52 g L^{-1}) and glycine (3.48 g L^{-1}) were on par with each other in biomass production in *Trametes* sp.

In *Phanerocheate* sp. the peptone recorded maximum growth of biomass (4.43 g L^{-1}) followed by ammonium sulphate (3.87 g L^{-1}) and glycine (3.11 g L^{-1}).

Effect of pH on Biomass Production of Isolated Fungi

In general, pH of the medium set at 5.0 and 6.0 supported higher biomass production in fungi. In *Trametes* sp. at pH 5 and 6 the biomass production was 2.99 g L^{-1} and 4.28 g L^{-1} respectively. In *Phanerochaete* sp at pH 5 and 6 the biomass production was 2.39 g L^{-1} and 3.91 g L^{-1} respectively (Table 19.3). At pH 8.0 the biomass production of *Trametes* sp. and *Phanerochaete* sp. recorded the lowest value of 1.97 and 1.91. Similarly at pH 9.0 the biomass production of *Trametes* sp. and *Phanerochaete* sp. recorded the lowest value of 1.62 and 1.73 g L^{-1} respectively.

Effect of Temperature on Biomass Production of Isolated Fungi

In general, 35°C appeared optimum for the growth of all the isolated fungi. *Trametes* sp. and *Phanerochaete* sp. at 35°C recorded 4.92 and 4.09 g L^{-1} biomass respectively (Table 19.4). At temperature of 15°C the biomass production of *Trametes* sp. and *Phanerochaete* sp. recorded the lowest value of 1.29 and 1.44 g L^{-1}. Similarly at temperature of 45°C the biomass production of *Trametes* sp. and *Phanerochaete* sp. recorded the lowest value of 1.17 and 2.07 g L^{-1} respectively.

Table 19.3: Effect of Different pH on Growth *Trametes* sp. and *Phanerochaete* sp.

pH	*Biomass Production (g l^{-1} dry weight)*	
	Trametes sp.	*Phanerochaete sp.*
5	2.99	2.39
6	4.28	3.91
7	2.44	2.31
8	1.97	1.91
9	1.62	1.73
SEd.	0.04	0.04
CD (0.05)	0.10	0.09

Table 19.4: Effect of Different Temperature on Growth *Trametes* sp. and *Phanerochaete* sp.

Temperature	*Biomass Production (g l^{-1} dry weight)*	
	Trametes sp.	*Phanerochaete sp.*
15	1.29	1.44
35	4.92	4.09
45	1.17	2.07
SEd.	0.03	0.02
CD (0.05)	0.06	0.05

Effects of Agitation on the Biomass Production of Isolated Fungi

Both the fungal cultures *Trametes* sp. (5.70 g L^{-1}) and *Phanerochaete* sp. (6.01 g L^{-1}) recorded higher biomass at static condition. Further increase in agitation tended to decrease fungal growth and mycellial pellets were formed (Table 19.5).

Table 19.5: Effect of Different Agitation on Growth *Trametes* sp. and *Phanerochaete* sp.

Agitation	*Biomass Production (g l^{-1} dry weight)*	
	Trametes sp.	*Phanerochaete sp.*
0 rpm	5.7	6.01
50 rpm	2.99	2.63
100 rpm	2.04	2.01
150 rpm	1.72	1.67
SEd.	0.03	0.03
CD (0.05)	0.07	0.10

In the present study *Trametes* sp. and *Phanerochaete* sp. were grown under eight different carbon and nitrogen sources at various levels of pH, temperature and shaking condition. The results showed that *Trametes* sp. yielded higher biomass when fructose and mannitol were used as carbon source and ammonium sulphate and sodium nitrate were used as nitrogen source at pH of 6.0 and temperature of

35°C under static condition. Similar reports were given by Pilat (1936) in *Polyporus versicolor.* In the case of *Phanerochaete* sp., the biomass production was higher when fructose was used as carbon sources and peptone and ammonium sulphate as nitrogen sources at pH 6.0 and temperature of 35°C under static condition. Similar results were reported by Sheng Hua wu (2000). Similarly Manorama and Neelakantan (1991) reported that pH 6.0 and a temperature of 35°C are favourable for good growth and enzyme activity of *Phanerochaete chrysosporium.*

References

Manorama and S. Neelakantan, 1991. Effect of different pH and temperature on the enzymatic activities of six ligocellulolytic fungi. In: *Abs XXXI Annual Conference of the Association of Microbiologists of India, Coimbatore*, pp. 44–45.

Mohapatra, P.K., L.N. Patnaik and G. Mista, 1990. Pollution due to textile industry: A case study. *Asian Env.*, 12: 51–66.

Pilat, 1936. Factors effecting growth of *Coriolus versicolor. Atl. Champ. Eur.*, 3: 261.

Pitt, J.I., 1979. The genus *Penicillum* and its telemorphic states. *Eupenicillum and Talaromyces*. Academic Publisher, New York.

Sheng Hua Wu, 2000. Six new species of Phanerochaete from Taiwan. *Botanical Bulletin of Academia Sinica*, 41: 65–174.

Chapter 20

Effect of Triazole on Growth and Chlorophyll Pigments in Radish

R. Sridharan,* A. Kishorekumar, R. Somasundaram, P. Manivannan, G.M. Alagu Lakshmanan, M. Gomathinayagam and R. Panneerselvam

Department of Botany, Annamalai University, Annamalainagar – 608 002, Tamil Nadu, India

ABSTRACT

The present study was conducted to investigate the effect of triazole compounds Triadimefon (10 mg L^{-1}) and Hexaconazole (5 mg L^{-1}) on radish. Each plants were treated with triadimefon and hexaconazole on 8th and 23rd DAS (Days After Sowing) in 30 cm plastic pot. The plants were harvested on 15th, 30th, 45th and 60th DAS for determining the effect of both triazoles on growth and chloroplast pigments. Both the triazole compounds increased the fresh and dry weight of tuber and pigments. However, both the treatments decreased the fresh and dry weight of shoot and leaf area.

Keywords: Triadimefon, Hexaconazole, Radish, Pigments, Growth.

Introduction

Radish is one of the important edible vegetable tuber crop cultivated in India and south east countries. Triazole compounds are mostly used as the systemic fungicides, also have plant growth regulating properties and effectively retard the shoot growth in monocotyledons and dicotyledons

---

* Corresponding Author.

(Fletcher *et al.*, 2000). Triazole compounds like triadimefon and hexaconazole have growth regulating properties and induced many morphological changes like reduction in shoot elongation, stimulation of rooting, inhibition of gibberellin biosynthesis, increased chlorophyll content, alter carbohydrate status and increased cytokinin synthesis (Davis *et al.*, 1988; Isumi *et al.*, 1988).

Triazole compounds have been shown to improve the yield of many root crops such as carrot, sugar beet, and potato (Davis and Curry, 1991) Triazole compounds influences hormone balance, photosynthetic rate, enzyme activities, lipid peroxidation and yield components in various crop plants (Zhou *et al.*, 1993; Ye *et al.*, 1995; Zhou and Ye, 1996). Triazole inhibit cytochrome P-450 mediated oxidative dimethylation reaction, including those which are necessary for the synthesis of ergosterol and the conversion of kaurene to kaurenoic acid in the gibberellin biosynthetic pathway (Rademacher *et al.*, 1987). Paclobutrazol induces a variety of morphological and biochemical responses in plants including reduction in shoot elongation, stimulation of rooting, increased cytokinin synthesis and a transient increase in ABA content (Fletcher and Hofstra, 1988).

The objective of the present study is to estimate the effect of triadimefon and hexaconazole on the growth, fresh and dry weight of shoot and tuber and photosynthetic pigments in *Raphanus sativus.* (Cv. Pusa chetki) during the growth and maturation of tubers.

Materials and Methods

The seeds of *Raphanus sativus* L. Cv. (Pusa Chetki) were obtained from Mahyco-hybrid seeds Co. Ltd, Maharashtra. The seeds were surface sterilized with 0.2 per cent $HgCl_2$ solution for 5 minutes with frequent shaking and thoroughly washed with tap water to remove $HgCl_2$. Two seeds were sown in each plastic pot of 30 cm diameter and 30 cm height containing 3 kg of soil mixture composed of red soil, sand and the farm yard manure (FYM) at 1 : 1 : 1 ratio. The seedlings were thinned to one per pot on 6th days after sowing. The pots were irrigated daily with tap water and care was taken to avoid leaching. The seedlings were treated with deionized water (control), 10 mg L^{-1} Triadimefon and 5 mg L^{-1} Hexaconazole solution alone per plant on 8th and 23rd days after sowing (DAS). Plants were harvested randomly on 15th, 30th, 45th and 60th DAS and separated into tuber and shoot and used for determining growth parameters. Total leaf area was determined using an electronic leaf area meter (Licor model L1-3100, Licoin USA). The dry weight was determined by drying the plant materials in an oven at 60°C until a constant dry weight was obtained. Chlorophyll were estimated by the method of Arnon (1949), carotenoid content was calculated using the formula of Kirk and Allen (1965). Analysis of variance (ANOVA) by the method outlined by Ridgman (1975).

Results and Discussion

Triazole treatment decreased the fresh weight of shoot significantly. Among the treatments triadimefon decreased it to a larger extent when compared to Hexaconazole (Table 20.1). Triazole treatments reduced the stem elongation, plant height and decreased the fresh weight of the shoot in bean (Asare-Boamah *et al.*, 1986), *Echingchola frumentacea* (Sankala *et al.*, 1992), grapevines (Hunter and Procter, 1992), citrus (Mehouach, *et al.*, 1996) *Prunus serotina* (Eliasson *et al.*, 1994) and Mango seedlings (Murti and Upreti 2005).

Triazole treatment decreased the dry weight of the shoot significantly in *Raphanus sativus.* Among the triazole treatment triadimefon decreased the shoot dry weight to a larger extent when compared to hexaconazole (Table 20.1). Triadimefon, triadimenol, paclobutrazol and S.3307 reduced the height and shoot dry weight in kentuchy blue grass (Kane and Smiley, 1983) and *Phaseolus vulgaris* (Fletcher *et al.*, 1986) and mango seedling (Murti and Upreti, 2005).

Table 20.1: Triadimefon and Hexaconazole Induced Variation in the Fresh Weight, Dry Weight and Leaf Area (Values are mean±SD of 3 replicates)

		Shoot				*Tuber*			
	DAS	*Control*	*TDM 10 mg L^{-1}*	*HEX 5 mg L^{-1}*	*LSD (P = 0.5)*	*Control*	*TDM 10 mg L^{-1}*	*HEX 5 mg L^{-1}*	*LSD (P = 0.5)*
	15	5.409	5.307	5.347	0.265	2.015	2.055	2.147	0.014
Fresh weight	30	21.210	19.950	20.289	0.835	40.100	44.258	44.380	0.147
gram plant^{-1} F.W	45	34.526	30.180	31.880	1.235	72.950	89.510	90.090	1.834
	60	41.000	35.000	36.780	1.146	125.000	163.170	163.63	3.467
	15	0.665	0.657	0.673	0.026	0.193	0.193	0.202	0.007
Dry weight	30	2.142	2.035	2.080	0.076	3.449	3.506	3.786	0.090
gram plant^{-1} D.W	45	3.384	2.979	3.124	0.123	5.676	6.623	6.675	0.128
	60	3.526	3.017	3.151	0.106	7.788	10.279	10.488	0.226
	15	76.94	75.90	74.73	2.86				
Leaf area	30	232.45	219.79	217.31	6.45				
cm^2 plant^{-1}	45	328.08	299.55	291.99	12.15				
	60	603.89	526.72	514.29	18.62				

Treatment with triazole compounds significantly increased the tuber weight of the radish. The fresh weight of the tuber increased with triazole treatments. Among the traizoles, hexaconazole treatment increased it to a higher level than triadimefon (Table 20.2). Triazole treated plants have exhibited thickened fleshy roots with increased root diameter and weight in apple and soybean (Steffens *et al.*, 1983; Sankhla *et al.*, 1986), Tomato (Berova *et al.*, 2000) and wheat (Berova *et al.*, 2002).

The dry weight of the tuber also increased by triazole treatment. Among the triazole, hexaconazole treatment significantly increased the dry weight of the tuber when compared to triadimefon (Table 20.1). The fresh and dry weight of the root were significantly higher in the triadimefon treated apple (Huang *et al.*, 1995) and mango seedlings (Murti and Upreti, 2005). Inhibited gibberellin biosynthesis and increased cytokinin and abscisic acid content induced by these triazoles might be the cause for decreased fresh and dry weight of shoot and increased fresh and dry weight of tuber growth in radish.

Both triadimefon and hexaconazole treatments inhibited the leaf area. Among these, hexaconazole treatment reduced the leaf area to a larger extent than triadimefon. Triadimefon reduced the leaf area in maize (Khalil *et al.*, 1990) and wheat (Gao *et al.*, 1988). Similar results were observed in tebuconazole treated *Barassica napus* (Child *et al.*, 1993).

Triadimefon and hexaconazole treatments increased the chlorophyll content to a larger extent when compared to control in radish. Among them hexaconazole treatment increased it to a larger level than triadimefon (Table 20.2). Increased chlorophyll content with triadimefon treatment was observed with uniconazole, LAB-150978, BAS-110 W treatments in *Cucumber* cotyledons (Thomas and Singh, 1995) and in paclobutrazol treated tomato leaves (Berova *et al.*, 2000).

The leaves of the triazole treated plants showed an increased carotenoid content when compared to control. Among these hexaconazole treatment showed higher level than triadimefon (Table 20.2).

Higher level of carotenoid content was observed in triadimefon treated cowpea (Gopi *et al.*, 1999), uniconazole treated wheat (Fletcher and Hofstra, 1998), tomato (Senaratna *et al.*, 1988) and dichlorobutrazole treated *Zea mays* (Khalil et *al.*, 1990).

Table 20.2: Triadimefon and Hexaconazole Induced Variation in the Total Chlorophyll, Chlorophyll-a, Chlorophyll-b and Carotenoid (Values are mean±SD of 3 replicates)

	DAS	*Leaf*			
		Control	*TDM 10 mgL⁻¹*	*HEX 5 mgL⁻¹*	*LSD (P = 0.5)*
	15	0.154	0.191	0.180	0.008
Total chlorophyll	30	0.207	0.234	0.233	0.007
mg g⁻¹.F.W	45	0.307	0.330	0.335	0.012
	60	0.404	0.440	0.452	0.014
	15	0.071	0.086	0.082	0.003
Chlorophyll-a	30	0.090	0.104	0.105	0.005
mg g⁻¹.F.W	45	0.145	0.159	0.161	0.008
	60	0.215	0.242	0.248	0.012
	15	0.082	0.104	0.098	0.003
Chlorophyll-b	30	0.117	0.130	0.128	0.004
mg g⁻¹.F.W	45	0.162	0.171	0.174	0.006
	60	0.189	0.198	0.204	0.009
	15	0.370	0.390	0.393	0.007
Carotenoid	30	0.510	0.520	0.536	0.005
mg g⁻¹.F.W	45	0.626	0.682	0.684	0.009
	60	0.653	0.731	0.738	0.013

References

Arnon, D.I., 1949. Copper enzymes in isolated chloroplasts, polyphenol oxidase in *Beta vulgaris* L. *Plant Physiol.*, 24: 1–15.

Asare-Boamah, N.K., G. Hofstra, R.A. Fletcher and E.B. Dumbroff, 1986. Triadimefon protects bean plants from water stress through its effects on abscisic acid. *Plant Cell Physiol.*, 27: 383–390.

Berova, M., Z. Zlatev and N. Stoeva, 2000. Physiological response and yield of paclobutrazol treated tomato plants (*Lycopersicon esculentum* Mill). *Plant Growth Regul.*, 39: 117–123.

Bervova, M., Z. Zlatev and N. Stoeva, 2002. Effect of paclobutrazol on wheat seedlings under low temperature stress. *Bulg. J. Plant Physiol.*, 28(1–2): 75–84.

Berova, M. and Z. Zlatev, 2000. Physiological response and yield of paclobutrazol treated tomato plants (*Lycopersicon esculentum* Mill.). *Plant Growth Regul.*, 30: 117–123.

Child, R.D., D.E. Evans, J. Allen and G.M. Arnold, 1993. Growth responses in oil seed rape (*Brassica napus* L.) to combined applications of the triazole chemicals triapenthenol and tebuconazole interactions with gibberellin. *Plant Growth Regul.*, 13: 203–212.

Davis, T.D. and E.A. Curry, 1991. Chemical regulation of vegetative growth. *Critic. Rev. Plant Sci.*, 10: 151–188.

Davis, T.D., G.L. Steffens and N. Sankhla, 1988. Triazole plant growth regulators. In: *Hort. Rev.*, (Ed.) J. Janick. Timber Press, Portland, *Oregon*, 10: 63–105.

Eliasson, M.K., C.A. Beyl and P.A. Barker, 1994. *In vitro* responses and acclimatization of *prunus serotina* with Paclobutrazol. *J. Plant Growth Regul.*, USA, 13: 137–142.

Fletcher, R.A. and G. Hofstra, 1988. Triazoles as potential plant protectants. In: *Sterol Biosynthesis Inhibitors*, (Eds.) Berg, D. and M. Plempel. Ellis Horwood Ltd., Cambridge, England, pp. 321–331.

Fletcher, R.A., A. Gilley, N. Sankhla and T.M. Davis, 2000. Triazoles as plant growth regulators and stress protectants. *Hort. Rev.*, John Wiley and Sons Inc., 24: 56–138.

Fletcher, R.A., G. Hofstra and J. Gao, 1986. Comparative fungitoxic and plant growth regulating properties of triazole derivatives. *Plant Cell Physiol.*, 27: 367–371.

Gao, J., G. Hofstra and R.A. Fletcher, 1988. Anatomical changes induced by triazoles in wheat seedlings. *Can. J. Bot.*, 66: 1178–1185.

Gopi, R., B.M. Sujatha, S.N. Rajan, L. Karikalan and R. Panneerselvam, 1999. Effect of triadimefon in the sodium chloride stressed cowpea (*Vigna unguiculata*) seedlings. *Indian J. Agri. Sci.*, 69: 743–745.

Huang, W.D., T. Shen, S.H. Han and S. Liu, 1995. Influence of paclobutrazol on photosynthesis rate and dry matter partitioning in the apple tree. *Plant Nutr.*, 18: 901–910.

Hunter, D.M. and J.T.A. Procter, 1992. Paclobutrazol affects growth and fruit composition of potted grapevines. *Hort. Sci.*, 27: 319–321.

Izumi, K., S. Nakagawa, M. Kobayashi, H. Oshio, A. Sakurai and N. Takahashi, 1988. Levels of IAA, cytokinins, ABA and ethylene in rice plants as affected by a gibberellin biosynthesis inhibitor, uniconazole. *Plant Cell Physiol.*, 29: 97–104.

Kane, R.T. and R.W. Smiley, 1983. Plant growth regulating effects of systemic fungicides applied to kentucky blue grass. *Agron. J.*, 75: 469–473.

Khalil, I.A., E.I. Mercer and Z.X. Wang, 1990. Effect of triazole fungicides on the growth, chlorophyllase pigments and sterol biosynthesis of maize (*Zea mays* L.). *Plant Sci.*, 66: 21–28.

Kirk, J.T.O. and R.L. Allen, 1965. Dependence of chloroplast pigment synthesis on protein synthesis: Effect of acsidione. *Biochem. Biophys. Res. Commun.*, 21: 532–530.

Mehouachi, J., F.R. Tadeo, S. Zaragoza, E. Primo-Millo and M. Talon, 1996. Effects of gibberellic acid and paclobutrazol on growth and carbohydrate accumulation in shoots and roots of citrus rootstock seedlings. *Hort. Sci.*, 71: 747–754.

Murti, G.S.R and K.K. Upreti, 2005. Paclobutrazol induced growth retardation of mango seedlings. *Indian Plant Physiol.*, 10: 9–13.

Rademacher, W., H. Fritsch, E. Graebe, H. Sauter and J. Jung, 1987. Tetcyclacis and triazole-type plant growth retardants: Their influence on the biosynthesis of gibberellins and other metabolic processes. *Pestic. Sci.*, 21: 241–252.

Ridgman, W.J., 1975. *Experimentation in Biology: An Introduction to Design and Analysis*. Thomson Litho Ltd., East Kilbridge, Scoland, pp. 81–100.

Sankhla, N., T.D. Davis, V.D. Jolley and A. Upadhyaya, 1986. Effect of paclobutrazol on the development of iron chlorosis in soybeans. *J. Plant Nutr.*, 9: 923–934.

Sankhla, N.A. Upadhyaya, T.D. Davis and D. Sankhla, 1992. Hydrogen peroxidase scavenging enzymes and antioxidants in *Echinochloa frumentacea* affected by triazole growth regulators. *Plant Growth Regul.*, 11: 441–442.

Senaratna, T., C.E. Mackay, B.D. McKersie and R.A. Fletcher, 1988. Uniconazole induced chilling tolerance in tomato and its relationship to antioxidant content. *J. Plant Physiol.*, 133: 56–61.

Steffens, G.L., S.Y. Wang, C.L. Steffens and T. Brennan, 1983. Influence of paclobutrazol (pp 333) on apple seedling growth and physiology. *Proc. Plant Growth Regul. Soc. Amer.*, 10: 195–205.

Thomas, R.M. and V.P. Singh, 1995. Effect of three triazole derivatives on mercury induced inhibition of chlorophyll and caroteniod accumulation in cucumber cotyledons. *Indian J. Plant Physiol.*, 38: 313–316.

Ye, Q.F., W.J. Zhou, W.F. Xi and J.Y. Fang, 1995. Effects of S-3307 on levels of endogenous (IAA, ABA and ZT) and some physiological of rape seedlings. *Acta. Agri. Zhejiang*, 7: 451–456.

Zhou, W.J., H.C. Shen, H.F. Xi and Q.F. Ye, 1993. Studies on the regulation mechanism of paclobutrazol to the growth of rape plant. *Acta Agri. Univ. Zhejian*, 19: 316–320.

Zhou, W.J. and Q.F. Ye, 1996. Physiological and yield effects of uniconazole on winter rape (*Brassica napus* L.). *J. Plant Growth Regul.*, 15: 69–73.

Chapter 21

Saccharification Studies by a Thermophilic Fungus *Sporotrichum thermophile* Isolated from Agriculture Waste of Bhopal

Rajhans Singh[1], Namita Singh[2], Anil Prakash[3] and Sarika Poonia[2]

[1]Maharana Partap Degree College, Kalor Road, Bhopal – 462 016

[2]Department of Biotechnology, Guru Jambheshwar University, Hisar – 125 001, Haryana

[3]Department of Biotechnology, Barkatullah University, Bhopal – 462 026, M.P., India

ABSTRACT

Thermophilic fungus has been reported from various habitat *viz.* municipal waste, natural compost, coal refuse, oil palm, and stored grains etc. Thermophilic fungus has been exploited for number of unique properties, like enzymes, antibiotic, amino acids production.

Cellulose is available in nature as plant residues. The cellulosic wastes cause environmental pollution, conservation of these wastes into useful products through microbes is in progress. Number of thermophilic fungi produces extra cellular enzymes, which diffuse and break the cellulose chain by saccharification.

In the present study attempt has been made to characterize and standardize the production of enzymes from *Sporotrichum thermophile* fungus isolated from agriculture waste. Following observations were recorded. High enzymatic activity was recorded between 12–15 day of incubation and a sharp decline in the activity after 15th days. 45° C temperatures shows better growth and production of enzyme and pH 6.8 showed better production of cellulose. Hence *S. thermophile* can be utilized for minimizing the environmental pollution by conservation of cellulosic waste into the useful products.

Keywords: *Cellulose, Endo-β-glucanase, Exo-β-glucanase, Sporotrichum thermophile.*

Introduction

Cellulose is most abundantly produced biopolymer in terrestrial environments. Each year fixation of CO_2 yields more than 10^{11} tons of dry plant material all over world (Schlesinger. 1991) and almost half of this consist of cellulose (Eriksson *et al.*, 1990). Cellulose is one of the major components of municipal waste and agro wastes which cause potent sources of environmental pollution. This potentially valuable source of food energy is largely unavailable to monogastric animals because of resistance of cellulose to digestive enzymes (Coutts and Smith, 1976). Cellulose almost never occurs alone in nature but is usually associated with plant substances. This association affects its natural degradation. Bacteria and fungal degradation of cellulose occur exocellularly either in association with outer cell envelope layer or extracellularly. Their enzymes help in conversion of cellulose into simpler saccharides. Previously mesophilic microorganisms were known to produce cellulose enzyme (Coutts and Smith, 1976) but now-a-days thermophilic microorganisms have received considerable attention for production of more amount (Singh *et al.*, 1990; Wase *et al.*, 1985) and variety of enzymes. This is largely due to the fact that enzyme produced by thermophiles exhibit optimal activity at high temperature that their mesophilic counterparts (Maheshwari *et al.*, 2000; Feldman *et al.*, 1988; Grajek, 1987; Hagerdal *et al.*, 1980).

The thermophilic microorganisms have adopted themselves to a wide array and ecological niche such as comport, soil, birds nest, wood chips, seed grams, volcano and hot springs etc (Maheshwari *et al.*, 1987). The resistance to survive at high temperature has modified the metabolism of thermophiles in favour of production of enzymes and other secondary metabolites of industrial importance (Feldman *et al.*, 1988; Bhat and Maheshwari, 1987).

Thermophilic fungi have become an experimental system for the basic and applied research due to their unique properties (Maheshwari *et al.*, 1987, 2000). These include growth and temperature optima, habitat relationship, enzymatic abilities, degradation and deterioration characteristics as well as possible allergenic reactions. They produce important extracellular enzymes such as xylanase, amylase, protease, lipase and cellulase. The production of thermostable enzymes at elevated temperature by thermophilic fungi is of economic importance (Feldman *et al.*, 1988). Some of these have already been exploited commercially (Chaudhuri *et al.*, 1988).

The existence of cellulolytic microorganism is well documented. Several efforts have been made by many workers to develop preproductive cellulose mutants of thermophilic fungi (Hayashida and Mo, 1986; Sen and Chakarbarty, 1984). Cellulose is a prominent agrowaste in each country and creating too much environmental pollution. To minimize the environmental pollution these cellulosic materials should be converted into useful product. Among thermophilic fungi cellulose system has been examined exhaustively. However, *Sporotrichum thermophile* has been studied in the most detail as it synthesizes a complete set of enzymes necessary for the break down of crystalline cellulose. Therefore, *Sporotrichum thermophile* appeared to be a useful source of cellulases. Keeping this in view, *Sporotrichum thermophile* was selected to study the cellulolytic potential.

The present study was carried with the objective to identify different environmental conditions for optimum production of cellulose enzymes by locally isolated thermophilic fungi *S. thermophile.*

Materials and Methods

Organism and Culture Conditions

The organism, *Sporotrichum thermophile*, a thermophilic fungi used for the enzymatic saccharification of cellulose, was isolated from local agriculture waste (Bhopal, M.P., India). Culture of

Sporotrichum thermophile was maintained on basal medium (Pattersson *et al.*, 1963). Slants were kept at 4ºC in cold room. Experiments were performed undersubmerged culture conditions.

Growth Parameters

The growth of *Sporotrichum thermophile* was estimated by measuring the extra cellular protein and mycelial protein. The extra cellular protein was measured by using Lowry's method (1951). Bovine Serum albumin was used as a standard mycelial protein was estimated by microbiuret method using Benedict's reagent (Herbert *et al.*, 1971). For studying growth behavior under different environmental conditions samples were taken after 3, 6, 9, 12, 15, 18 and 21 days incubation under different environmental conditions.

Enzymes Assay of Cellulose Enzymes

Cellulose activity was assayed viscometrically using Ostwald Fenske Viscometer (Reese *et al.*, 1950). The percentage loss in viscosity was calculated according to the following formula:

$$\% \text{ loss in viscosity} = \frac{(ET_o - ET_t)}{(ET_o - ET_w)} \times 100$$

where,

ET_o: Efflux time at zero hour

ET_t: Efflux time at time t

ET_w: Efflux time of distilled water

Endo-β-glucanase (EC. 3.2.1.4)

The Endo-β-glucanase was asseyed by the method of Sumner and Somer (1944) in terms of reducing sugar (R.S.) released from the substrate carboxyl methyl cellulose (CMC). 1 ml of cold CMC reagent was added to 1 ml of enzyme sample incubated at 50ºC for 30 mins and then cooled to 4ºC. Then amount of reducing sugar produced assayed by DNS reagent (Miller, 1959).

Exo-β-glucanase (E.C. 3.2.1.9.1)

The Exo-β-glucanase was assayed by the method suggested by Mendels and Weber (1969) by analyzing reducing sugars (R.S.) with Dinitrosalicyclic acid reagent 50 mg or cotton fibers in 1 mM citrate phosphate buffer pH-5 and 1 ml of crude enzyme preparation were incubated at 45ºC for 24 hrs. Then reducing sugar was assayed by DNS reagent (Miller, 1959).

Results and Discussion

It is a well known fact that incubation period greatly influences the growth and production of the organism. To find optimum period of incubation cultures were kept for 3 to 15 days under submerged conditions. The optimum conditions of incubation for growth vary from organism to organism (Bhat and Maheshwari, 1987). The maximum growth in our experiment in terms of extracellular protein (105.4 µg/ml) and mycelial protein (98.2 µg/ml) was observed on 15th day of incubation after which the growth sharply declines (Figure 21.1). This finding was in contrast to Bhat and Maheshwari (1987) who reported maximum growth of organism on 4th day and Coutts and Smith (1976) reported that growth of organism increases with increase in incubation period.

The mycelial dry weight maximum was on 18th day of incubation (0.92 gm/70 ml) (Figure 21.1). It increases with increase in period of incubation and than remained approx stationary with culture age. Similar observation was reported by Bhat and Maheshwari (1987).

Bhat and Maheshwari (1987) reported that growth of *Trichoderma ressei* (mesophilic strain) was excellent in comparison to *S. thermophile* (thermophilic fungus). This was in contrast to our results that *S. thermophile* grow very well and to result of Maheshwari *et al.* (2000) who found that *S. thermophile* grew at five times the rate of *T. ressei* and to Tansey (1971) results.

Cellulolytic system mainly consists of three components, exoglucanase, endoglucanase and β-glucosidase that act in synergy to degrade cellulose (Ryu and Mandels, 1980; Grajek, 1987; Beguin, 1990; Gilbert and Hazlewood, 1993; Nidetzky and Steiner, 1993). Endoglucanase, exogluctnase and β-glucosidase may form ternary complex with substrate (Moloney, 1985). It is hypothesized that in *S. thermophile*, β-glucosidase physically associates with endo and exoglucanase to form a catalytically more active cellulose complex (Maheshwari *et al.*, 2000). The different level of celluloses produced by *S. thermophile* could be because of difference in induction by same substrate (Macris and Galiotou-Panayotou, 1986). Bhat and Maheshwari (1987) and Bagga *et al.* (1990) have reported that cellulases were not synthesized at the same time and same rate. Cellulose productivity also varies among strains of *S. thermophile* (Maheshwari *et al.*, 2000; Oberson *et al.*, 1992). The maximum production of endo-β-glucanase (40.6 μg R.S./ml) and exo-β-glucanase (31.2 μg R.S./ml) was on the same 12th day of incubation. These findings are not agreeable with the earlier concept of Bhat and Maheshwari (1987). Coutts and Smith (1976) reported that endo-β-glucanase appeared in early stages of growth. Similar observations were made by Neudoerffer and Smith (1970). However present study the production of both enzymes are on the same day. Production of endo and exoglucanase enzyme increased with increase in incubation period upto 12th day after which it declines. But the production of endo-β-glucanase is higher than the exoglucanase. This result was agreeable to some extent with Coutts and Smith (1976).

The maximum growth was on 15th day and maximum enzyme produced was on 12th day which is in actively growing period. This result was same to that of Maheshwari *et al.* (2000) that in *S. thermophile* endo and exoglucanase were typically formed during active growth. The percentage viscosity loss was maximum (67.5 per cent) on 12th day of incubation (Figure 21.4).

Growth and the production of enzymes by *Sporotrichum thermophile* was greatly affected by the hydrogen ion concentration of the synthetic medium. Exo-β-glucanase and endo-β-glucanase can tolerate a wide range of pH, the maximum production was at 6.5 pH (41.25 μg R.S./ml and 31.5 μg R.S./ml) (Figure 21.5). The percentage viscosity loss is maximum at pH 6.5 (69.5 per cent) and it declines to approx 31 per cent on moving from pH 6.5 to 8.5. Both exo and endoglucanase activity decline with increase in pH from 6.5. In contrast to our result many author reported that the cellulose production was observed at pH 6.0 by *Sporotrichum thermophile* (Bhat and Maheshwari, 1987). Singh *et al.* (1990) have also reported pH 6.0 as optimum for cellulase production by immobilized cells of *Sporotrichum thermophile.* Grajek (1987) has reported that the optimum pH for extracellular β-glucosidase production for his isolate of *Sporotrichum thermophile* from wood waste material was around 6.5 pH. In contrast to our results Coutts and Smith (1976) reported that his organism *Sporotrichum thermophile* produces exo and endoglucanases more efficiently at acidic pH. Similar results were observed by Rao (1977) that initial pH of 5.6 was found to be most favorable for obtaining high yield of cellulase.

For better growth of organism the basal medium was adjusted in the pH range from 3.5–8.5. The maximum growth was found when pH of the medium was adjusted to 6.5 prior to autoclaving

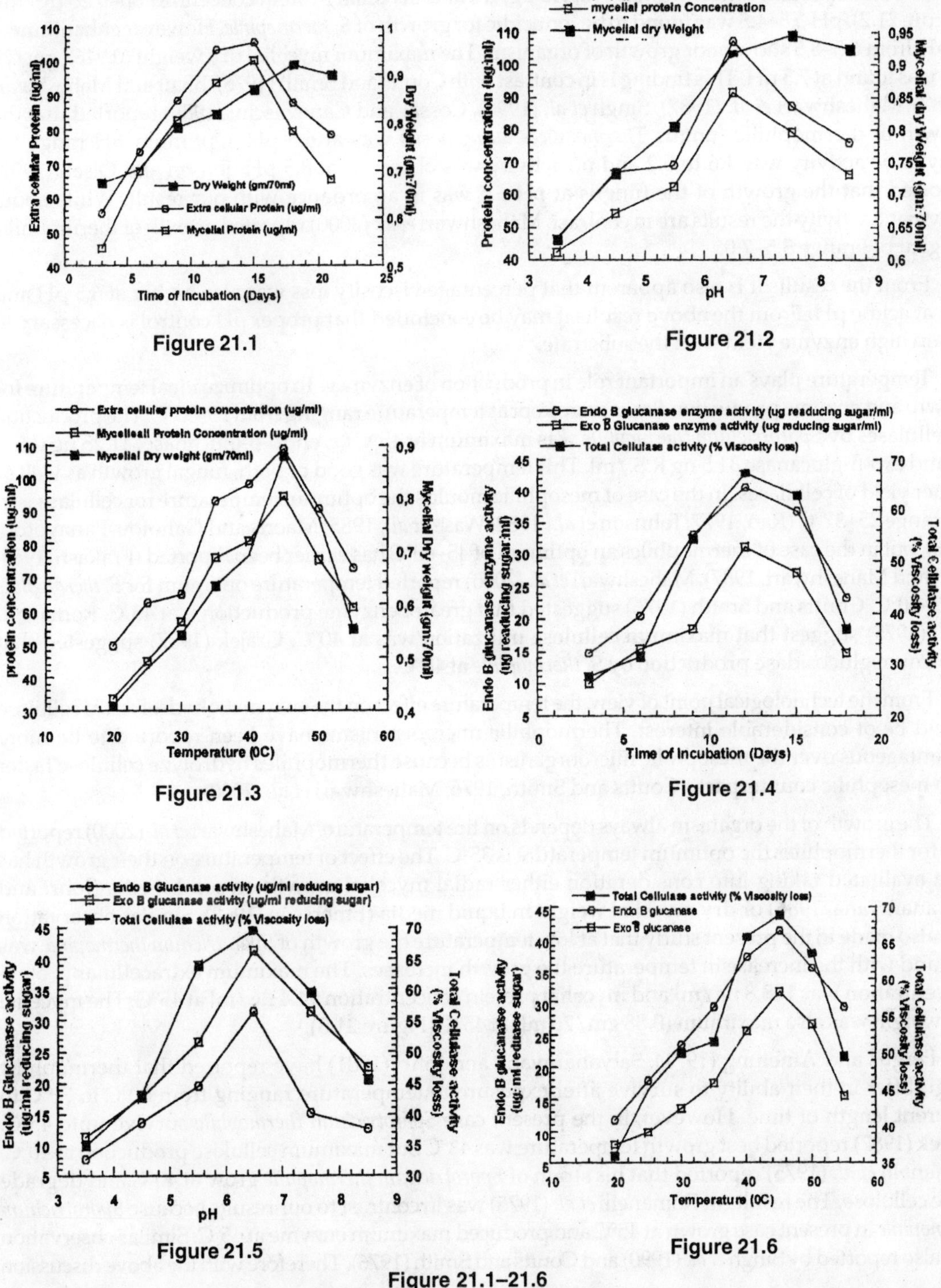

Figure 21.1–21.6

(extracellular protein concentration– 107.42 μg/ml and mycelial protein concentration–91.5 μg/ml) (Figure 21.2). pH 3.5–4.5 was found to be too acidic for growth of *S. thermophile*. However enhancement in pH from 6.5–8.5 show poor growth of organism. The maximum mycelial dry weight (0.9468 gm/70 ml) was found at 7.5 pH. This finding is in contrast with Coutts and Smith (1976), Bhatt and Maheshwari (1987), Maheshwari *et al.* (1987), Singh *et al.* (1990). Cossar and Canevascini (1986) reported that the growth of thermophilic fungus *Thermomyces lanuginosus* was at 6.5 pH. Optimum pH range for enzymatic activity was 4.6 to 5.2 and no activity was observed at 8.5 pH. Jensen and Olsen (1992) reported that the growth of the fungus at pH 6.5 was in accordance with our result. While about enzymatic activity the results are in contrast. Maheshwari *et al.* (2000) reported growth of thermophilic fungi at pH range 5.5–7.0.

From the results it is also apparent that percentage viscosity loss was also higher at 6.5 pH and low at acidic pH. From the above results it may be concluded that proper pH control is necessary to obtain high enzyme yield with the substrate.

Temperature plays an important role in production of enzymes. To optimize ideal temperature for growth and enzyme production flasks were kept at temperature ranging from 20–55°C. The production of cellulases by *Sporotrichum thermophile* was maximum at 45°C *i.e.* endo-β-glucanase, 41.25 μg R.S./ml and exo-β-glucanase 31.5 μg R.S./ml. This temperature was good or both, fungal growth as well as higher yield of cellulases. In the case of mesophilic moulds the optimum temperature for cellulase is in the range 25–37°C (Rao, 1977; Johnson *et al.*, 1982; Wase *et al.*, 1985; Macris and Galiotou-Panayotou, 1986), but in the case of thermophiles an optimum of 45–60°C has earlier been reported (Prakash, 1984; Bhat and Maheshwari, 1987). Maheshwari *et al.* (2000) reported temperature optimum for *S. thermophile* at 45–50°C. Coutts and Smith (1976) suggested that greater enzyme production is at 45°C. Romanelli *et al.* (1975) suggest that maximum cellulose utilization was at 40°C. Grajek (1987) suggested that maximum glucosidase production by *S. thermophile* at 40°C.

From the technological point of view, the temperature effect on the enzymatic hydrolysis of cellulose would be of considerable interest. Thermophilic microorganisms have been reported to be more advantageous over the mesophilic microorganisms because thermophiles hydrolyze cellulose faster than mesophilic counterpart, (Coutts and Smith, 1976; Maheshwari *et al.*, 2000).

The growth of the organism always depends on tire temperature. Maheshwari *et al.* (2000) reported that for thermophiles the optimum temperature is 35°C. The effect of temperature on their growth has been evaluated taking into consideration either radial mycelial growth on solid media (Johri and Satyanarayana, 1986) or dry mycelial weight in liquid media (Emersson, 1968). Similar observation was also made in the present study that at low temperature the growth of *Sporotrichum thermophile* was low and with the increase in temperature the growth increases. The maximum extracellular protein concentration was 108.8 μg/ml and mycelial protein concentration 94.4 μg/ml at 45°C. The mycelial dry weight was also maximum (0.88 gm/70 ml) at 45°C (Figure 21.3).

Fergus and Amelung (1971), Satyanarayana and Johri (1981) have reported that thermophilic fungi differ in their ability to survive after exposure to temperature ranging from 60°C to 70°C to different length of time. However, in the present case *Sporotrichum thermophile* survive upto 45°C. Grajek (1987) reported best growth temperature was 43°C but maximum cellulose production at 40°C. Romanelli *et al.* (1975) reported that his strain of *Sporotrichum thermophile* grow at 40°C and degrade more cellulose. The results of Romanelli *et al.* (1975) was in contrast to our results because *Sporotrichum thermophile* in present case grown at 45°C and produced maximum enzyme at 45°C. Similar observation was also reported by Singh *et al.* (1990) and Coutts and Smith (1976). Therefore with the above discussion it may be concluded that temperature plays a very important role in growth as well as in the production of metabolites.

The results obtained in this experiment were very useful and encouraging, *Sporotrichum thermophile* is one of the most important organism which produce thermostable enzymes in nature.

References

Bagga, P.S., Sandhu, D.K. and Sharma, S., 1990. Purification and characterization of cellulolytic enzymes produced by *Aspergillus nidulans. Journal of Applied Bacteriology*, 68: 61–68.

Beguin, P., 1990. Molecular biology of cellulose degradation. *Ann. Rev. Microbiol.*, 44: 219–228.

Bhat, K.M. and Maheshwari, R., 1987. *Sporotrichum thermophile:* Growth, cellulose degradation, and cellulase activity. *Appl. Environ. Microbiol.*, 53(9): 2175–2182.

Chaudhuri, S., Thakur, I.S., Goel, R. and Johri, B.N., 1988. Purification and characterization of two thermostable xylanases from melanocarbus albomyces. *Biochem. Int.*, 17: 563–575.

Cossar, D. and Canevascini, G., 1986. Cellulose enzyme production during continuous culture growth of *Sporotricum* (*Chrysoporium*) *thermophile. Appl. Microbiology Biotechnology*, 24: 306–310.

Coutts, A.D. and Smith, R.E., 1976. Factors influencing the production of celluloses by *Sporotrichum thermophile. Appl. and Environ. Microbiol.*, 31: 819–825.

Emersson, R., 1968. Thermophiles. In: *The Fungi*, (Eds.) Ainsworth, G.C. and Sursman, A.S. Academic Press, New York, 3: 105–128.

Eriksson, K.E.L., Blanchette, R.A. and Ander, P., 1990. *Microbial and Enzymatic Degradation of Wood and Wood Components*. Springer-Verlag, Berlin, 407 pp.

Feldman, K.A., Lovett, J.S. and Tsao, G.T., 1988. Isolation of the cellulose enzymes from the thermophilic fungus *Thermoascus aurantiacus* and regulation of enzyme production. *Enzyme Microb. Technol.*, 10: 202–271.

Fergus, C.L. and Amelung, R.M., 1971. The heat resistance of some thermophilic fungi on mushroom compost. *Ibid*, 68: 675–679.

Gilbert, H.J. and Hazlewood, G.P., 1993. Bacterial celluloses and xylanases. *Journal of General Microbiology*, 139: 187–194.

Grajek, W., 1987. Comparative studies on the production of cellulases by thermophilic fungi in submerged and solid state fermentation. *Appl. Microbiol. Biotechnol.*, 26: 126–129.

Hagerdal, B., Ferchak, J.O. and Pye, E.K., 1930. Saccharification of cellulase by cellulolytic enzyme system of *Thermomonospora* sp. I. stability of cellulolytic activities with respect to time, temperature and pH. *Biotech. Bioeng.*, 22: 1515–1526.

Hayashida, S. and Mo, K., 1986. Production and characteristics of avicel-disintegrating endoglucanase from a protease-negative *Humicola grisea* var. *thermoidea* mutant. *Applied and Environ. Microbiol.*, 51(5): 1041–1046.

Herbert, D., Phipps, P.J. and Strange, R.E., 1971. Chemical analysis of microbial cells. In: *Method in Microbiology*, 5B: 209–344.

Jensen, B. and Olsen, J., 1992. Physio-chemical properties of a purified α-amylase from the thermophilic fungus *Thermamyces lanugenuses. Enzyme Microb. Technol.*, 14: 112–116.

Johnson, Eric A., Saka, J.M., Halliwell, G., Madia, A. and Demain, A.L., 1982. Saccharification of complex cellulosic substrate by the cellulose system from *Clostridium thermocelium. Appl. Environment Microbiol.*, 43(5): 1125–1132.

Johri, B.N. and Satyanarayana, T., 1986. Thermophilic moulds perspective in basic and applied research. *Indian Rev. Life Sci.*, 6: 75–100.

Lowry, O.H., Rosebrough, N.J., Farr, A.L. and Randall, R.J., 1951. Protein measurement with the folin phenol reagent. *Journal of Biological Chemistry*, 193: 265–275.

Macris, B.J. and Galiotou-Panayotou, M., 1986. Enhanced cellobiohydrolase production from *Aspergillus ustus* and *Trichoderma harzianum. Enzyme Microbiol Technology*, 8: 141–144.

Maheshwari, R., Bharadwaj, G. and Bhat, M.K., 2000. Thermophilic fungi: Their physiology and enzymes. *Microbiology and Molecular Biology Reviews*, 64(3): 461–488.

Maheshwari, R., Kamalam, P.T. and Balasubramanyam, P.V., 1987. The biogeography of thermophilic fungi. *Curr. Sci.*, 56: 151–155.

Mendels, M. and Weber, J., 1969. The production of cellulases. In: *Cellulases and their Applications.* Advances in chemistry series, (Ed.) F. Gould. American Chemical Society, Washington, 95: 391–413.

Miller, G.L., 1959. Use of dinitrosalicilic reagent for determination of reducing groups. *Anal. Chem.*, 31: 426–428.

Moloney, A.P., McCrae, S.I., Wood, T.M. and Coughlan, M.P., 1985. Isolation and characterization of 1,4-β-D-glucan glucanohydrolases of *Talaromyces emersonii. Biochem. J.*, 225: 365–374.

Neudoerffer, T.S. and Smith, R.E., 1970. An evaluation of fungal enzymes for the solubilization of wheat bean constituents. *Can J. Microbiol.*, 16: 139–146.

Nidetzky, B. and Steiner, W., 1993. A new approach for modeling cellulase cellulose adsorption and the kinetics of the enzymatic hydrolysis of microcystalline cellulose. *Biotechnology and Bioengineering*, 42: 469–479.

Oberson, J., Binz, T., Fracheboud, D. and Canevascini, G., 1992. Comparative investigation of cellulose-degrading enzyme systems produced by different strains of *Myceliophthora thermophila* (Apinis) v. Oorschot. *Enzyme. Microb. Technol.*, 14: 303– 312.

Pettersson, B., Cowling, E.B. and Porath, J., 1963. Studies on cellulolytic enzymes. I. Isolation of a low molecular weight cellulose from *Polypores versicolor. Biochem. Biphys. Acta*, 67: 1–8.

Prakash, A., 1984. Antagonistic attributes of thermophilic fungi and thermophilism. *Ph.D. Thesis*, Bhopal University, Bhopal.

Rao, R.R., 1977. Studies on cellulose production by an Aspergillus strain. *Indian Journal of Microbiology*, 17(3): 123–128.

Reese, E.T., Siu, R.G.H. and Levinson, H.S., 1950. The biological degradation of soluble cellulose derivatives and its relationship to the mechanism of cellulose hydrolysis. *Journal of Bacteriology*, 59: 485–497.

Romanelli, R.A., Houston, C.W. and Barnett, S.M., 1975. Studies on thermophilic cellulolytic fungi. *Appl. Microbiology*, 30: 276–281.

Ryu, D.D.Y. and Mandels, M., 1980. Cellulases: Biosynthesis and application. *Enzyme Microb. Technol.*, 2: 91–102.

Satyanarayana, T. and Johri, B.N., 1981. Lipolytic activity of thermophilic fungi of paddy straw compost. *Current Sci.*, 50: 682–688.

Schlesinger, W.H., 1991. *Biogeochemistry: An Analysis of Global Change*. Academic, San Diego, 443 pp.

Sen, S. and Chakraborthy, S.L., 1984. Isolation of cellulose hyperproductive mutants of *Myceliophthora thermophila*. In: *Proc. 7th In. Biotechnol. Symp.*, New Delhi, 2: 604–605.

Singh, A., Goel, R. and Johri, B.N., 1990. Production of cellulolytic enzymes by immobilized *Sporotrichum thermophile. Enzyme Microb. Technol.*, 12: 464–468.

Sumner, J.B. and Somer, G.F., 1944. *Laboratory Experiments in Biological Chemistry*. Academic Press, Inc., New York.

Tansey, M.R., 1971. Agar-diffusion assay of cellulolytic ability of thermophilic fungi. *Arch. Microbiol.*, 77: 1–11.

Wase, D.A. J., McManamey, W.J., Raymahasay, S. and Vaid, A.K., 1985. Comparison between cellulose production by *Aspergillus fumigatus* IMI 255901. *Enzyme Microb. Technol.*, 7: 225–229.

Chapter 22

Fly Ash-soil Amendment Technique: A Harbinger to Enhanced Vegetable Production in Jharkhand

N. Tripathi[1*], P.K. Mishra[3], R.S. Singh[1] and P.S.M. Tripathi[3]

[1]Central Mining Research Institute, Barwa Road, Dhanbad – 826 001, Jharkhand, India

[2]V.B. University, Hazaribagh, Jharkhand, India

[3]Central Fuel Research Institute, Dhanbad – 828 001, Jharkhand, India

ABSTRACT

The present article discusses the prospects of utilization of bulk quantities of fly ash in agriculture sector to get enhanced yields of vegetable crops in the degraded and nutrient-deficient, less fertile, acidic, red, sandy soils. After making such soil nutrient enriched and more productive by fly ash amendment technique, investigations were carried out to make appraisal of the performance/growth pattern of the vegetables, such as beans (*Phaseolus vulgaris*–family Leguminosae), tomato (*Lycopersicum esculentum*–family Solanaceae), carrot (*Daucus carota*–family Umbelliferae) and radish (*Raphanus sativus*–family Cruciferae) on the amended soil. The experiments were carried out in pots, both in control and fly ash mixed soil, fly ash dose being 2.5 per cent, 5 per cent, 10 per cent, and 15 per cent by weight, in addition to the usual application of recommended doses of NPK fertilizer. The physico-chemical properties, such as pH, EC, porosity, bulk density (ED), water holding capacity (WHC), cation exchange capacity (CEC), dehydrogenase activity and texture, of the soil after mixing with fly ash were found to improve greatly, making the amended soil richer in essential nutrient elements and more fertile. The comparison of morphometric data and biochemical properties, such as chlorophyll, carbohydrate and protein contents of these crops grown in control and amended red soil has evinced much better performance and higher yields. The fly ash-soil amendment technology will thus go a long way in greatly enhancing and improving the agricultural production and economy of the state.

* E-mail: nymphaea7@yahoo.com.

Introduction

Sobriqueted as the mineral bowl of India, the newly created state of Jharkhand accounts for about 48 per cent of the country's coal production. Opening of 50 new coal mines in Jharkhand area is on the anvil to achieve the target productin of 417 million tons of coal per year. Side by side, to meet the present as well as future energy requirements and to bridge the gap between the demand and supply of power in the state, establishment of many more big coal-fired thermal power plants as also several captive power generation plants is well in the offing. As can be well imagined, the coming of several new Coal-fired power plants along with the existing operational power plants in the state will simultaneously give rise to another serious problem, *i.e.*, generation of unmanageable quantities of fly ash (current production of fly ash alone being ca. 110 million tonnes per annum) [C.F.R.I. Report (2002); Central Electricity Generating Board (1973)], for the storage and disposal of which quite a big area of precious agricultural land will be occupied near the power plants, making the available cultivable land in the state shrink. It has already assumed alarming proportions, which calls for taking urgent remedial measures for proper and bulk gainful utilization of coal ashes, fly ash in particular, in an eco-friendly manner. Particularly, with reference to fly ash utilization in agriculture, it has been found that fly ash can be used as a soil amendment (Davis, *et al.*, 2002) and soil-coal ash admixture (20-30 per cent) is optimum in terms of soil health, plant growth, fertility status and yield, where heavy metals in crop produce were within the permissible limits (Mandal, *et al.*, 1997, 98, 99). The alkaline fly ash could be used as a substitute of lime for amending acidic soils and increase the availability of P, K, Ca, Mg, Zn, Cu and Co besides improving soil physico-chemical properties. The results indicated the prospect of safe disposal and utilization of fly ash in agriculture for retaining productivity of problem soils, reduce the usage of costly chemical fertilizer, bring greater economy in cultivation and minimize environmental problems (Mitra *et al.*, 2003).

In the present article the effect of amendment of red lateritic soil with different doses of fly ash on the growth and yields of some vegetable crops, like beans, tomato, carrot and radish has been discussed.

Materials and Methods

Collection and Characterization of Samples

The soil samples from Sudamdih area of Dhanbad (Jharkhand, India) and fly ash samples from DPPL, Durgapur (West Bengal, India) were collected and analyzed for various physico-chemical properties, following prescribed standard methods.

The physical properties, *viz.* colour, texture, BD, porosity, WHC of fly ash from DPPL, Durgapur and soil sample from Dhanbad with and without the additions of fly ash were determined following the prescribed methods (Piper, 1950 and Black, 1965).

Chemical properties, *viz.* pH, electrical conductivity, total and available elements (N, P, and S) of fly ash and soil were also estimated following the methods as described by Black (1965), Jackson (1973) and Tandon (1995).

Pot Experiment

Pot experiments under green house conditions were done using acidic soil (pH 5.0) for growing vegetable crops with and without different doses of fly ash. After proper mixing of fly ash with the soil with pre-decided doses, *i.e.*, 0 (control), 50, 100, 200 and 300 t/ha (equivalent to 0.0, 2.5, 5.0, 10.0, 15.0 per cent fly ash w/w, respectively), 4 kg each of amended soils was added in their respective pots. Recommended doses of NPK fertilizers were applied in both control and fly ash amended soil. The pots were properly lined with the polythene sheets to check the leaching of trace and heavy metals, if any, from fly ash-soil amended system under green house conditions. Four crops, *i.e.*, beans, tomato, radish and carrot were grown in pots with different doses of fly ash to study whether there is any carry-over of toxic metals present in fly ash, such as copper, zinc, manganese, iron, lead, nickel, cobalt, arsenic, mercury, chromium and cadmium by these vegetable crops. The crops grown with the fly ash were harvested and their yield data (both on harvest and oven-dried at 60°C basis) and biochemical parameters were recorded. The vegetables grown with and without fly ash were also analyzed for trace and heavy metals.

Biochemical Analysis of the Vegetables Grown

Biochemical characterization, *viz.*, chlorophyll, carbohydrate and protein contents of the vegetables was done following Dogra and John (1978).

Analysis of Trace and Heavy Metals

The digestion of soil, fly ash and roots of different crops was done in an Analytical Microwave System (Prolabo Microwave Oven) using different "Suprapure" acids for the analysis of micro nutrients (Cu, Zn, Fe, Mn), trace elements and heavy metals, such as Pb, Cr, Ni, Co, As, Hg, Cd. The trace elements in various digested samples were analyzed by Atomic Absorption Spectrometry technique (Schimadzu, Japan; Model AAS-680), but As and Hg were analyzed by another Spectrometer (Unicorn make SP-2900 AAS), using hydride vapour generator.

Results and Discussion

The analytical results of soil samples before the start of experiment and after harvest of each crop are shown, in respect of various physico-chemical and biochemical parameters in Tables 22.1–22.5 while the crop yield data of the crops grown are presented in Table 22.6. The results of trace and heavy metal concentrations are given in Tables 22.7 and 22.8.

Physical Properties of Soil

The comparative physico-chemical characteristics of the fly ash and soil show (Tables 22.1 and 22.2) that the soil is acidic (pH = 5.01), while the fly ash is alkaline (pH = 8.92) in nature. Bulk density (BD), water holding capacity and porosity values of the fly ash are 0.81 g/cc, 54.34 per cent and 65.81 per cent, respectively, while those of the soil are 1.63 g/cc, 30.92 per cent and 37.31 per cent, respectively. The alkaline fly ash promoted the nutrient supplying capacity of the soil by raising pH level from 5.01 to 5.48 in soil.

Table 22.1: Initial Physical Characterization of Soil and Fly Ash

Parameter	*Soil*	*Fly Ash*
pH	5.01	8.92
Electrical Conductivity (d s/m)	0.05	0.12
Bulk Density (g/cc)	1.63	0.81
Water Holding Capacity (per cent)	30.92	54.35
Porosity (per cent)	37.31	65.81
Texture		
Sand (per cent)	66.10	33.50
Silt (per cent)	21.60	50.10
Clay (per cent)	12.30	16.40
Textural Class	Sandy loam	Silty loam

Table 22.2: Initial Chemical Characterization of Soil and Fly Ash

Parameter	Soil		Fly Ash
	Total	Available	Total
Nitrogen (per cent)	0.0515	0.0120	BDL
Phosphorus (per cent)	0.0140	0.00095	0.29
Potassium (per cent)	0.77	0.0098	0.84
Calcium (per cent)	0.86	0.0024	1.09
Magnesium (per cent)	0.56	0.00179	0.79
Sulphur (per cent)	0.052	0.0045	0.026
Zinc (per cent)	44.10	2.10	71.72
Copper (per cent)	42.90	2.34	78.20
Iron (per cent)	229	45.8	293
Manganese (per cent)	284.40	17.0	212.80
Lead (per cent)	22.60	3.70	52.0
Nickel (per cent)	228.30	0.58	134.0
Chromium (per cent)	179.0	BDL	225.0
Cadmium (per cent)	6.90	BDL	10.60
Arsenic (per cent)	0.56	BDL	0.59
Mercury (per cent)	BDL	BDL	0.01
Cobalt (per cent)	32.05	BDL	51.20

The results of the physical characteristics of soil and fly ash indicate that due to its alkaline nature, fly ash can be suitably applied to the acidic soils to neutralize their acidity, thereby acting as a liming agent for the acidic soils. According to Martens, *et al.* (1976), the beneficial effects of coal ash on plant growth may result from a combination of pH and nutrient addition. Similarly, water holding capacity (WHC) and porosity of the soil are significantly improved by fly ash addition into the soil. Chang *et al.* (1977) reported that an addition of Fly Ash (8 per cent by weight) increase the water holding capacity of soil. They also reported that soil hydraulic conductivity improved at lower rates of fly ash application but deteriorated when the rate of fly ash amendment exceeded 20 per cent in calcareous soils and 10 per cent in acidic soils. This improvement in water holding capacity is beneficial to the plants especially under rainfed agriculture. Thus fly ash can be potentially used and is quite effective to act as an efficient soil modifier/conditioner.

The physico-chemical properties of soil were improved when chemical fertilizer was supplemented with fly ash. It resulted into the decreased bulk density and increased organic carbon and pH of soil besides increasing the available nutrients like N, P, K, Ca, Mg, Zn, Cu and Co. The reduction in bulk density of amended soil was because of lower bulk density of fly ash (0.81 g/cc) as compared to that of soil (1.63 g/cc). Page *et al.* (1979, 1980) reported that Fly Ash amendment to a variety of agricultural soils tend to decrease the bulk density. Optimum bulk density in turn improves the soil porosity, the workability of the soil, the root penetration and the moisture retention capacity of the soil.

The potential benefits of fly ash addition to soil include improved structure for fine- and coarse-textured soil (Martens, 1971; Erickson *et al.*, 1987; Ghodrati *et al.*, 1994). Fly Ash addition @ 70 t/ha has been reported to alter the texture of sandy and clayey soil to loamy one (Fail and Wochock, 1977). Capp

(1978) also suggested that both sandy and clayey soils tend to become loamy in texture after fly ash addition. Sharma *et al.* (1989) reported the average silt content in U.S. fly ash to be about 63.2 per cent, while in India this content ranges from about 16 per cent (IIT, Kharagpur) to 45 per cent (VAS, Raichur). In the present study, the silt content in fly ash is 50 per cent.

Major and Secondary Nutrients in Soil

The results (Table 22.3) reveal that the application of maximum dose of fly ash results in the enrichment of the available major and secondary nutrients (N, P, K, S, Ca and Mg) status of the soil, which has contributed to the improvement of soil fertility, most probably owing to higher retention of the fertilizer ions applied through the application of inorganic fertilizers in the presence of fly ash. Further, the results show that at the highest dose of the fly ash and the maximum yield at these doses, the final status of the major nutrients in the soil is proportionately increased in accordance with their contents in the soil and in fly ash. The presence of essential plant nutrients in fly ash has also been reported by several other workers (Ashokan, *et al.*, 1995; Neelima, *et al.*, 1995).

Table 22.3: Physico-chemical Characteristics of Soil after Fly Ash Amendment

Parameter	*Fly ash-amended Soil*
pH	5.48
EC (d s/m)	0.03
BD (g/cc)	1.52
WHC (per cent)	34.82
Porosity (per cent)	42.35
Mechanical composition	
Sand (per cent)	61.95
Silt (per cent)	25.30
Clay (per cent)	12.75
Textural Class	Sandy loam
Nitrogen (per cent)	0.0156
Phosphorus (per cent)	0.00108
Potassium (per cent)	0.01004
Calcium (per cent)	0.00288
Magnesium (per cent)	0.00206
Sulphur (per cent)	0.00420
Zinc (ppm)	2.36
Copper (ppm)	2.70
Iron (ppm)	44.10
Manganese (ppm)	15.62
Lead (ppm)	3.66
Nickel (ppm)	0.61
Chromium (ppm)	BDL
Cadmium (ppm)	BDL
Arsenic (ppm)	BDL
Mercury (ppm)	BDL
Cobalt (ppm)	0.05

Trace and Heavy Metals in Soil

Results indicated that (Table 22.3) the application of maximum dose of fly ash does not influence at all the available status of trace elements and heavy metals such as Pb, Ni, Se, Cr, Cd, As, Hg and Co in the soil system after harvest of each crop. This may be due to the presence of heavy metals in alkaline ashes in oxide forms.

Biochemical Characterization of Soil

The results of biochemical analysis of vegetable samples, in terms of carbohydrate, chlorophyll and protein contents have shown a marked difference in the crops grown in amended soil and those grown without fly ash (Tables 22.4 and 22.5). The crops grown with maximum fly ash dose have shown a marked increase in chlorophyll, carbohydrate and protein contents over those found in control. The percentage increase in chlorophyll content of beans, tomato, radish and carrot was 18.2, 3.84, 27.7 and 17.41, respectively. The carbohydrate and protein contents in all the four crops increased respectively by 11.76 per cent, 13.88 per cent, 14.28 per cent, 29.31 per cent and 15.7 per cent, 15.49 per cent, 8.56 per cent, 18.87 per cent. Biochemical characteristics may be altered and improved due to the improved physico-chemical properties and fertility status of soil after fly ash amendment.

Table 22.4: Biochemical Characteristics of Vegetable Crops Before and After Fly Ash Amendment (±1SE)

Parameters (%)	*Beans*				*LSD*	*Tomato*				*LSD*
	Control	T_1	T_2	T_3		*Control*	T_1	T_2	T_3	
Chlorophyll	3.44 ±0.69	3.62 ±0.88	4.10 ±0.86	4.21 ±0.93	0.49	3.50 ±0.48	3.52 ±0.67	3.62 ±0.39	3.64 ±0.66	0.75
Carbohydrate	6.0 ±1.2	6.2 ±1.14	6.4 ±1.26	6.8 ±1.18	0.62	6.2 ±1.4	6.6 ±1.33	7.0 ±1.16	7.2 ±1.8	0.82
Protein	3.81 ±0.39	4.21 ±0.46	4.33 ±0.52	4.52 ±0.73	0.56	2.89 ±0.16	3.10 ±0.34	3.21 ±0.6	3.42 ±0.48	0.67

LSD values are at 5 per cent level.

Table 22.5: Biochemical Characteristics of Root Crops Before and After Fly Ash Amendment (%1SE)

Parameters (%)	*Radish*				*LSD*	*Carrot*				*LSD*
	Control	T_1	T_2	T_3		*Control*	T_1	T_2	T_3	
Chlorophyll	3.21 ±0.56	3.64 ±0.38	4.10 ±0.62	4.44 ±0.86	0.54	2.75 ±0.16	2.81 ±0.18	3.21 ±0.26	3.33 ±0.33	0.67
Carbohydrate	7.2 ±1.24	7.4 ±1.52	8.2 ±1.21	8.4 ±1.3	0.49	8.2 ±1.2	9.4 ±1.32	10.2 ±1.32	11.6 ±1.24	0.46
Protein	3.31 ±0.64	3.41 ±0.38	3.52 ±0.48	3.62 ±0.53	0.62	3.61 ±0.55	3.82 ±0.63	4.31 ±0.65	4.45 ±0.48	0.71

LSD values are at 5 per cent level.

Effect of Treatment on Crop Yield

An increase of 10–30 per cent in the yield of the crops grown in the soil amended with different doses of fly ash (Table 22.6), without any apparent adverse effect can be envisaged due to the improved physico-chemical, biochemical properties and fertility status of soil after the addition of fly ash in soil. Regional Research Laboratory (RRL) Bhopal reported that on an average in comparison to control, there was around 50–60 per cent more yield of Brinjal, around 45 per cent more yield of potato and pea, around 40 per cent more yield of Tomato and around 29 per cent more yield of cabbage in fly ash treated plot when Fly Ash was applied @25 per cent of soil (Kumar *et al.*, @ www.tifac.org.in/news/flyagr.orgl.htm). Selvam and Mahadevan (2002) have also reported the increased seed germination rate in tomato sown on fly ash amended soil. Nevertheless, the penetration of the root crops was also found to be better in fly ash-amended soil than in control. Taylor, *et al.* (1988) and Pichtel, *et al.* (1994) have also reported the deeper root penetration in fly ash amended soil.

Table 22.6: Crop Yield After Amendment with Different Doses of Fly Ash (on harvest and oven dry basis)

Yield (g/pot)	*Radish (after 4 months)*	*Carrot (after 4 months)*	*Beans (after 4 months)*	*Tomato (after 4 months)*
Harvest basis				
Control	124.6	89.6	22	120
$^*T_1$	137.4	100.8	28	135
$^*T_2$	145.0	108.8	33	175
$^*T_3$	156.8	120.9	50	315
$^*T_4$	174.6	141.6	52.8	540
CD (5%)	1.6	0.8	0.6	1.4
Oven dry (60°C) basis				
Control	6.59	8.10	5.24	4.85
$^*T_1$	7.34	9.22	6.82	6.22
$^*T_2$	7.55	9.92	8.35	10.54
$^*T_3$	8.32	10.9	9.26	12.57
$^*T_4$	9.22	12.5	10.98	20.0

*: T_1, T_2, T_3, T_4 are the treatments at different doses of fly ash (2.5 per cent, 5 per cent, 10.0 per cent and 15 per cent, respectively).

The concentration of major and secondary plant nutrients (N, P, K, S, Ca and Mg) also increased progressively with the increasing dose of fly ash in comparison to that in the control. Micronutrient (such as Cu, Zn, Mn, Fe) concentrations also increased in all the crops grown with different doses of fly ash application over control. Nevertheless, the concentration of trace and heavy metals, such as Cu, Zn, Mn, Fe, Cr, Co, Pb, Cd, As, Hg and Ni is non-significant in comparison to control (Tables 22.7 and 22.8), albeit the concentration of trace and heavy metals was increased in all the vegetable crops through the fly ash application. Although some of the heavy metals, such as Cr, Co, Pb and Ni have shown a per cent increase, yet, it is almost insignificant as the results reported are less than 1 ppm in all the cases. These results are within the normal range generally found in crop produce and are in agreement with those suggested by Srivastava *et al.* (1996). Sims *et al.* (1994) have also reported that the uptake of trace and heavy metals by the plants grown with fly ash at higher dose (20 per cent by weight of soil) is quite insignificant.

Table 22.7: Trace and Heavy Metal Concentration in Root Crops Grown on Fly ash-amended Soil

Elements	Radish					Carrot				
	Control	50 t/ha	100 t/ha	200 t/ha	300 t/ha	Control	50 t/ha	100 t/ha	200 t/ha	300 t/ha
N (per cent)	2.28	2.25	2.31	2.32	2.34	1.57	1.59	1.61	1.62	1.63
P (per cent)	0.15	0.18	0.19	0.22	0.25	0.08	0.10	0.12	0.14	0.16
K (per cent)	2.85	2.87	2.89	2.93	2.96	2.38	2.40	2.42	2.45	2.47
S (per cent)	0.15	0.18	0.20	0.23	0.23	0.17	0.19	0.21	0.23	0.25
Ca (per cent)	2.17	2.21	2.25	2.31	2.34	1.18	1.20	1.21	1.23	1.25
Mg (per cent)	0.19	0.21	0.24	0.25	0.28	0.19	0.21	0.22	0.25	0.26
Cu (ppm)	2.47	2.52	2.56	2.60	2.63	3.75	3.79	3.81	3.83	3.86
Zn (ppm)	10.6	10.8	11.1	11.6	11.9	18.9	19.1	19.5	19.3	19.6
Mn (ppm)	30.8	31.2	30.6	30.8	31.2	52.6	53.1	53.2	53.6	54.0
Fe (ppm)	52.4	52.6	53.0	53.2	53.5	93.7	94.2	94.2	94.3	94.4
Cr (ppm)	0.38	0.35	0.42	0.46	0.49	0.66	0.68	0.71	0.76	0.78
Co (ppm)	0.25	0.27	0.34	0.37	0.40	0.72	0.75	0.77	0.80	0.82
Pb (ppm)	0.15	0.21	0.23	0.28	0.32	0.18	0.16	0.21	0.23	0.25
Cd (ppm)	BDL	BDL	BDL	BDL	BDL	BDL	BDL	BDL	BDL	BDL
As (ppm)	BDL	BDL	BDL	BDL	BDL	BDL	BDL	BDL	BDL	BDL
Hg (ppm)	BDL	BDL	BDL	BDL	BDL	BDL	BDL	BDL	BDL	BDL
Ni (ppm)	0.65	0.71	0.76	0.72	0.75	1.78	1.80	1.81	1.83	1.86

Therefore, this study evinces that in spite of the increasing concentration of trace and heavy metals in fly ash, the uptake of these metals by vegetable crops, grown with different doses of fly ash under green house conditions, is insignificant.

Statistical Analysis

The statistical analysis was done using the statistical software SPSSwin (1993). There is a significant difference in the physico-chemical properties of control and those of amended soil. The most significant difference has been found at 15 per cent amendment ($P < 0.05$). The biochemical properties of the vegetables grown in fly ash amended soil have shown a significant increase ($P < 0.05$) as compared to those of the control; the greatest difference found at 15 per cent amendment.

Conclusion

Fly Ash has potential for use in agriculture because it contains almost all macro as well as micronutrients except organic carbon and nitrogen. It can be used in conjunction with the recommended doses of chemical fertilizer for different crops to increase the yield of various agricultural crops. Although, Fly Ash contains moderate quantities of trace and heavy metals, radioactive elements, but their concentrations are too low to have any harmful effect and their effects on groundwater, soil health and uptake by plants are probably negligible. Thus, the soil-fly ash amendment improves various physico-chemical properties and biochemical properties of soil making rather less productive lands more fertile, without having any adverse environmental impact. Therefore, it is imperative that concomitant with the plans for more coal mining and power generation, blue print should also be carefully prepared for the environmentally-safe management and gainful utilization of fly ash in different ways.

Table 22.8: Trace and Heavy Metal Concentration in Bean and Tomato Crops Grown on Fly Ash-amended Soil

Elements	Beans					Tomato				
	Control	50 t/ha	100 t/ha	200 t/ha	300 t/ha	Control	50 t/ha	100 t/ha	200 t/ha	300 t/ha
N (per cent)	1.15	1.12	1.16	1.14	1.22	1.06	1.1	1.12	1.14	1.17
P (per cent)	0.11	0.16	0.19	0.24	0.28	0.06	0.11	0.13	0.14	0.18
K (per cent)	2.27	2.31	2.38	2.46	2.53	2.19	2.21	2.26	2.3	2.33
S (per cent)	0.11	0.13	0.16	0.2	0.21	0.12	0.15	0.16	0.19	0.22
Ca (per cent)	2.09	2.11	2.16	2.19	2.23	1.1	1.13	1.18	1.19	1.22
Mg (per cent)	0.16	0.18	0.21	0.24	0.25	0.17	0.19	0.2	0.22	0.25
Cu (ppm)	2.38	2.41	2.44	2.47	2.48	2.86	2.87	2.89	2.92	2.96
Zn (ppm)	9.99	10.0	10.14	10.17	10.21	18.54	18.59	18.62	18.67	18.71
Mn (ppm)	28.88	28.9	28.95	28.98	30.04	51.74	51.76	51.77	51.81	51.84
Fe (ppm)	51.54	51.56	51.57	51.59	60.0	89.86	89.9	89.92	89.95	89.99
Cr (ppm)	0.26	0.27	0.31	0.35	0.39	0.61	0.62	0.65	0.69	0.72
Co (ppm)	0.17	0.18	0.22	0.24	0.27	0.68	0.69	0.73	0.77	0.78
Pb (ppm)	0.11	0.13	0.16	0.19	0.23	0.12	0.16	0.19	0.20	0.24
Cd (ppm)	BDL	BDL	BDL	BDL	BDL	BDL	BDL	BDL	BDL	BDL
As (ppm)	BDL	BDL	BDL	BDL	BDL	BDL	BDL	BDL	BDL	BDL
Hg (ppm)	BDL	BDL	BDL	BDL	BDL	BDL	BDL	BDL	BDL	BDL
Ni (ppm)	0.59	0.61	0.63	0.65	0.68	1.65	1.68	1.69	1.72	1.78

Acknowledgement

Ministry of Environment and Forests (MOEF), New Delhi and Council of Scientific and Industrial Research (CSIR), New Delhi are acknowledged.

References

Ashokan, P., Saxena, M., Bose, S.K.J. and Khazenchi, A.C., 1995. *Proceedings of Workshop on Fly Ash Management in the State of Orissa*, RRL, Bhubaneshwar, April 11, p. 64–75.

Black, C.A. (Ed.), 1965. *Methods of Soil Analysis*, Part 1 and 2. American Society of Agronomy Inc., Publisher, Wisconsin, USA.

Capp, J.P., 1978. Power plant fly ash utilization for land reclamation in the Eastern United States. In: *Reclamation of Drastically Disturbed Lands*, (Eds.) Schaller, F.W. and P. Sutton. Sol. Sci. Soc. of Am., Madison, WI, pp. 339.

C.F.R.I. (Central Fuel Research Institute), 2002. Utilization of fly ash in agriculture, at Bakreshwar Thermal Power Project, Bakreshwar (West Bengal). C.F.R.I. Dhanbad (India) Report No. TR/ C.F.R.I./3.03/2001–2002.

Carlson, C.L. and Adriano, D.C.J., 1993. *Environ. Qual.*, 22: 227–247.

Central Electricity Generating Board, 1973. The uptake of minerals by plants growing on pulverized fuel ash. CEGB, London, U.K. Report No. SSD.MID./R.2/73, pp. 30.

Chang, A.C., Lund, L.J., Page, A.L. and Warneke, J.E., 1977. Physical properties of fly ash amended soils. *J. Environ. Qual.*, 6(3): 267.

Coal Abstracts Nos., 1992. 9207635 and 9207645, April.

Daniels, W. Lee, Stewart, B., Haering, Kathryn and Zipper, Carl, 2002. Virginia Tech Publication Number 460–134.

Dogra, J.V.V. and John, M.S.H., 1978. *A Handbook of Experiments in Botany*. Bharati Bhawan Publishers.

Elseewi, A.A. Bingham, F.T. and Page, A.L. 1980. *J. Env. Qual.*, 7: 69–73.

Fail, J.L. and Wochok, Z.S., 1977. Soyabean growth on fly ash amended strip mine soils. *Plant Soil*, 48: 473.

Indian Institute of Technology, Kharagpur.

Jackson, M.L., 1967. *Soil Chemical Analysis*. Prentice Hall of India Pvt. Ltd., Delhi.

Kumar, V., Goswami, G. and Zacharia, K.A. (www.tifac.org.in/news/flyagrl.htm). *Fly Ash Use in Agriculture: Issues and Concerns*.

Mandal, S. and Saxena, M., 1998. Response of soyabean grown on vertisols amended with coal ash. *Chem. Environ. Res.*, 7(3&4): 309.

Mandal, S. and Saxena, M., 1997 Role of fly ash in the improvement of black cotton soil. *J. Env. Studies and Policy*, 2(1): 29–38.

Mandal, S. and Saxena, M., 1999. Fly ash: A boon for clay and sandy soil. *J. Env. Res.*, 9(1): 19–24.

Martens, D.C., 1971. Availability of plant nutrients in fly ash. *Bumines Report of Inv.*, p. 15–18

Martens, D.C. and Beham, B.R., 1976. *Proceedings of 4th Intl. Symp. on Ash Utilization* (St. Lois, MO) 24–25 March, p. 657–664.

Mittra, B.N., Karmakar, S., Swain, D.K. and Ghosh, B.C., 2003. Fly ash: A potential source of soil amendment and a component of integrated plant nutrient supply system. In: *Proceedings of International Fly Ash Utilization Symposium*, Centre for Applied Energy Research, University of Kentucky, Paper No. 28.

Page, A.L., Elseewi, A.A. and Straughan, I.R., 1979. Physical and chemical properties of fly ash from coal-fired plants with reference to environmental impacts. *Residue Rev.*, 7: 83.

Page, A.L., Elseewi, A.A., Lund, L.J., Bradford, G.R., Mattigod, S., Chang, A.C. and Bingham, F.T., 1980. *Consequences of Trace Element Enrichment of Soils and Vegetation from the Combustion of Fuels Used in Power Generation*. University of California, Riverside, pp. 158.

Pichtel, J.R., Dick, W.A. and Sutton, P., 1994. *J. Environ. Quality*, 23: 766.

Piper, C.S., 1950. *Soil and Plant Analysis*. University of Adelaide, Adelaide, Australia.

Reddy, Neelima M., Khandaul, S., Tripathy, A. and Sahu, P.K., 1995. *Proceedings of Workshop on Fly Ash Management in the State of Orissa*, RRL, Bhubaneswar, April 11, 1995, p.76–87.

Selvam, A. and Mahadevan, A., 2002. Effect of ash pond soil and amendments on the growth and arbuscular mycorrhizal colonization of *Allium cepa* and germination of *Arachis hypogaea*, *Lycopersicon esculentum*, and *Vigna mungo* seeds. *Soil and Sediment Contamination (formerly Journal of Soil Contamination)*, 2(5): 673–686.

Sharma, S. *et al.*, 1989. Fly ash dynamics in soil-water systems. *Critical Reviews in Environmental Control*, 19(3): 251–275.

Sims, J.T., Vasilas, B.L. and M. Ghodrati, M., 1994. Soil hydraulic properties and elemental leaching due to amendment of soil by fly ash addition. *Water Air and Soil Pollution*, 81: 349–361.

SPSSwin, 1993.

Srivastava, P.C. and Gupta U.C., 1996. *Trace Elements in Crop Production*. Oxford and IBH Publishing Co. Pvt. Ltd., New Delhi.

Tandon, L. (Ed.), 1995. *Methods of Analysis of Soils, Plants, Water and Fertilizers*. Fertilizer Development and Consultation Organization, New Delhi.

Taylor E.M. Jr. and Schuam G.E., 1988. Fly ash and lime amendment of acidic coal spoil aid revegetation, *J. Env. Qual.*, 17(1): 120–124.

University of Agricultural Sciences, Raichur.

Chapter 23

Noise-induced Hearing Loss (NIHL) Among Flourmill Workers: A Case Study of Jalgaon City, Maharashtra

S.B. Attarde and S.T. Ingle

School of Environmental Sciences, North Maharashtra University, Post Box 80, Jalgaon – 425 001, Maharashtra

ABSTRACT

The most well known effect of noise at work is hearing loss. The flourmill workers are exposed to continuous noise during their work and hence are at high risk of noise-induced hearing loss (NIHL). The noise levels were monitored in the flourmills situated in Jalgaon city to evaluate the average noise environment. The study shows that the flourmill workers are exposed to the noise above 85 dB. The self-reported hearing health survey and pure tone audiometry were used to evaluate hearing status of the workers. The binaural average hearing thresholds were determined for low frequency (250, 500, 1000 and 1500 Hz), mid frequency (1000, 1500, 2000, 3000 and 4000 Hz.) and high frequency (3000, 4000, 6000 and 8000 Hz.) ranges of human hearing. The self-report indicated the hearing impairment in 64 per cent workers. The hearing loss (shift in hearing threshold above 25 dB) as assessed by audiometry, was observed in more than 72 per cent workers and was more pronounced for high frequency range. The self-report and audiometry showed good agreement with each other on the basis of sensitivity, specificity, false positive and false negative for the frequency ranges studied. Mitigation measures such as use of hearing protection and acoustic design of flourmill shades are necessary to reduce the impacts of exposure.

Keywords: Flourmill workers, Noise monitoring, Noise-induced hearing loss, Audiometry.

Introduction

Flourmill is an important small-scale industry in urban as well as rural areas. The health impacts of workplace environment are not studied among the flourmill workers. At most of the flourmills, the mill owners themselves are the workers hence the pollution control and occupational safety rules are not rigorously followed. The flourmills are not constructed by taking in to considerations of acoustic designs. The flourmill workers work at an average of 8 hours per day without wearing any ear protective device and thus get routinely exposed to a constant mill noise during their duty period. As a result most of the flour mill workers loose a lot of their hearing. The noise does not give any acute effect or pains and the effects of noise are so gradual that they are ignored for many years until the hearing loss becomes severe. The most dangerous opinion, which is very common among people, is that a person can be accustomed to noise. This belief is totally wrong and accustomed to noise is an evidence of loosing hearing abilities. Another dangerous opinion is that noise level is accepted as non-hazardous unless it makes a clink in the ear. It is true that high noise causes such clinks but it is not true that sounds without clinks are below the harmful levels (Celen and Am, 2003).

The human ear is very sensitive and delicate instrument and intense noise levels as encountered in many industries and workplaces can cause temporary as well as permanent hearing loss. The causal relationship between workplace noise and hearing loss has been observed for centuries (Franks, 1998). Noise not only affects hearing but also affects physiological and psychological well-being of people. The physiological changes include changes in heart beat rates, blood pressures, increased metabolic rates etc. The non-auditory effects include annoyance, speech disturbance, sleep disturbance, reduced working efficiency, headache and fatigue. It also gives tinnitus, a ringing sound in the ear. A typical health effect of noise is noise-induced hearing loss (NIHL). Hearing loss can be temporary following noise exposure. However if exposure persists the loss can become permanent over time. The development of NIHL is affected by many factors, such as individual sensitivity, noise level, duration of exposure, noise characteristics, the proportion of impulse noise and the effectiveness of noise protection determines the magnitude of damage (Henderson *et al.*, 1993). About 85 dBA and above of noise level can endanger hearing if it continues for longer time (Celen and Arn, 2003).

Monitoring of the workplace noise levels is essential to establish the causative relationship between workplace noise and the noise induced hearing loss. The grinder of the flourmills produces continuous sound, which is sufficient to impart hearing loss on continuous exposure among the flourmill workers. The workers have to stand very close to the mill to feed the mill grains and collect the flour and hence expected to receive high noise dose during entire duty period. Most of the flourmill workers reported annoyance due to their workplace noise and most of them have had moderate to severe hearing loss. The hearing loss in flourmill workers was both unilateral as well as bilateral. The objectives of the study were to measure the noise levels in the flourmills, measurement of noise dose, which the workers of the flourmill receive, and to assess the impacts of noise exposure on hearing thresholds of the mill workers. It was also aimed to check the agreement between self-reported hearing loss data and the audiometric test data. To provide recommendations for noise reduction and health protection of workers was an extreme objective.

Materials and Methods

The present study was performed in Jalgaon city, which is the major urban center in north Maharashtra region with population of about 5 lakhs with the status of district place. As detected by preliminary survey, about 75 commercial flourmills are running on an average of 8 hours per day in the city. Out of which the workers working in 60 flourmills were studied for their hearing status. The

10 flourmill workers having age above 55 years were excluded from the study due to the possibility of having age induced hearing loss (Leong and Laortanakul, 2003) thereby reducing the significant number of samples to 50. A control group was the population selected from isolated communities, residing away from the main area of the city and was not much exposed to continuous, heavy noise in their routine life. The persons above age 55 are not included in control group as they may have age induced hearing loss *i.e.*, presbyacusis. The persons having well recognized hearing impairment and other illness were also excluded as controls.

The noise environment was monitored in 10 representative flourmills in the city. The selected parameters like equivalent noise levels (Leq). Maximum noise level (Lmax). maximum sound pressure level (MaxP) and 8-hour per cent dose were measured using Bruel and Kjear noise dose meter type-4436 (made in Denmark). The noise dose meter was attached to the body of worker as described in the instrument manual and allowed to remain there throughout the whole work shift (Ingle *et al.*, 2005). The dose meter in this position can measure the personal noise dose to which the worker is exposed though he moves here and there during his work. The dose meter also measured the distribution of noise levels during the measurement period at 5 dB intervals. As a part of preliminary survey a general questionnaire related with the noise annoyances such as hearing difficulty, irritation, poor concentration, anxiety, sleep disturbance, speech disturbance, headache, and tinnitus was get filled in from the workers. The self-reported hearing loss data was prepared using a standard questionnaire designed by 1977-National Health Interview Survey (NHIS), USA (Ries, 1982). The questionnaire was divided in to two parts-screening question and self-rating. The first question, called screening question was whether the worker have any hearing difficulty? If the answer was yes then the worker was asked to describe self-rating of his hearing in each ear as good, at least trouble, at lot of trouble and deafness. This has yielded the independent rating of hearing in each ear. The subjects who reported hearing difficulty in screening and also on self-rating scale were classified as having self-reported hearing loss whereas the subjects who reported hearing difficulty in screening but reported good hearing in both the ears on self rating were classified as having no hearing loss (Gomez *et al.*, 2001).

The audiometric screen test was carried out in a closed room for flourmill workers and controls with minimum background noise using a calibrated pure tone audiometer (Elkon make). The audiometry was carried out for samples before the start of work with the rest period of about 8 hours after the previous duty schedule and after the sleep at the morning for controls. Audiometry was conducted by following the HSE guidelines (Health and Safety Executive, 1995) as closely as possible to avoid temporary threshold shift (TTS). Thresholds were measured by air conduction only (Toppila *et al.*, 2001) as the noise-induced hearing loss occurs due to damage to cochlea, hairy cells of inner ear and it is sensory neural type hearing loss. The pure tone audiometer was capable of generating nine frequencies, *viz.*, 250, 500, 1000, 1500, 2000, 3000, 4000, 6000 and 8000 Hz. The audiometric testing of left and right ears was carried out separately (Leong and Laortanakul, 2003). The three threshold averages were calculated: a binaural low frequency average (250, 500, 1500 and 2000 Hz), a binaural mid frequency average (1000, 1500, 2000, 3000 and 4000 Hz) and a binaural high frequency average (3000, 4000, 6000 and 8000 Hz) to evaluate the hearing status of the workers (Ingle *et al.*, 2005). The agreement between the self-reported hearing loss and hearing impairment determined by audiometry was measured by sensitivity, specificity, false positive and false negative (Fleiss, 1981). The statistical analysis was performed using simple computer program for calculating averages and standard deviations.

Results and Discussions

Epidemiological and laboratory studies involving workers exposed to occupational noise and general population (including children) living in noisy areas around airports, industries, noisy streets, indicates that noise may have both temporary as well as permanent physiological impacts on health of people (WHO, 2000). Research has correlated the exposure to environmental noise with physiological changes in blood pressure; sleep disturbance and other stress related disorders. Worldwide noise induced hearing impairment is the most common and irreversible occupational hazard. There is general agreement that the noise levels bellow 70 dB do not cause hearing damage. Noise induced hearing impairment is accompanied by loudness perception (loudness recruitment), distortion (paracusis) and a disturbing ringing in the ear (tinnitus). Most of the workers with developing hearing loss try to adapt it but preventive measures are not searched by them, hence it intensifies with exposure period which ultimately gives deafness having serious sociological and psychological consequences.

The average noise environment for flourmill is given in Table 23.1. The average equivalent continuous sound levels (Leq) represents the mean energy of sound levels averaged over the measurement period, that is, it is the continuous steady noise level which would have the same acoustic energy as the actual, fluctuating noise levels measured over same period of time. It was found that the average noise levels for the flourmill workplace were 88.2 dB (control 44.8 dB). The Central Pollution Control Board (CPCB) of India has set up a noise exposure limit of 90 dB(A) for industrial workers over an 8 hour working day (Pandya, 2003). OSHA recommends that noise levels averaging more than 90 dB (A) over an 8-hour period at the workplace should be reduced and the employees exposed to noise levels averaging more than 85 dB must be included in hearing conservation program. Lmax is the maximum sound pressure level for the measurement period and was obtained as 97.7 for the averaged flourmill workplace environment. The maximum peak level sound pressure (MaxP) for the average measurement was 126.7. The most important parameter of concern with noise induced health problems is 8-hour per cent dose, which is the dose extrapolated to a period of 8 hours. It was 258.8 for the flourmill workers, which was much higher than the control (11.30). Any dose per cent value of more than 100 (accumulated during a working day) suggests that the worker is being exposed to too much noise (Bruel and Kjear noise dose meter type-4436 manual, 1999). The dose values for the flourmill workers were fluctuated to the greater extent during the individual measurements (SD±75.39), which shows the considerable variation of noise levels within the flourmills. The average noise level distribution of flourmill is shown in Figure 23.1, from which it is clear that the noise levels in the flourmill fluctuated between 40–105 dB during the study period. The noise distribution during sampling period shows that 75.8 per cent noise levels ranged between 70–85 dB. The 14.4 per cent noise levels were within the range of 40–65 dB and 9.8 per cent were within 85–100 dB range.

Table 23.1: Average Workplace Noise Environment (noise levels) and Range of Exposure (noise dose) for Flourmill Workers and Controls (n = 10)

Parameter	*Workers ± SD*	*Controls ± SD*
Leq	88.2±4.39	44.8±3.33
Lmax	105.9±7.00	97.7±7.36
MaxP	127.6±6.17	126.7±9.10
8-h per cent dose	258.8±75.39	13.1±8.52

Adapting the WHO definition of health a state of complete physical, mental and social well-being and not merely the absence of disease or infirmity "the extra aural effects of noise, the impairments of

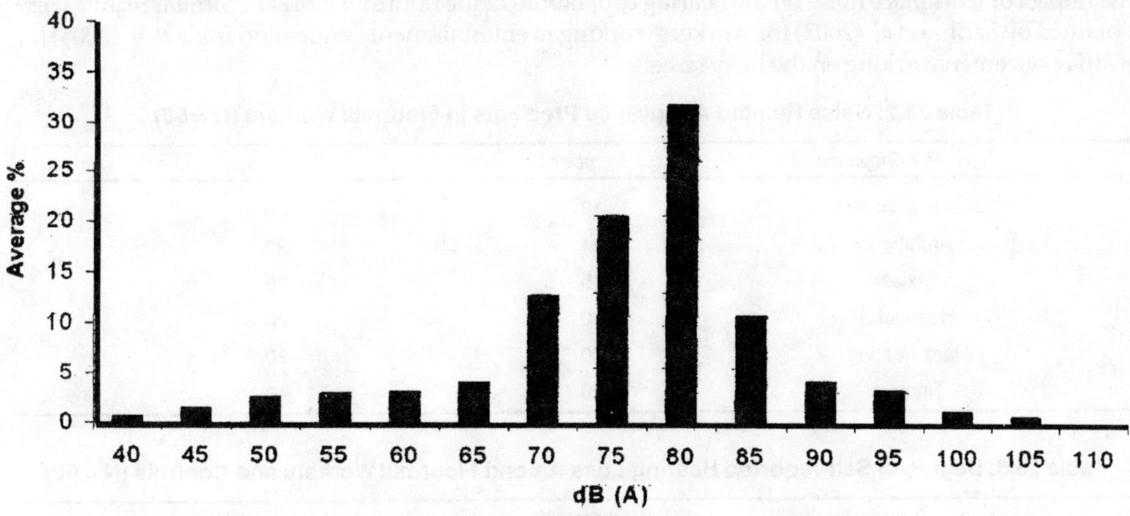

Figure 23.1: Average Noise Distribution for Flourmill Workplace

rest and sleep" of communication psychosocial well-being, and performance are clearly classified as health effects (Porter *et al.*, 2000). The analysis of general questionnaire revealed that the workers are suffering from various types of annoyance effects due to noise. It was observed that most of the workers suffered for one or other difficulty due to noise exposure (Table 23.2). Higher number of workers suffered from sleep disorders, anxiety and tinnitus. Few of them complained for headache, irritation and hard hearing. The self-reported hearing loss among the workers as listed in Table 23.3, was 64 per cent. Among controls 10 per cent persons reported hearing difficulties which might be due to exposure to anthropogenic noisy activities 8 per cent workers and 2 per cent controls reported good hearing in both the ears on self-rating scale but reported hearing difficulty on screening question which might be due to lack of understanding of the questionnaire. The persons having hearing threshold average greater than 25 dB were classified as hearing impaired (Gomez *et al.*, 2001). The audiometric average thresholds of workers revealed that 72 per cent workers in low frequency, 76 per cent in mid frequency and 82 per cent in high frequency range have increased hearing thresholds above 25 dB. The respective values for controls were 8 per cent in low frequency, 8 per cent in mid frequency and 10 per cent in high frequency range (Table 23.4), which may be due to exposure to anthropogenic noise generated by domestic appliances and exposure to leisure time noisy activities like listening music with loud sound or due to personal errors in detection of these hearing thresholds. In the average threshold range of 26 to 35 dB (mild hearing loss), 30 per cent workers have hearing impairment in low frequency region while, 26 and 20 per cent workers have impairment in mid and high frequency ranges respectively. The corresponding values for controls were 8, 8 and 10 per cent for low, mid and high frequency ranges respectively, which are much less as compared to the workers. The greatest impact of noise for workers was observed for the hearing threshold range of 36 to 50 dB (moderate hearing loss). The percentage of workers falling in this range was 20, 26 and 26 per cent for the low, mid and high frequency ranges respectively. The hearing threshold range > 50 dB (moderately severe hearing loss) includes 22,24 and 36 per cent workers in low, mid and high frequency ranges. The controls were not found any hearing loss in moderate and moderately severe ranges. The average hearing thresholds for controls were much less as compared to the samples, which clearly indicates the impact of workplace noise on the hearing capabilities of the flourmill workers. Similar results were obtained by Sadhra *et al.* (2002) for workers working in entertainment venues and Ingle *et al.* (2005) for traffic policemen working on the busy streets.

Table 23.2: Noise Related Annoyance Problems in Flourmill Workers (N = 50)

Type of Annoyance	*Number*	*Percentage*
Sleep disorders	13	26
Irritation	11	22
Anxiety	15	30
Headache	10	20
Hard hearing	10	20
Tinnitus	20	40

Table 23.3: Degree of Self-reported Hearing Loss Among Flourmill Workers and Controls (N = 50)

Type	*Workers (%)*	*Controls (%)*
Screening Question		
No hearing difficulty	36	90
Hearing difficulty	64	10
Self-rating		
Good hearing*	08	02
Unilateral at least trouble	14	04
Unilateral at lot of trouble	14	04
Bilateral at least trouble	12	00
Bilateral at lot of trouble	10	00
Unilateral deafness	04	00
Bilateral deafness	02	00

*: These participants reported hearing difficulty on screening question but reported good hearing in both the ears on self-rating scale.

The comparison between self-report and audiometry showed good agreement between the two methods (Table 23.6). Sensitivity is the ability of self report to detect the presence of hearing loss when hearing impairment is detected by audiometry. The sensitivity of the self-report was found to be good for both samples as well as controls for all the three frequency ranges and the values ranged between 44–68 per cent. Specificity is the ability of self-report to detect the absence of hearing loss when it is absent on audiometry. The self-report was found to be very good in terms of specificity also and the values were within the range of 42–74 per cent for samples and controls for the three frequency ranges. False positive is the per cent proportion of participants who reported presence of hearing loss and did not have hearing loss by audiometry and false negative are those who reported the presence of hearing loss while it was not detected by an audiometry. The per cent false positive cases were between 14 and 20 per cent which may be due to underestimation of hearing capability while per cent false negative were between 12 and 24 per cent for workers and controls and may be due to overestimation of self hearing capability. Similar results were obtained by (Nandahl, 1998 and Gomez *et al.*, 2001). This suggests that the self-report method may be used as a quick epidemiological survey method to detect the hearing status in larger populations.

Table 23.4: Degree of Hearing Impairment Based on Audiometric Threshold Average for Fourmill Workers and Controls (N = 50)

Category	*Threshold Average (dB HL)**					
	Binaural Low Frequency (%)		*Binaural Mid Frequency (%)*		*Binaural High Frequency (%)*	
	Workers	*Controls*	*Workers*	*Controls*	*Workers*	*Controls*
No impairment < 25 dB HL	28	92	24	92	18	90
Impairment > 25	72	08	76	08	82	10
> 25 – ≤ 35	30	08	26	08	20	10
> 35 – ≤ 50	20	00	26	00	26	00
> 50	22	00	24	00	36	00

* Low frequency is the average threshold at 250, 500, 1000 and 1500 Hz, mid frequency is the average threshold at 1000, 1500, 2000, 3000 and 4000 Hz and high frequency is the average threshold at 3000, 4000, 6000 and 8000 Hz.

Table 23.5: Age Group Wise Distribution of Average Hearing Thresholds Among Flourmill Workers and Controls (N = 50)

Age Group (Number)	*Average Threshold Hearing (dB HL)**					
	Binaural Low Frequency (%)		*Binaural Mid Frequency (%)*		*Binaural High Frequency (%)*	
	Workers	*Controls*	*Workers*	*Controls*	*Workers*	*Controls*
25 – ≤ 35 Years (15)	26.3	19.0	32.7	23.5	30.7	22.3
> 35 – ≤ 50 (20)	33.5	25.7	39.6	25.3	43.1	24.3
> 50 (15)	37.3	27.0	41.7	28.3	54.3	27.8

* Low frequency is the average threshold at 250, 500, 1000 and 1500 Hz, mid frequency is the average threshold at 1000, 1500, 2000, 3000 and 4000 Hz and high frequency is the average threshold at 3000, 4000, 6000 and 8000 Hz.

The age-group wise distribution of average hearing thresholds is represented in Table 23.5, which shows that the hearing threshold increases with the age in the workers. This increase is more pronounced in workers as compared to controls, which may be due to continuous exposure to mill noise during duty period. It is observed that the increase in hearing threshold is more for the high frequency region as compared to low and mid frequency regions. The mild hearing loss was observed in some cases of controls group. This may be due to the anthropogenic exposure to noise (domestic noise and leisure time activity noise) or the personal error in detection of these hearing thresholds. The real situation at the workplace and in the non-occupational environment is determined by the simultaneous influence of different agents that modify the responses of the organism to noise (Griefahn, 2000). From the above study we have concluded that the continuous exposure to higher noise levels imparts hearing impairment in the exposed individuals and the hearing loss spreads more towards the high frequency side. Noise induced hearing loss continues to be one of the most common occupational disorders. Some ear protecting devices such as earplugs and earmuffs are necessary for the fourmill workers during their work. Also, taking acoustic parameters in to considerations the designing of fourmill shades should be performed to reduce the impact of mill noise on workers.

Table 23.6: Measure of Agreement Between the Presence or Absence of Self-reported Hearing Loss and Hearing Impairment Defined by Audiometric Threshold Averages for Flourmill Workers and Control (N = 50)

Age Group	*Threshold Average (dB HL)**					
	Binaural Low Frequency (%)		*Binaural Mid Frequency (%)*		*Binaural High Frequency (%)*	
	Workers	*Controls*	*Workers*	*Controls*	*Workers*	*Controls*
Sensitivity	68	64	66	68	44	48
Specificity	42	48	70	74	64	68
False positive	20	16	16	14	20	16
False negative	24	22	18	12	18	16

* Low frequency is the average threshold at 250, 500, 1000 and 1500 Hz, mid frequency is the average threshold at 1000, 1500, 2000, 3000 and 4000 Hz and high frequency is the average threshold at 3000, 4000, 6000 and 8000 Hz.

References

Bruel and Kjaer Limitd, 1999. *Noise Dose Meter Type*-4436 manual, Denmark, pp.53.

Celen, I.H. and Arn, S., 2003. Noise levels of agricultural tractors. *Pakistan Journal of Biological Sciences*, 6(19): 1706–1711.

Gomez, M.I., Hawang, S., Sobotova, L., Stark, A.D. and May, J.J., 2001. A comparison of self-reported hearing loss and audiometry in a Cohort of New York farmers. *Journal of Speech, Languages and Hearing Research*, 44: 1201–1208.

Griefahn, B., 2000: Noise-induced extra-aural effects. *J. Acoust. Soc. Jpn.*, 21(E): 307–317.

Ingle, S.T., Pachpande, B.G., Wagh, N.D. and Attarde, S.B., 2005. Noise exposure and hearing loss among the traffic policemen working at busy streets of Jalgaon urban center. *Transportation Research*, 10(D): 69–75.

Koweir, M.H. and Jamil, A.T.M., 2003. Noise pollution in textile, printing and publishing industries in Saudi Arabia. *Environmental Monitoring and Assessment*, 83: 103–111.

Leong, S.T. and Laortanakul, P., 2003. Monitoring and assessment of daily exposure of roadside workers to traffic noise levels in an Asian city: A case study of Bangkok streets. *Environmental Monitoring and Assessment*, 85: 69–85.

McBride, D.I. and Williams, S., 2001. Audiometric notch as sign of noise induced hearing loss. *Occupational Environmental Medicine*, 58: 46–51.

Pandya, G.H., 2003. Assessment of traffic noise and its impact on the community. *Indian Journal of Environmental Studies*, 60(6): 595–602.

Sadhra, S., Jackson, C.A., Ryder, T. and Brown, M.J., 2002. Noise exposure and hearing loss among students employees working at university entertainment venues. *Ann. Occup. Hyg.*, 46(5): 455–463.

Toppila, E., Pyykko, I. and Stark, J., 2001. Age and noise induced hearing loss. *Scand. Audiology*, 30(4): 236–244.

Chapter 24

Allelopathic Potential of *Vicia sativa* L. on Germination, Growth and Dry Matter Production in Wheat

A. Rahman and Musarrat Jabeen
Department of Botany, R.D. & D.J. College, Munger – 811 201, Bihar, India

ABSTRACT

Allelopathic effect of *Vicia sativa* was studied in terms of effect of aqueous extracts from radicles and plumules of the plant on seed germination and seedling growth in Wheat. Reduction in the percentage of seed germination and growth of radicle and plumule of the seedling were recorded in all treatments. The degree of inhibition was dependant on the concentration of the extract. Similar inhibitory effects was also recorded in the form of reduction in the dry weight of the seedling. Radicle growth was more severely inhibited in comparison to plumule and proved phytotoxic to the seedling. Such differential effect might indicate the presence of higher concentration of growth inhibitors in the radicle than the plumule.

Keywords: Allelopathic, Vicia sativa, Inhibition, Growth reduction, Wheat.

Introduction

Chemical compounds have been reported to be exudated by plants and their inhibitory effect have also been observed in many cases. The ecological significance of such compounds lie in their ability to effect species composition, rate of succession and plant productivity. These are evidences showing liberation of chemicals by volatilization from aerial parts, exudation from roots, leaching from plants and their residues by rain or by decomposition of residues (Rice, 1987). Such chemicals which impose allelochemics released into the environment by a species is directly responsible for the survival as well as dominance of that species and reduction or even elimination of associated plant species.

Vicia sativa L. is a dominant weed associated with wheat crop of Munger (24°30′ North Latitude, 86°30′ East Longitude and 45 m above sea level). It causes considerable damage to the crop. Very little work has been done on the aspect of allelopathic studies on this weed except by Sinha (1993). However, a considerable work on this aspect has been done on the other weed (Sen *et al.*, 1969; Srivastava, 1969; Chatterjee, 1975; Lines and Fournier, 1979; Mukherjee and Sahai, 1985; Acharia and Sinha, 1992; Dhawan *et al.*, 1998; Rahman, 1998, 2002).

Considering the above fact endeavour has been made to study the allelopathic effect of aqueous extract of root and Shoot of *Vicia sativa* on the seed germination, radicle and Plumule elongation and dry weight of wheat seedling.

Materials and Methods

Root and shoot of *V. sativa* were collected from R.D. & D.J. college campus, Munger, in January 2004. Extracts were prepared by crushing 10 gm of these organs in a mixture with 100 ml of distilled water. The mixture was kept for 72 h before filtering in Whatman (No. 1) filter paper and the filtrate brought to 100 ml with addition of water.

This constitute the stock solution from which desired concentration (5 per cent, 10 per cent, 15 per cent and 20 per cent) were obtained by way of dilution. Each bioassay consisted of placing 25 wheat seeds in a sterile petridishes (11 cm dia) which contained one thickness of filter paper and 10ml test solution for each treatment. Controls were maintained with equal quantity of distilled water. Seeds were allowed to germinate under room temperature (20± 2°C).

Seed germination and linear growth of radicle and plumule were measured after 7 days. These seedlings were dried at 80°C for 72 h and then weighted (Konis, 1940). The above data have been subjected to statistical analysis by students 't' test of significance.

Results and Discussion

The process of seed germination was observed to be affected by the treatment of extracts. The germination percentage showed decline along with the increase in the concentration of weed extract (Table 24.1). Root extract was more inhibitory as compared to shoot extract, as at 20 per cent concentration germination percentage was recorded to be only 36 per cent in case of root extract as compared 45 per cent in shoot extract treated seeds. Decrease in the percentage of germination due to presence of inhibitory substance in the treatment of *V. sativa* extract. On the other hand 96 per cent seed germination was recorded in the controlled condition. Thus, the decrease in the percentage of seed germination with the increase in the concentration of aqueous extract, as observed with presented work, is in conformity with the observation of Lines and Fournier (1979), Singh and Sen (1982), Acharia and Sinha (1992), Rahman (2002).

Growth performance of the seedlings as recorded from the measurement of length radicle and plumule and also by the dry weight of the seedlings showed that inhibition was appeared in all concentration of the extracts (Table 24.1). Treatment with both root and shoot extracts showed considerable decrease in the root and shoot length. The radicle growth was affected more as compared to the plumule. Such affect might be correlated with the fact that toxic chemicals present in the extracts have also affected the process of growth of seedlings in wheat (Afaq and Sinha, 1970); Sharma and Lavania, 1978).

Table 24.1: Effect of Aqueous Extract of *V. sativa* on Seed Germination (%) Linear Growth (cm) and Dry Weight (mg) of Radicles and Plumules of Wheat (*T. aestivum*) after 7 Days (mean±S.D.)

Plant Extract	Cone (%)	Germination (%)	Linear Growth (cm)		Dry Weight (mg)	
			Radicle	Plumule	Radicle	Plumule
Control	0	96	6.90±1.30	8.88±1.51	1.56±0.04	0.85±0.01
Root Extract	5	79	7.11±1.01	7.47±1.73	1.41±0.06	0.73±0.03
	10	60	6.39+1.83	7.23+1.77	1.37+0.05	0.60+0.01
	15	42	6.17±0.99	6.77±1.88	1.37±0.04	0.60±0.03
	20	36	6.02±1.47	6.99±1.04	1.02±0.03	0.59±0.01
Shoot Extract	5	88	6.18±1.46	8.16±1.28	1.54±0.03	0.71±0.02
	10	87	5.12±0.74	8.00±1.19	1.53±0.05	0.70±0.02
	15	60	4.35±0.83	6.00±0.97	1.52±0.06	0.61±0.01
	20	45	4.21±0.46	5.85±1.40	1.32±0.03	0.60±0.01

*: All differences significant at 1 per cent level.

Studies on the effect of extracts of *V. sativa* organ on the dry weight of radicle and plumule of the seedling of wheat showed that the observations were in similar lines as in the case of studies on the germination and seedling growth (Table 24.1). Reduction in dry weight of seedling (Radicle and plumule) with the increase in the concentration of the extract was suggestive of the fact that allelopathic chemicals present in *V. sativa* were causing reduction in growth by reducing the dry matter content in the plant. It indicated the effect of extracts of the weed at biosynthet level. The degree of action was same as in case of growth of the seedlings.

Negative correlation values supported the above fact. Though Del Moral and Cates (1971) had observed that the inhibition potential of any vegetation in an ecosystem is of paramount importance in judging the community of dynamics of the system. Thus, it might indicate the effect of leachate from *V. sativa* affecting the growth of wheat in the natural population. This might be also one of the major reason for better growth of these weeds during initial stage of growth in rabi fields.

Acknowledgement

The authors are thankful of Dr. (Professor) S.S. Acharia, University Professor, Department of Botany, RD and DJ College, Munger for his valuable encouragement.

References

Acharia, S.S. and Sinha, A.K., 1992. Allelopathic effect of *Circium arvense* on the seed germination and seedling growth of wheat. *Geobios*, 19: 187–188.

Afaq, S.H. and Sinha, A.S., 1970. Effect of stem extract of *Portulaca grandiflora* on germination and growth of *Cicer arietinum* seed germination. *Sci. and Cult.*, 16: 402–404.

Chatterjee, U.N., 1975. Some aspects of plant-plant chemical interactions. *Indian J. Plant Physiol.*, 18: 91–96.

Del Moral, R. and Cates, R.G., 1971. Allelopathic potential of the dominant vegetation of Western Washington. *Ecology,* 52: 1030–1037.

Dhawan, S.R., Dhawan, P. and Gupta, S.K., 1998. Allelopathic interaction between *Parthemium* and some vegetable plants: Studies on seed germination. *Adv. Plant Sci.,* 11(2): 101–104

Konis, E., 1940. On germination inhibitors II: On the action of Germination inhibiting substances in the tomato fruit. *Palestinian Journal of Botany,* Jerusalem, Ser., 11: 6–27

Lines, M.N. and Fournier, C.A., 1979. Allelopathic effect of *Cupressus lusitamica* on germination of weed seeds. *Rev. Biol. Trop.,* 27: 223–230

Mukherjee, U. and Sahai, R., 1985. Allelopathic effect of *Indigofera enneaphylla* on seed germination and seedling establishment of *Alysocarpus monilifer. Indian Bot. Soc.,* 64: 169–171

Rahman, A., 1998. Allelopathic potential of *Parthemium hysterophorus* Linn. on germination, growth and dry matter production in *Cassia sophera* L. *Bionature,* 18(1): 17–20.

Rahman, A., 2002. Allelopathic potential of *Parthemium hysterophorus* on three weedy species of *Cassia. J. Ecotoxicological Environ. and Monit.,* 12(4): 309–313.

Rice, E.L., 1987. Allelopathy: An overview. In: *Allelochemicals: Role in Agriculture and Forestry.* (Ed.) R.C. Walker. American Chemical Soc., Washington, 2: 8–22.

Sen, D.N., Dhawan, D.D. and Sharma, K.D., 1969. Preliminary studies on the influence of certain weeds and germination and growth of *Pennisetum typhoides,* Rich. *Proc. Indian Acad. Sci.,* 70: 111–117.

Sharma, B.K. and Lavania, G.A., 1978. Effect of some hormones as seed germination and early seedling growth of *Vicia* sp. *Geobios,* 5(4): 176–178.

Singh, S.C. and Sen, D.N., 1982. Observation on seedling allelopathy on weeds. *Curr. Sci.,* 51: 45–46.

Sinha, A.K., 1993. Cytological and ecological studies in important leguminous weeds of Rabi crop of Munger. *Ph.D. Thesis,* pp. 108–137.

Srivastava, A.K., 1969. Effect of seed extract of *Heliotropium eichwaldi* stead on seed germination of *Phaseolus aureus,* Roxb. *Curr. Sci.,* 38: 440–441.

Chapter 25

The Role of Abiotic Factors in the Distributional Patterns of Acarina and Collembola in the Sub-tropical Forest Ecosystem of Manipur

Ch. Chitrapati and Th. Binoy Singh
Department of Life Sciences, Manipur University, Canchipur – 795 003, Imphal

ABSTRACT

The present investigation was conducted to determine the distributional pattern of Acarina and Collembola the two most dominant soil microarthopods group in the subtropical forest ecosystems of Manipur (24°45′N and 93°55′E). The soil samples were collected with the help of soil corer of 19.64 cm^2 in surface area and microarthopods were extracted by modified Tullgren funnel. Acarina and Collembola were recorded the most dominant group in both the study sites contributing 20–65 per cent to the total soil microarthopods. Maximum population density was recorded during rainy season and decline with the onset of winter season. The population density showed decreasing trend with the increase of soil depth in both the study sites. Among the four suborders of Acarina recorded, Cryptostigmata was the most dominant and Astigmata was the least. Nine genera each of Cryptostigmata and Collembola were identified. Microclimatic conditions in different layers of soil, primarily soil moisture content, soil temperature and soil organic carbon content exhibited a significant positive correlation ($P < 0.01$) with the changes in the population density of Acarina and Collembola. Sorensens Quotient of Community Similarity (Q/S) of oribatid and collembolan community analysed between the two study sites in different soil layers ranged from 25–90 per cent.

Keywords: Soil microarthropods, Collembola, Acarina, Abiotic factors, Community similarity.

Introduction

The soil is a physically and chemically diverse environment that is used by a wide variety of invertebrate animals and microorganisms. The species differ in the distribution of their activities in space, time and biological function. In the present study soil microarthropod would mean the microarthopods fauna in soil in general and Acarina and Collembola in particular. Soil microarthopods play a key role in regulation of microbial population, decomposition of organic matter and cycling of nutrients within the soils (Coleman and Crossley, 1996; Tian *et al.*, 1998). The abiotic factors are important in determining the soil faunal population pattern and species composition. Fluctuations in temperature and moisture are less secure than they are at and above the ground surface and the amplitude of these fluctuations decreases with depth in the soil (Geiger, 1959). Under Manipur climatic conditions except studies made by Singh *et al.* (1998), no work has been reported so far. Gradual soil degradation particularly due to slash and burnt method has led to the drastic changes in the population dynamics of the soil microarthopods and thereby either directly or indirectly affecting the soil ecosystem. With this objective the present study carried out in the sub-topical forest ecosystem of Manipur to know the distributional patterns of soil microarthopods under the different abiotic factors.

Materials and Methods

The present investigation was carried out from March 2000 to February 2001 in both the protected forest (Koirengei reserve forest) and degraded sites (Khonghampat area). A total of 324 soil samples were collected with the help of cylindrical soil corer of 19,64 cm^2 in surface area at three layers *i.e.* 0–5, 5–10 and 10–15 cm during the study period. The soil microarthropods were extracted using modified Tullgren funnel at a constant temperature of 40°C (±2°C) for 6–7 days depending on he moisture content of soil samples. Soil temperature, the soil moisture contents and the soil organic carbon (Walkley and Black method) were also analysed. Community similarity was analysed by the Coefficient of Similarity (CCS) of Sorensen (1948) also known as Quotient of Similarity (Q/S) as follows:

$$\frac{Q}{S} = \frac{2C}{S_1 + S_2} \times 100$$

where,

C: Number of species common to both the communities

S_1: Number of species in community 1 (protected site)

S_2: Number of species in community 2 (degraded site)

Results and Discussion

The total population density of Acarina and Collembola recorded higher in the protected forest site as compared to the degraded study site (Table 25.1) both annually and vertically which may be attributed to high soil moisture content, high organic carbon content (Figures 25.1 and 25.2) and close canopy in the study site 1. Vegetation itself does not directly influences Collembola but may affect them indirectly through microclimate, soil fertility, soil profile, microflora and soil porosity (Hagvar, 1982). The population density of Acarina exhibited highest aggregation of 38 per cent and 41 per cent in the uppermost soil layer *i.e.* at 0–5 cm in the protected forest and degraded sites respectively and least population density of Acarina was found at 10–15 cm soil layer contributing only 28 per cent in both the study sites. Collembola was also recorded maximum at the upper soil layer and decreased with depths. This may be attributed to adequate living space, favourable moisture conditions, aeration rates and rich accumulation of humus soil in the upper soil layer (Wallwork, 1970).

Table 25.1: Showing the Total Population Density of Acarina and Collembola in the Protected Forest and Degraded Ecosystem (Number±S.E.) × 10^2 m^{-2}

Soil Microarthropod	*Protected Forest*		*Degraded Ecosystem*
	Soil Layers (cm)	*Number±S.E.*	*Number±S.E.*
Acarina	0–5	718±36.05	495±19.20
	5–10	657±20.05	372±20.33
	10–15	530±34.97	339±25.69
	Total	**1905±91.07**	**1206±65.22**
Collembola	0–5	241±12.68	166±4.95
	5–10	190±3.32	148±3.8
	10–15	171±7.66	143±6.20
	Total	**602±23.66**	**457±14.15**

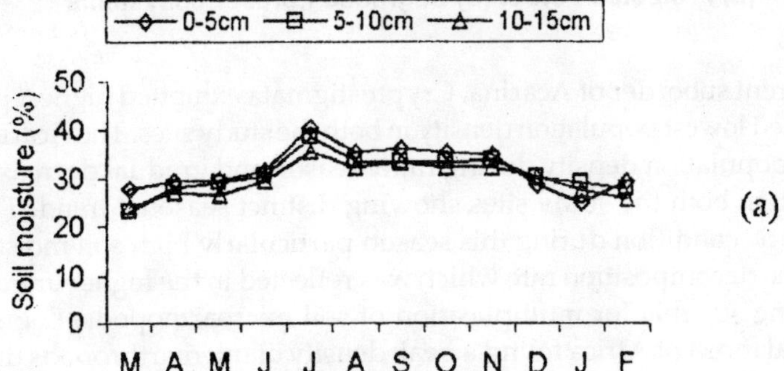

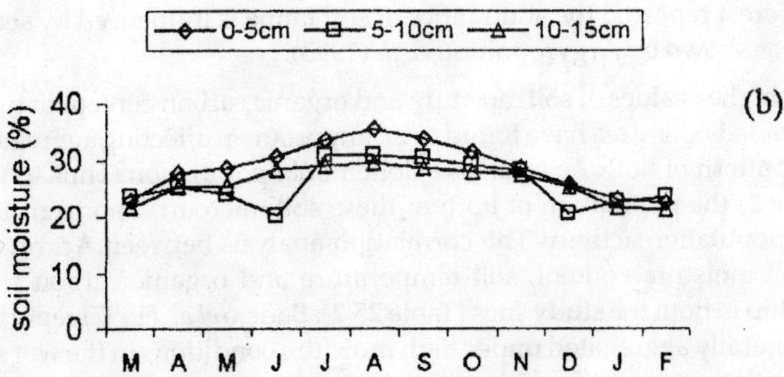

Figure 25.1: Monthly Variation of Soil Moisture (%) at Different Soil Layers (cm) in the (*a*) Protected Forest (*b*) Degraded Forest Ecosystems

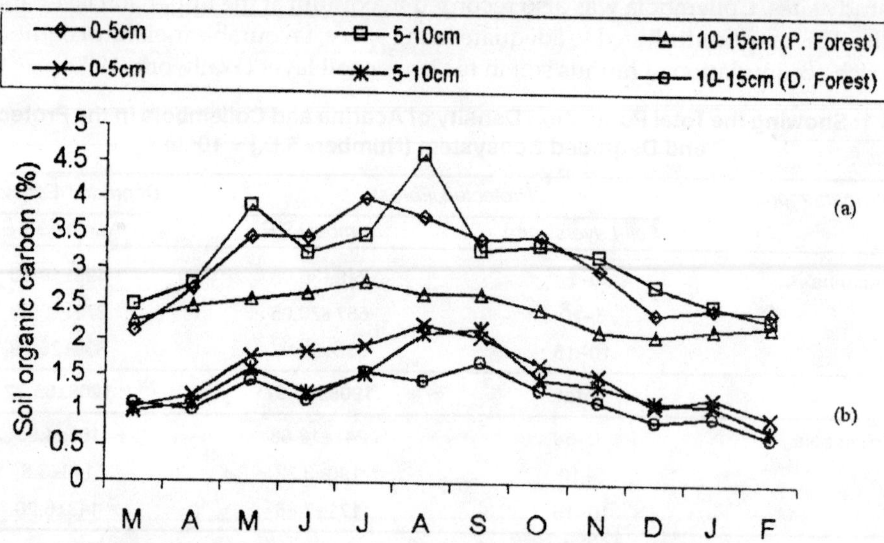

Figure 25.2: Monthly Variation of Soil Organic Carbon (%) at Different Soil Layers (cm) in the (*a*) Protected Forest (*b*) Degraded Forest Ecosystems

Among the different suborder of Acarina, Cryptostigmata exhibited highest population density and Astigmata occupied lowest population density in both the study sites. The Acarina and Collembola exhibited maximum population density during rainy season and gradual decreasing trend towards the dry winter season in both the study sites showing distinct seasonal trend. This may be due to congenial microclimatic condition during this season particularly high soil moisture content had a favourable affect in the decomposition rate which was reflected in the higher organic carbon content of the soil thus making suitable for multiplication of soil microarthopods, Loots and Ryke (1966) working in the tropical forest of Africa found a peak density of microarthropods during late summer and minimum during dry winter season. Yadava and Singh (1988) while worl:ing in the subtropical forest ecosystem of Shiroy Hill of Manipur in North Eastern India also exhibited higher population density of soil microarthopods during rainy season which gradually decreased with the approach of winter season. These findings are in agreement with the present study. Sharon *et al.* (2001) while working in an Oak forest reported the abundance of soil fauna is influenced by season and not only the upper leaf layer as shown by Argyropoulou *et al.* (1993).

Comparatively higher values of soil moisture and organic carbon content in the protected study site than that of degraded open area have found to be important in affecting aggregation behaviour as well as fluctuation pattern of both Acarina and Collembola population. Thus in the degraded and open study site (Site 2) the population of both of these soil microarthopods group resulted in the dispersed and less population activity. The correlation analysis between Acarina and Collembola population with soil moisture content, soil temperature and organic carbon showed a positive significant relationship in both the study sites (Table 25.2). Badejo *et al.* (1997) reported that Mites and Collembola were generally aggregated under high moisture conditions in the wet season which was also in accordance with the present finding.

Table 25.2: Correlationship Between Acarina and Collembola with Soil Moisture and Temperature in the Protected and Degraded Study Site

Variables	*Soil Layers (cm)*	*r*	*df*	*Y*	*t*	*p*
			Acarina			
Protected Forest						
Soil moisture (%)	0–5	0.8660	10	–8.41+0.45x	5.4732	< 0.01
	5–10	0.6719	10	–6.81+0.39x	2.8670	> 0.05
	10–15	0.7040	10	–6.39+0.36x	3.32	< 0.01
Soil temperature (°C)	0–5	0.8440	10	–1.03+0.35x	4.976	< 0.01
	5–10	0.807	10	–1.54+0.36x	4.315	< 0.01
	10–15	0.785	10	–11.94+5.48x	3.937	< 0.01
Soil organic carbon (%)	0–5	0.801	10	–3.13+2.93x	7.062	< 0.01
	5–10	0.860	10	–8.55+4.69x	5.318	< 0.01
	10–15	0.789	10	–8.08+14.12x	4.060	< 0.01
Degraded Ecosystem						
Soil moisture (%)	0–5	0.828	10	–3.79+0.27x	4.67	< 0.01
	5–10	0.673	10	–3.42+0.25x	2.875	< 0.01
	10-15	0.559	10	–2.34+0.20x	2.133	< 0.05
Soil temperature (°C)	0–5	0.879	10	–0.49+0.23x	8.005	< 0.01
	5–10	0.915	10	–1.45+0.23x	7.197	< 0.01
	10–15	0.911	10	–1.38+0.21x	6.954	< 0.01
Soil organic carbon (%)	0–5	0.799	10	–0.005+2.68x	4.198	< 0.01
	5–10	0.500	10	1.71+0.93x	1.824	> 0.05
	10–15	0.649	10	0.12+2.50x	2.697	< 0.05
			Collembola			
Protected Forest						
Soil moisture (%)	0–5	0.596	10	–2.81+0.15x	2.339	< 0.05
	5–10	0.514	10	–1.29+0.09x	1.897	> 0.05
	10–15	0.627	10	–1.91+0.11x	2.859	< 0.05
Soil temperature (°C)	0–5	0.655	10	–0.64+0.14x	2.738	< 0.05
	5–10	0.628	10	–0.083+0.085	2.551	< 0.05
	10–15	0.627	10	–0.17+0.07x	2.543	< 0.05
Soil organic carbon (%)	0–5	0.678	10	–1.76+1.21x	2.914	< 0.05
	5–10	0.539	10	0.14+0.53x	2.022	> 0.05
	10–15	0.629	10	–2.19+1.48x	2.556	< 0.05
Degraded Ecosystem						
Soil moisture (%)	0–5	0.880	10	–1.35+0.09x	5.793	< 0.01
	5–10	0.858	10	–1.52+0.103x	5.291	< 0.01
	10–15	0.580	10	–0.55+0.066x	3.165	< 0.05
Soil temperature (°C)	0–5	0.766	10	0.08+0.06x	3.765	< 0.01
	5–10	0.819	10	–0.15+0.68x	4.518	< 0.01
	10–15	0.576	10	0.17+0.05x	6.954	< 0.01
Soil organic carbon (%)	0–5	0.670	10	–0.12+0.92x	2.853	< 0.05
	5–10	0.711	10	0.25+0.72x	3.196	< 0.01
	10–15	0.440	10	0.61+0.49x	1.550	> 0.05

In the present investigation the community analysis have been confined to the individuals belonging to only nine genera each of Cryptostigmata (Acarina) and Collembola. Oribatid belonging to nine genera *viz., Chaunoproctus* sp., *Scheloribates* sp., *Zetorchestes* sp., *Rhysotritia* sp., *Haplophorella* sp., *Oppia* sp., *Xylobates* sp., *Galumna* sp. and *Euphthiracus* sp. and nine genera of Collembola *viz., Sminthurinus* sp., *Isotoma* sp., *Sinella* sp., *Folsomia* sp., *Lepidocyrtus* sp., *Onychiurus* sp., *Xenylla* sp. and *Tullbergia* sp. were analysed to show the similarity between the two study sites.

The value of Sorensen Quotient of Oribatid and Collembola showed high similarity between the two study sites (Table 25.3) which was regulated by different abiotic factors. Some genera such as *Chaunoproctus* sp., *Scheloribates* sp., *Oppia* sp. and *Xenylla* sp. were commonly available throughout the seasons and soil layers and exhibited high similarities between the two study sites. Sharon *et al.* (2001) pointed out that every taxon has a geographical range which fluctuates through times and becomes discontinuous and though individuals were distributed within the range of various points at any given time, they may not be available at the same points. But Addison *et al.* (2003) reported that the community structure of Collembola in old-growth forests differed significantly from those of younger forests and could still be differentiated from those of mature forests 80–102 years after clearcutting. However, the results of the present study showed that Collembola and Acarina population tends to be similar in diversity and richness so long the microhabitat factors are similar and least difference in the elevation of the study site could not influence much in their distribution pattern. As it was revealed from the investigation in the protected site with congenial abiotic factors particularly soil moisture played important role in determining population distribution both horizontally as well as vertically. In the degraded site, population response, associated with clear cutting and fire, resulted in a low population density as compared to the protected study site. Both Collembola and Acarina abundance preferred humus rich soils as well as soil moisture content which was clearly indicated from the high Verhoef *et al.* (1988) and reported soil moisture as the most important environmental factor acting upon soil arthropods distribution. Thus the present study revealed that the microhabitat factors particularly soil moisture and soil organic matter content tended to respond more rather than the macrohabitat factors in the distribution pattern of Acarina and Collembola population both horizontally as well as vertically.

Table 25.3: Showing Community Similarity of Oribatid and Collembola in Different Seasons and Soil Layers

Soil Layer (cm)	*Season*	*Sorensen's Coefficient Q/S (%)*	
		Oribatid	*Collembola*
	Summer	66.70	54.54
0–5	Rainy	80.00	54.54
	Winter	72.72	66.67
	Summer	54.54	60.00
5–10	Rainy	66.70	60.00
	Winter	25.00	50.00
	Summer	75.00	57.14
10–15	Rainy	88.80	50.00
	Winter	40.00	28.57

References

Addison, J.A., Trofymow, J.A. and Marshall, V.G., 2003. Abundance, species diversity and community structure of Collembola in successional coastal temperate forests on Vancouver Island, Canada. *Applied Soil Ecology*, 24: 233–246.

Argyropoulou, M.D., M.D. Asikidis, G.D. Latrou and G.P. Stamou, 1993. Colonization patterns of decomposing litter in a maquis ecosystem. *European Journal of Soil Biology*, 29: 183–191.

Badejo, M.A., T.O. Obilade and B.A. Oblubakin, 1997. Spatial distribution and abundance of mites and springtails under different temperature and moisture regimes in a tropical rain forest floor. *Tropical Ecology*, 38(1): 31–38.

Coleman, D.C. and D.A. Crossley, 1996. *Fundamentals of Soil Ecology*. Academic Press, New York, USA.

Geiger, R., 1959. *The Climate Near the Ground*. Harvard University Press, Cambridge, Mass, USA.

Hagvar, S., 1982. Collembola in Norwegian coniferous forest soils: Relations to plant communities and soil fertility. *Pedobiologia*, 24: 255–296.

Loots, G.C. and P.A.J. Ryke, 1966. A comparative quantitative study of the microarthopods in different types of pasture soils. *Zool. Afr.*, 21:157–197.

Prabhoo, N.R., 1976. Soil microarthropods of virgin forest and adjoining tea fields in the Western Ghats in Kerala. *Oriental Insects*, 10(33): 435–442.

Sharon, R., G. Degani and M. Warburg, 2001. Comparing the soil macrofauna in two Oak wood forests: Does community structure differ under similar ambient conditions? *Pedobiologia*, 45: 355–366.

Singh, T.B. and P.S. Yadava, 1998. Seasonal fluctuations of oribatid mites in a subtropical forest ecosystem of Manipur, North Eastern India. *International Journal of Ecology and Environmental Sciences*, 24: 923–129.

Sorensen, T., 1948. A method of establishing groups of equal amplitude in plant sociology based on the similarity of species content and its application to analysis of the vegetation Danish commons. *Biol. Skr.*, 5: 1–34.

Tian, G., C.O. Adejuyigbe, G.O. Adeoye and B.T. Khang, 1998. Role of soil microarthopods in leaf decomposition and N release under various land-use practices in the humid tropics. *Pedobiologia*, 42: 33–42.

Verhoef, H.A., Prast, J.E. and Verweij, R.A., 1988. Relative importance of fungi and algae in the diet and nitrogen nutrition of *Orchesella cinta* (L.) and *Tomocerus minor* (Lubbock). *Funct. Ecol.*, 2: 195–201.

Wallwork, J.A., 1970. *Ecology of Soil Animals.* Mc-Graw Hill Publishing Company Limited.

Yadava, P.S. and E.J. Singh, 1988. Some aspects of ecology of oak forest in Shiroy Hills, Manipur (North-Eastern India). *International Journal of Ecology and Environmental Sciences*, 14: 111.

Chapter 26

Influence of Ecological Factors to the Agroecosystem of Paddy in Imphal Valley: Special Reference to Biomass

S.R. Singh[1], P. Rukamani Devi[1], N.B. Devi[1], W.K. Devi[1] and N.S. Devi[2]
[1]Post Graduate Studies Centre, HRDRI, Canchipur –795 003, Imphal
[2]Botany Department, Imphal College, Imphal, Manipur

ABSTRACT

Agroecosystem, an environment of indispensable for sustainability in food production and its changing would be consequent to changes in vast cereal ecosystem have been investigated on paddy fields at Imphal Valley on two varieties *viz.*, Leimaphou (CV_1) and Phoudongba (CV_2) at different phases of crop growth during two consecutive kharif seasons of 2001 and 2002 and correlate to the meteorological factors. The dry weight, biomass in dry, biomass in wet, have remarkable influenced by environmental factors. The meteorological factors have close correlation to the growth rate and increment rate of yield components, in the wetland environment of Imphal valley.

Keywords: *Agroecosystem, Leimaphou, Phoudongba, Wetland, Biomass, Agro-climatic zone, Biological efficiency.*

Introduction

It is well known fact that the urgent need of ecology is to look deep into the study of agroecosystem as its knowledge have been based on ecological principles in food production or agriculture as well as keeping environment clean and healthy. In Manipur such work have not yet been taken up even by

Department of Agriculture. For this purpose, the present work have been undertaken on the following objectives:

1. Determine the impact of ecological parameters to the biomass of test crop.
2. To analyse the consequence of ecological factors to the rate of increment of dry biomass and wet biomass.

Materials and Methods

The agro-ecological investigation on paddy field was conducted in farmers field during two kharif crop seasons *viz.*, 2001 and 2002 at "Kitna Panung Loukon" situated in the Imphal East District of Manipur (24°5′ to 25°16′ N Latitude and 93°15′ E Longitude). The corresponding meteorological parameters were recorded from the Tulihal Airport, Imphal and ICAR, Lamphelpat, Imphal. The investigation was conducted on 3 phases *viz.*, vegetative, reproductive and ripening phases of crop growth (IRRI Research, 2001; Sorangi *et al.*, 2001 and Uphoff, 2001) with determination of biomass in dry weight, biomass in wet. Determination of above parameters an area of 1 sq. m. in the farmer's field was taken as sample unit and such units of 20 numbers/ha were selected as required to complete the observation for the period of investigation. Experimental observations were recorded for one hundred forty nine days in two seasons for Leimaphou (CV_1) and Phoudongba (CV_2), a high status high fertilizer response rice variety *in situ* from 1st June to 26th October for consecutive years of 2001 and 2002. The field was fertilized @ 100 kg N, 25.8 kg P and 50 kg K/ha. The soil of the site was clay-loam in texture and having 206.75 kg, 32.7 kg and 181.6 kg of available N, P and K respectively with a pH of 6.7. Biomass in dry and biomass in wet were taken at one observation per four days interval in vegetative phase, one observation per seven days interval in reproductive and one observation per ten days in ripening phases with the help of balance and physical counting.

Results and Discussion

Table 26.1 revealed that biomass in dry and wet during vegetative phase of rice cultivars CV_1 and CV_2 during 2001 and 2002. Biomass in wet for CV_1 during 2001 was observed as 8 mg, 16 mg, 115 mg, 150 mg, 200 mg, 1270 mg and 1370 mg on 1st, 4th, 8th, 12th, 16th, 20th, 24th, and 28th days. Biomass in dry weight were recorded as 2 mg, 5 mg, 10 mg, 16 mg, 21 mg, 24 mg, 260 mg, and 350 mg respectively. In CV_2 during 2001 biomass in wet were recorded as 17 mg, 100 mg, 150 mg, 230 mg, 260 mg, 1280 mg and 1380 mg on 1st, 4th, 8th, 12th, 16th, 20th, 24th and 28th days respectively. Biomass in wet for CV_1 (2002) on 1st, 4th, 8th, 12th, 16th, 20th, 24th and 28th days were recorded as 8 mg, 16.5 mg, 110 mg, 140 mg, 208 mg, 250 mg, 1270 mg and 1400 mg respectively. Dry biomass for CV_1 in 2002 was observed as 2 mg, 5 mg, 10 mg, 15 mg, 21.5 mg, 24 mg, 260 mg, and 405 mg on 1st, 4th, 8th, 12th, 16th, 20th, 24th and 28th days. Dry biomass for CV_2 in 2002 was accorded as 2 mg, 5 mg, 7 mg, 9 mg, 14 mg, 24 mg, 270 mg, and 380 mg on 1st, 4th, 8th, 12th, 16th, 20th, 24th and 28th days. Biomass in wet for CV_2 in 2002 was observed as 7 mg, 18 mg, 95 mg, 120 mg, 160 mg, 270 mg, 1300 mg, and 1400 mg on 1st, 4th, 8th, 12th, 16th, 20th, 24th and 28th days respectively.

Regarding biomass in dry, attainment of dry biomass increases progressively with time and attainment of two maximum dry biomass for CV_1 was observed on 24th and 28th June with values of 260 mg and 350 mg respectively in 2001 (Table 26.1 and Figure 26.1). Likewise in CV_2 two maximum attainment of dry biomass were accorded on 24th and 28th June with values of 259 mg and 370 mg respectively in 2001 [Figure 26.1(a)]. The period coincides with meteorological parameters of temperatures 30.1°C in maximum and 22.4°C in minimum, relative humidity 72 per cent–70 per cent and rainfall 1.2 mm on 24th June and 31.6°C in maximum temperature and 21.8°C in minimum

Table 26.1: Determination of Biomass (increment value of dry and wet biomass) During Vegetative Phase of Rice Varieties CV_1 and CV_2 in 2001–2002

Date of Observation	Dry Biomass (mg)		Biomass in Wet (mg)		Increment Value of Dry Biomass (mg)		Increment Value of Wet Biomass (mg)		Max. Temp. (°C)	Min. Temp. (°C)	Mean Temp. (°C)	RH Mor. (%)	RH Eve. (%)	RH Mean (%)	Rainfall (mm)
	CV_1	CV_2	CV_1	CV_2	CV_1	CV_2	CV_1	CV_2							
Y_1 (2002)															
1st June	2	2	8	8	–	–	–	–	31	19.2	25.1	83	72	77.5	17.9
4th June	5	5	16	17	3	3	8	9	26.9	20.6	23.75	91	90	90.5	8.4
8th June	10	8	115	100	5	3	99	83	27.5	21.1	24.3	94	75	84.5	18.4
12th June	16	14	150	150	6	6	35	50	30.3	20.1	25.2	84	75	79.5	10.4
16th June	21	22	200	230	5	8	50	80	27.4	21.3	24.35	82	82	87.5	24.3
20th June	24	24	245	260	3	2	45	30	28.7	21	24.85	80	62	171.5	3.4
24th June	260	250	1270	1280	236	226	1025	1020	30.1	22.4	26.25	72	70	71	1.2
28th June	350	370	1370	1380	90	120	100	100	31.6	21.8	26.7	73	65	69	1.2
Y_2 (2002)															
1st June	2	2	8	7	–	–	–	–	30.5	19.1	24.8	90	72	81	Nil
4th June	5	5	16.5	18	3	3	8.5	11	30.6	19.9	25.25	70	66	68	Nil
8th June	10	7	110	95	5	2	93.5	77	30.6	19.5	25.05	68	66	67	Nil
12th June	16	9	140	120	5	2	30	25	28.3	20.7	24.5	95	82	88.5	41.5
16th June	21.5	14	208	160	6.5	5	68	40	30.7	21.6	26.15	90	86	88	4.8
20th June	24	24	250	270	2.5	10	42	110	31.5	21.5	26.5	80	65	72.5	5.3
24th June	260	270	1270	1300	1226	246	1020	1030	28.6	21.5	25.05	83	81	82	10.8
28th June	405	380	1400	1400	145	110	130	100	30.1	22.4	26.25	79	63	71	0.1

CV_1: Laimaphou; CV_2: Phoudongba.

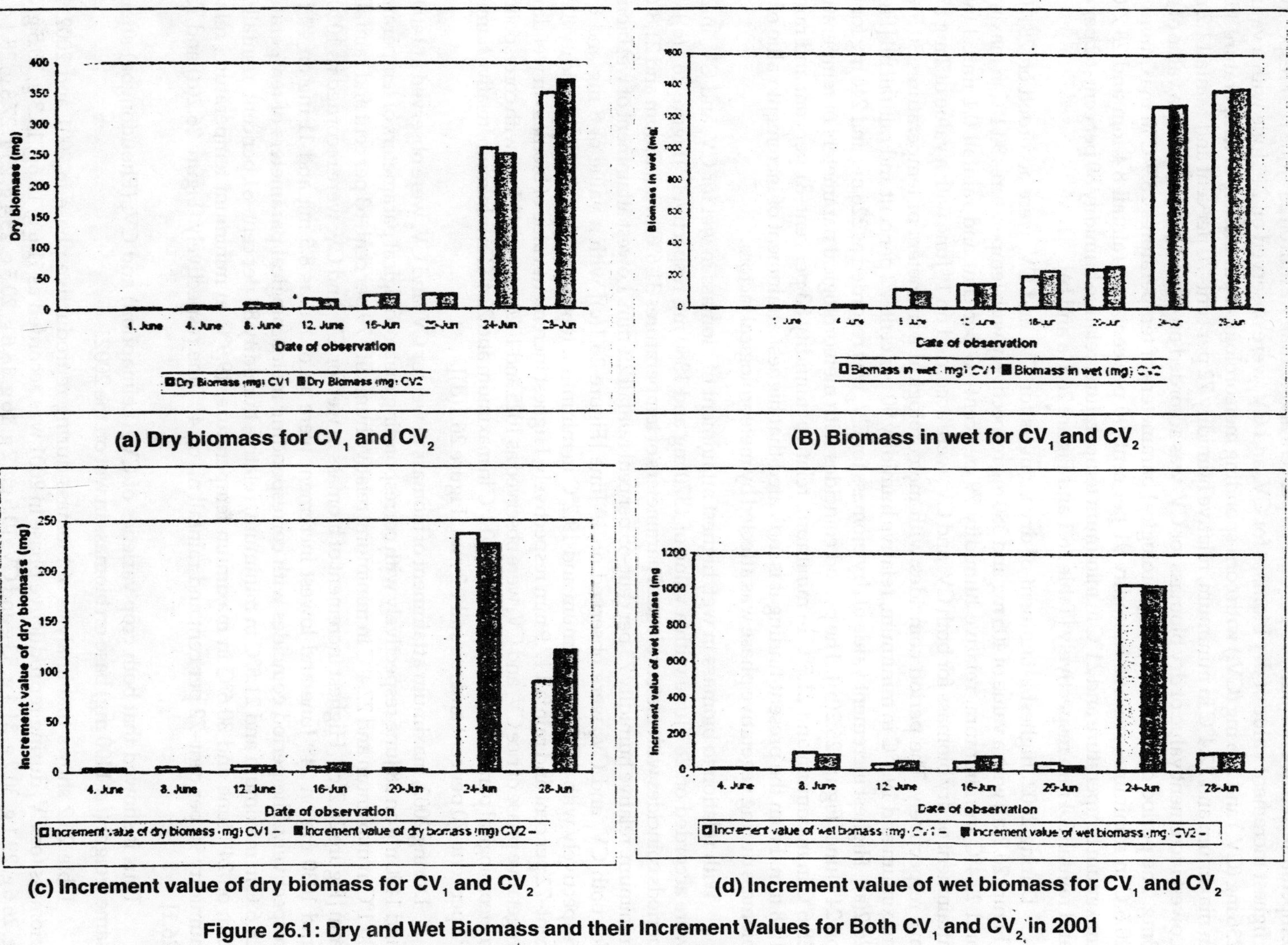

(a) Dry biomass for CV_1 and CV_2

(B) Biomass in wet for CV_1 and CV_2

(c) Increment value of dry biomass for CV_1 and CV_2

(d) Increment value of wet biomass for CV_1 and CV_2

Figure 26.1: Dry and Wet Biomass and their Increment Values for Both CV_1 and CV_2 in 2001

temperature, relative humidity 73 per cent–65 per cent and rainfall 1.2 mm on 28th June 2001 respectively. Highest increment value of dry biomass for CV_2 and CV_1 were observed on 24th June with a value of 236 mg (CV_1) and 226 mg (CV_2) with corresponding meteorological parameters of temperature 30.1°C in maximum and 22.4°C in minimum, relative humidity 72 per cent–70 per cent and rainfall 1.2 mm. Lowest increment value of dry biomass for CV_1 was accorded on 4th and 20th June with a value of only 3mg. The period coincides the meteorological parameters of temperatures 26.9°C in maximum and 20.6°C in minimum, relative humidity 91 per cent–90 per cent and rainfall 8.4 mm and 28.7°C in maximum temperature and 21°C in minimum temperature, relative humidity 80 per cent–62 per cent and rainfall 3.4 mm respectively [Table 26.1 and Figure 26.1(a and b)].

During 2002, highest attainment of dry biomass for CV_1 and CV_2 were accorded on 28th June [Figure 26.2(a)] with a value of 405 mg and 380 mg respectively with temperatures 30.1°C in maximum and 22.4°C in minimum, relative humidity 79 per cent–63 per cent and rainfall 0.1 mm. Lowest attainment of dry biomass for both CV_1 and CV_2 were accorded on 1st June with a value of 2 mg and 2 mg respectively. The period coincides with meteorological parameters of temperatures 30.5°C in maximum and 19.1°C in minimum, relative humidity 90 per cent–72 per cent and rainfall nil [Figure 26.2(a)]. Highest increment value of dry biomass for CV_1 was recorded as 226 mg and 246 mg for CV_2 on 24th June [Figure 26.2(b)]. The period coincides with meteorological parameters of temperatures 28.6°C in maximum and 21.5°C in minimum, relative humidity 83 per cent–81 per cent and rainfall 10.8 mm. From the present finding it is indicated that highest attainment of increment value of dry biomass during vegetative phase was affected by meteorological factors.

With relation to biomass in wet, highest attainment of biomass in wet for CV_1 and CV_2 in 2001 were accorded on 28th June with a value of 1370 mg and 1380 mg respectively [Figure 26.1(c)]. The periods coincides with meteorological parameters of temperatures 31.6°C in maximum and 21.8°C.in minimum, relative humidity 73 per cent–65 per cent rainfall 2.1 mm. Lowest attainment of wet biomass for both CV_1 and CV_2 were observed on 1st June [Figure 26.1(c)] with a value of 8 mg and 8 mg respectively with 31°C in maximum and 19.2°C in minimum temperature, relative humidity 83 per cent–72 per cent and rainfall 17.9 mm respectively. Highest increment value of biomass in wet during vegetative phase of rice CV_1 and CV_2 were observed as 1025 and 1020 mg on 24th June with corresponding meteorological parameters of temperature 30.1°C in maximum and 22.4°C in minimum, relative humidity 72 per cent–70 per cent and rainfall 1.2 mm [Figure 26.1(d)].

During 2002, maximum attainment of biomass in wet for CV_1 and CV_2 were observed at 1400 mg and 1400 mg on 28 June respectively with corresponding meteorological parameters of temperatures 30.1°C in maximum and 22.4°C in minimum, relative humidity 79 per cent–63 per cent and rainfall 0.1 mm [Figure 26.2(c)]. Highest increment of biomass in wet for CV_1 and CV_2 were recorded ns 1020 mg and 1030 mg on 24th June and lowest increment were accorded as 8.5 mg and 11 mg on 4th June respectively. The period coincides with corresponding meteorological parameters of temperatures 28.6°C in maximum and 21.5°C in minimum, relative humidity 83 per cent–81 per cent, rainfall 10.8 mm on 24th June and 30.6°C in maximum temperature, 19.9°C in minimum temperature, relative humidity 90 per cent–72 per cent and rainfall nil on 4th June respectively [Figure 26.2(d) and Table 26.1].

Thus it showed that both crop varieties of CV_1 (Leimaphou) and CV_2 (Phoudongba) attained same weight (*i.e.* 1400 mg) base on biomass in wet on the 2002.

Table 26.2 shows that analysis of biomass during reproductive phase in 2001 and 2002. Dry biomass for CV_1 during reproductive phase in 2001 was accorded as 11.6 g, 15.6g, 18.3 g, 20.5 g, 23.1 g, 26.6 g, 29.1 g, 30.2 g, 31.4 g, 33 g and CV_2 10.5 g, 12 g, 16 g, 18.6 g, 20.5 g, 23.4 g, 25 g, 27.3 g, 28.1 g and

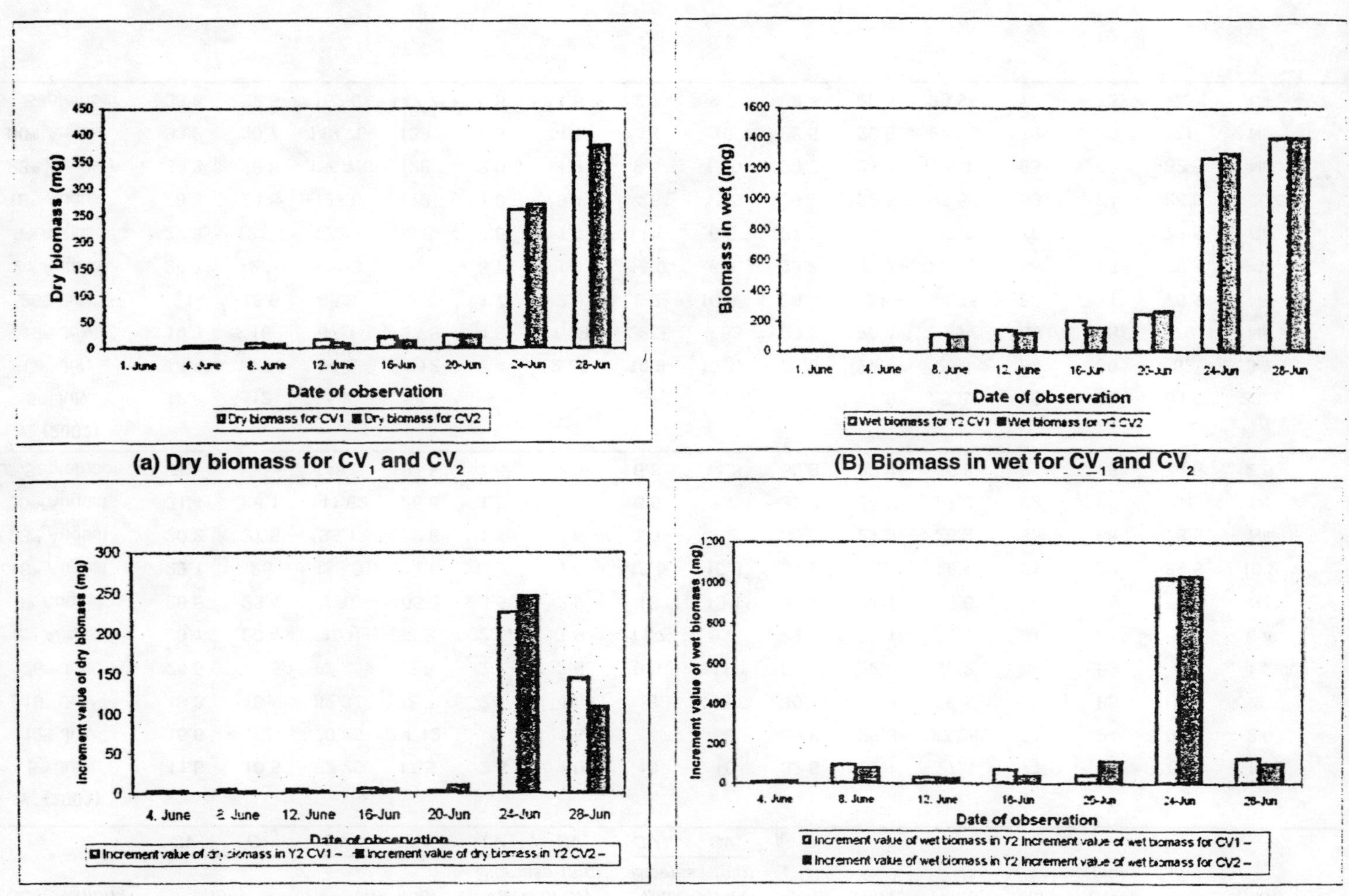

(a) Dry biomass for CV_1 and CV_2

(B) Biomass in wet for CV_1 and CV_2

(c) Increment value of dry biomass for CV_1 and CV_2

(d) Increment value of wet biomass for CV_1 and CV_2

Figure 26.2: Dry and Wet Biomass and their Increment Values for Both CV_1 and CV_2 in 2002

Table 26.2: Analysis of Biomass (increment value of dry and wet biomass) During Reproductive Phase of Rice Varieties CV_1 and CV_2 in 2001–2002

Date of Observation	Dry Biomass (mg)		Biomass in Wet (mg)		Increment Value of Dry Biomass (mg)		Increment Value of Wet Biomass (mg)		Max. Temp. (°C)	Min. Temp. (°C)	Mean Temp. (°C)	RH Mor. (%)	RH Eve. (%)	RH Mean (%)	Rainfall (mm)
	CV_1	CV_2	CV_1	CV_2	CV_1	CV_2	CV_1	CV_2							
Y_1 (2001)															
5th July	11.6	10.5	52.25	47.5	2.5	1.0	10	10	32.5	22.1	27.3	71	64	675	Nil
12th July	15.6	12	70.25	54.12	4	1.5	16	66	30.8	23.3	27.05	91	94	92.5	2.6
19th July	18.3	18.6	82.37	72.7	2.7	4.0	18	118.5	30.1	22.9	26.5	87	85	86	9
26th July	20.5	20.5	92.25	83.8	2.2	2.6	9.8	111.1	29.8	22.6	26.2	84	82	83	1.2
2nd August	23.1	20.4	104	92.2	2.6	1.9	11.7	8.4	29.1	22.1	25.6	91	81	86	6.4
9th August	26.6	23.4	120	105.3	3.5	2.9	16	13.1	32.1	21.1	26.6	78	78	78	Nil
16th August	29.1	25	132.5	115.6	12.5	1.6	12.5	10.3	31.2	22	26.6	91	88	89.5	10.2
23rd August	30.2	27.3	135.12	122.8	1.1	2.6	2.3	7.2	30.5	21.9	26.2	85	68	76.5	Nil
30th August	31.4	28.1	141.62	126.5	1.2	1.2	6.5	3.7	30.3	24.3	27.3	82	80	81	Nil
6th September	33	130.6	148.5	137.8	1.6	2.5	6.8	11.3	30.8	22.4	26.6	81	78	79.5	2.5
Y_2 (2002)															
5th July	12.5	12	54	54	–	–	–	–	27.7	22.8	25.21	86	83	84.5	23.3
12th July	14.4	14.25	64.8	64.12	2.4	2.25	10.8	10.12	27.3	22.9	25.1	84	80	82	3.7
19th July	19.3	16	87.1	72.75	4.9	1.79	22.3	8.63	29.6	20.4	25	85	73	79	Nil
26th July	21	18.6	95.3	83.7	1.7	2.6	8.2	10.95	28.5	21	24.75	82	71	76.5	Nil
2nd August	24.7	20	111.3	90	3.7	1.4	16.0	6.3	29.8	22.7	26.25	86	73	79.5	4
9th August	27.3	121.4	122.8	96.5	2.6	1.4	11.5	6.5	31.5	23.7	27.6	76	71	73.5	Nil
16th August	28.3	24.8	127.6	112	1.0	3.4	4.8	15.5	28.8	22.5	25.65	93	84	88.5	20
23rd August	30.3	26.6	136.75	128	2.0	1.8	9.1	16.0	29.7	21.1	25.4	83	82	82.5	Nil
30th August	33.2	30.7	149.75	138	2.9	4.1	13	10	28.5	20.6	24.55	72	70	71	Nil
6th September	33.8	32.5	152.5	147.7	0.6	1.8	2.7	9.7	29.4	22.5	25.95	77	73	75	Nil

30.6 g respectively on 5th July, 12th July, 19th July, 26th July, 2nd August, 9th August, 16th August, 23rd August, 30th August and 6th September. On these observation period biomass in wet for CV_1 during 2001 was accorded as 52.25 g, 70.25 g, 82.37 g, 92.25 g, 104 g, 120 g, 132.5 g. 135.12 g, 141.62 g, and 148.5 g for CV_2 it was recorded as 47.5 g, 54.12 g, 72.7 g, 83.8 g, 92.2 g, 105.3 g, 115.6 g, 122.8 g, 126.5 g and 137.8 g respectively.

Dry biomass of CV_1 in 2002 was observed as 12.5 g, 14.4 g, 19.3 g, 21 g, 24.7 g, 27.3 g, 28.3 g, 30.3 g, 33.2 g, and 33.8 g in CV_2 it was accorded as 12 g, 14.25 g, 16 g, 18.6 g, 20 g, 21.4 g, 24.8 g, 26.6 g, 30.7 g, and 32.5 g on 5th July, 12th July, 19th July, 26th July, 2nd August, 9th August, 16th August, 23rd August, 30th August and 6th September respectively. Biomass in wet for CV_1 during 2002 was observed as 54 g, 64.8 g, 87.1 g, 95.3 g, 111.3 g, 122.8 g, 127.6 g, 136.75 g, 149.75 g, 152.5 g, and in CV_2 it was observed as 54 g, 64.12 g, 72.75 g, 83.7 g, 90 g, 96.5 g, 112 g, 128 g, 138 g and 147.7 g on 10 consecutive observation periods *i.e.* 5th July, 12th July, 19th July, 26th July, 2nd August, 9th August, 16th August, 23rd August, 30th August and 6th September respectively (Table 26.2).

Regarding biomass in dry, highest and lowest increment of dry biomass for CV_1 in 2001 was observed with the values of 4.0 (*i.e.* 11.6 g to 15.6 g) on 12th July and 1.1 g (*i.e.* 20.1 g to 30.2 g) on 23rd August respectively. The periods coincides with meteorological parameters of temperatures 30.8°C in maximum and 23.3°C in minimum, relative humidity 91 per cent–94 per cent, rainfall 2–6 mm and 30.5°C in maximum temperature, 21.9°C in minimum temperature, relative humidity 85 per cent–68 per cent and rainfall nil. In case of CV_2 during 2001, highest and lowest increment of biomass in dry were recorded as 4 g (*i.e.* 12 g to 16 g) 19th July and 1.0 g on 5th July respectively. The periods coincides with meteorological parameters of temperatures 30.1°C in maximum and 22.9°C in minimum, relative humidity 87 per cent–85 per cent, rainfall 9 mm on 19th July and 32.5°C in maximum temperature, 22.1°C in minimum temperature, relative humidity 71 per cent–64 per cent and rainfall nil on 5th July.

In 2002, highest rate of increment of biomass in dry for CV_1 was observed as 4.9 g (14.4 g to 19.3 g) on 19th July with corresponding meteorological parameters 20.4°C in maximum and 22.5°C in minimum, relative humidity 77 per cent–73 per cent and rainfall nil. In CV_2 highest rate of increment of biomass in dry was recorded as 4.1 g on 30th August with corresponding meteorological parameters of temperature 28.5°C in maximum and 20.6°C in minimum, relative humidity 72 per cent–70 per cent and rainfall nil. The lowest rate of increment of biomass in dry for CV_2 in 2002 was accorded as 1.4 g on 2nd August. The period coincides with meteorological parameters of temperatures 29.8°C in maximum, relative humidity 86 per cent–73 per cent and rainfall 4.0 mm. It is clear that the present investigation recorded the rate of increment of biomass in dry as high as 4 g and 4.9 g during 2001 and 2002 in CV_1 whereas in CV_2 the values were 4 g and 4.1 g during 2001 and 2002 respectively.

With relation to rate of increment of biomass in wet for CV_1 in 2001, it recorded as 18.0 g as highest on 19th July with corresponding meteorological parameters of temperatures 30.1°C in maximum and 22.9°C in minimum, relative humidity 87 per cent–85 per cent and rainfall 3 mm. Lowest increment of biomass in wet for CV_1 in 2001 was observed as 2.6 g on 23rd August with corresponding meteorological factors or temperatures 30.3°C in maximum and 24.3°C in minimum, relative humidity 82 per cent–80 per cent and rainfall nil. For CV_2 in 2001, the highest rate of increment of wet biomass was accorded as 18.5 g on 19th July and corresponding meteorological parameters of similar to that of CV_1. But lowest rate of increment of biomass in wet was recorded as 3.7 g on 30th August with corresponding meteorological parameters 30.3°C in maximum and 24.3°C in minimum, relative humidity 82 per cent–80 per cent and rainfall nil. During 2002, highest rate of increment of biomass in wet for CV_1 was accorded as 22.3 g on 19th July with corresponding meteorological parameters of temperature 29.6°C in maximum and 20.4°C in minimum, relative humidity 85 per cent–73 per cent and rainfall nil.

Lowest rate of increment of biomass in wet was observed as 2.7 g on 6th September with corresponding meteorological parameters of temperature 29.4°C in maximum and 22.5°C in minimum relative humidity 77 per cent–73 per cent and rainfall nil.

In CV_2 highest rate of increment of biomass in wet was observed as 16.0 g on 23rd August. The period coincides with meteorological parameters of temperatures 29.7°C in maximum and 21.1°C in minimum, relative humidity 83 per cent–82 per cent and rainfall nil. Lowest increment rate of biomass in wet for CV_2 in 2002 was recorded as 6.3 g on 2nd August with corresponding meteorological parameters of temperatures 29.8°C in maximum and 22.7°C in minimum, relative humidity 86 per cent–73 per cent and rainfall 4 mm. From the present analysis it is evident that rate of increment of biomass in wet is higher in 2002 than 2001 in CV_1 whereas in CV_2 the values were reverse as 18.5 g and 16.0 g during 2001 and 2002 respectively. The present finding emphasized the dry and wet biomass as an important entities towards the growth eventually to yield and yield parameters (William, 1946; Thorne and Evam, 1964 and Thompson, 1988).

Table 26.3 revealed that biomass of rice CV_1 and CV_2 during ripening phase in 2001 and 2002. Biomass of CV_1 in 2001 was recorded as 34 g, 35 g, 36 g, 36.6 g and 40 g and CV_2 as 31.5 g, 33 g, 34.5 g, 35 g and 35.13 g on 16th September. 26th September, 6th October, 16th October, and 26th October respectively. Biomass in wet for CV_1 in 2001 was recorded as 153 g, 160 g, 163 g, 165 g, 175 g, and in CV_2 it was accorded as 143.7 g, 150.3 g, 158.1 g, 157.3 g and 158.12 g on five consecutive observation dates (*i.e.* 16th September, 26th September, 6th October, 16th October and 26th October). Biomass for CV_1 and CV_2 in 2002 was observed as 36.38 g, 37.8 g, 39.1 g, 40 g, 40.2 g, and 33.9 g, 35 g, 35.6 g, 36 g, 36.25 g on five consecutive periods (*i.e.* 16th September, 26th September, 6th October, 16th October, and 26th October). Attainment of biomass in wet for CV_1 was observed as 163.75 g, 170.25 g, 176.25 g, 180 g and 181.25 g and in CV_2 it was recorded as 152.62 g, 158 g, 160.25 g, 162.5 g and 163.12 g on five consecutive observation dates (*i.e.* 16th September, 26th September, 6th October, 16th October and 26th October) respectively.

Regarding the rate of increment of dry biomass during ripening phase, highest increment value of dry biomass in CV_1 was observed as 3.4 g on 26th October (2001) with corresponding meteorological parameters of temperatures 30.1°C in maximum, 15°C in minimum, relative humidity 78 per cent–71 per cent and rainfall nil. In case of CV_2 in 2001, highest increment value of dry biomass was recorded as 1.5 g on 26th September. The period coincides with meteorological parameters of temperatures 29.5°C in maximum and 21°C in minimum, relative humidity 85 per cent–75 per cent and rainfall nil. During 2002, highest increment of biomass in dry for CV_1 was accorded as 2.58 g on 16th September with corresponding meteorological parameters of temperatures 29°C in maximum and 22.8°C in minimum, relative humidity 81 per cent–75 per cent and rainfall 1.1 mm. In case of CV highest rate of increment of dry biomass was observed as 6.5 g on 26th September. The period coincides with meteorological parameters of temperatures 29.6°C in maximum and 20.4°C in minimum, relative humidity 85 per cent–90 per cent and rainfall nil (Table 26.3). From the present analysis it is clear that rate of increment of biomass in dry was as high as 3.4 g and 2.58 g for CV_1 in 2001 and 2002 and 1.5 g and 6.5 g for CV_2 in 2001 and 2002 respectively.

During ripening phase, highest rate of increment of biomass in wet for CV_1 was observed as 10 g on 26th October with corresponding meteorological parameters of temperatures 30.1°C in maximum and 15°C in minimum, relative humidity 78 per cent–71 per cent and rainfall nil. In CV_2 during 2001, highest rate of increment of wet biomass was accorded as 8.0 g on 6th October. The period coincides with meteorological parameters of temperatures 28.5°C in maximum and 20.5°C in minimum, relative humidity 78–72 per cent and rainfall nil. During 2002, highest rate of increment of biomass in wet for

Table 26.3: Determination of Biomass during Ripening Phase of Rice Varieties CV_1 and CV_2 in 2001–2002

Date of Observation	Dry Biomass (mg)		Biomass in Wet (mg)		Increment Value of Dry Biomass (mg)		Increment Value of Wet Biomass (mg)		Max. Temp. (°C)	Min. Temp. (°C)	Mean Temp. (°C)	RH Mor. (%)	RH Eve. (%)	RH Mean (%)	Rainfall (mm)
	CV_1	CV_2	CV_1	CV_2	CV_1	CV_2	CV_1	CV_2							
Y_1 (2001)															
16th September	34	31.5	153	143.7	1.0	0.9	4.5	5.9	26	20.6	23.3	85	75	80	Nil
26th September	35	33	160	150.3	1.0	1.5	7.0	6.6	29.5	21	25.25	85	75	80	Nil
6th October	36	34.5	163	158.3	1.0	1.5	3.0	8.0	28.5	20.5	24.5	78	72	75	Nil
16th October	36.6	35	165	157.3	0.6	0.5	2.0	1.0	30.2	19.6	24.9	88	85	86.5	Nil
26th October	40	35.13	173	15.12	3.4	0.13	10.0	0.9	30.1	15	22.55	78	71	74.5	Nil
Y_2 (2002)															
16th September	36.38	33.9	163.75	152.62	2.58	1.4	11.25	4.92	29	22.8	25.9	81	75	78	1.1
26th September	37.8	35	170.25	158	1.42	6.5	6.5	5.38	29.6	20.4	25	85	80	82.5	Nil
6th October	39.1	35.6	176.25	160.25	1.3	0.6	6.0	2.25	30.8	20.6	25.7	80	76	78	Nil
16th October	40•	36	180	162.5	0.9	0.4	3.75	2.25	22.1	18.3	20.2	96	91	93.5	5.2
26th October	40.2	36.25	181.25	163.12	0.2	0.25	1.2	0.625	27.7	13.4	20.55	80	68	74	Nil

CV_1 was observed with a value of 11.25 g on 16th September with corresponding meteorological parameters of temperatures 29.6°C in maximum and 22.8°C in minimum, relative humidity 81 per cent–75 per cent and rainfall 1.1 mm. In CV_2 in 2002 highest rate of increment of biomass in wet was accorded as 5.38 g on 26th September. The period coincides with meteorological parameters of temperatures 29°C in maximum, 20.4°C in minimum, relative humidity 85 per cent–80 per cent and rainfall nil. It is clear that the present investigation recorded the biomass in wet as height as 10.0 g and 11.25 g during 2001 and 2002 in CV_1 respectively whereas in CV_2, the value were 8.0 g and 5.38 g during 2001 and 2002 respectively. The present finding indicates the impact or varieties to biomass in wet, the key characteristics of exponent to the yield and yield factors with their micro and macro environment. Huang and Lvr (2000), Venkateswarlu and Prasad (1981) reported biomass in wet was highly affected by meteorological parameters. Similar reports support the present finding in various crops and from different parts of the country (Landsberg, 1975; Malsushima, 1966).

From the present analysis it is noteworthy to determine the agroecosystem or high yielding rice type in different agro-climatic zones of the state which will help in understanding the acclimatizational qualities, environmental impact and developmental factors, consequently to more production and maximum utilization of histological efficiencies with environmental conservation for the betterment of human kind conservation.

References

Huang, J.J. and Lvr, H.S., 2000. The influence of temperature during grain filling stage on grain quality in rice (*Oryza sativa* L.). I. Effects of temperature on yield components, milling quality and grain physiochemical properties. *J. Agric. Assoc., China,* 1(4): 370–389.

IRRI, Research, 2001. *Climate Change and Rice Growth Aust. Grain,* 10(6): 20.

Landsberg, J.J., 1975. Temperature effects and plant response. In: *Progress in Biometeorology, Division C Progress in Plant Biometeorology,* (Ed.) L.P. Smith. Swets and Zeitlinger, Amsterdam, pp. 860107.

Matsushima, S., 1966. *Crop Science in Rice: Theory of Yield Determination and its Application.* Fuji Publication, Tokyo, 374 pp.

Sorangi, S.K., Lenka, D., Jena, S.N. and Gulati, J.M.L., 2001. Modelling summer rice grain yield using growth and environmental parameters. *Crop Res.,* 21(13): 274–277.

Thompson, L.M., 1988. Effects of changes in climate and weather variability on the yield of corn and soybeans. *Journals of Production Agriculture,* 1: 20–27.

Thorne, G.N. and Evam, A.F., 1964. Influence of tops and roots on net assimilation rate of sugar beet and spinach but and grafts between them. *Ann. Bot.,* 28: 499–508.

Uphoff, N., 2001. The system of rice intensification agro-ecological opportunities for small. *Leisa,* 17(4): 15–16.

Venkataswarlu, B. and Prasad A.S.R., 1981. Harvest index and biomass: Criteria for selecting rice plants with high yielding ability. *Indian J. Plant Physio.,* 25: 149–157.

Williams, R.F., 1946. The physiology of plant growth with special reference to the concept of net assimilation rate. *Ann. Bot.,* 10: 41–72.

Chapter 27

Effect of Different Layouts and Mulches on Green Yield of Turmeric cv. Salem

V.T. Yamgar, M.S. Shirke and P.S. Patil
Agricultural Research Station, K. Digraj, Dist. Sangli (M.S.), India

ABSTRACT

An experiment was conducted during the year 1999–2002, to study the effect of different layouts alongwith mulches on growth attributes and green yield of turmeric. The data of experiment during 2001–2002 as third year of experiment revealed that the ridges and furrow method of planting with soybean straw mulch @ 20 tonnes/ha produced taller plants (188.4 cm) and maximum average number (1.8) of tillers per clump, per cent recovery (21) and green yield (280.24 qt/ha). However, the same method of planting without any mulch recorded less (113.02 cm) heighted plants and lower yields (276.88 qt/ha) as compare to same method with polythene mulch (372.92 qt/ha). The pooled data revealed that the treatment, ridges and furrow method with soybean straw mulch found best as it has recorded maximum green yield (268.44 qt/ha). The differences in yield due to various treatments was found statistically significant.

Keywords: Turmeric, Mulch, Straw, Layouts, Green yield, Polythene.

Introduction

Turmeric is an important commercial crop grown over 147000 hectare in India (Mishra *et al.*, 2000). In India, Maharashtra, Karnataka, Andhra Pradesh, Tamil Nadu, Orissa and Bihar are major growing states. In western Maharashtra the crop is grown in various soil. However, alluvial soil with good organic matter status in the pH range of 5.00 to 7.50 is optimum for the crop, it cannot withstand water logging. Heavy clay, gravelly, stony soils, are unsuitable for the crop because of its interference with development of rhizomes.

Spacing and plant density has also been found to influence significantly the growth attributes, dry matter accumulation and thereby the rhizome yield. Shaikh and Munges (1998) reported the differential growth attributes and yields of potato as affected by ridges and furrow and flat bed method of planting. According to them the planting system may contributes the availability of loose and friable soil at the base of plant and may have stored higher amount of soil moisture in addition to proper root aeration resulting luxuriant growth.

Mulching of turmeric with dry leaves was reported to be beneficial in several states (Jha *et al.*, 1994; Hussain *et al.*, 1969; Mohanty *et al.*, 1991 and Aiyadurai, 1966). Mulching treatments hasten germination by a week and allow less weeds thereby resulting in higher rhizome yield than non-mulched treatments. In Kerala, mulching with green leaves at the rate of 15 tonnes/ha immediately after planting is recommended for turmeric.

In western Maharashtra the crop is generally planted in hot summer May–June during which it is essential to conserve moisture and lower down the temperature at root zone of plant. Positive relation in effect of depth of planting which may vary according to planting method and mulching on germination, growth and yield of fresh rhizome is reported (Mishra *et al.*, 2000). Hence, the experiment was conducted to evaluate the effects of different layouts and mulching on yield and yield attributes of turmeric under rainfed condition.

Materials and Methods

The experiment was laidout in the randomised block design in medium black soils at Agricultural Research Station K. Digraj. During the Kharif 1999–2002, the experiment consisted six treatments including check (Table 27.1). The field was laidout as per treatments, the healthy rhizomes of variety, 'Salem' were planted during June as per the layouts in total four replications. The soil fertility before experiment was found as the available NPK were 218, 8.42 and 524 kg/ha, pH 8.2 and EC (0.30 ds/m). The available organic carbon was at 0.52 per cent (poor). The mulches were applied after planting. Each plot of gross ($5.40 \times 4.50\ m^2$) and net ($4.80 \times 3.75\ m^2$) size was provided with FYM @ 25 tonnes/ha. The recommended dose of fertilizer (200: 100: 100 kg NPK/ha) was applied. Cultural operations were carried out as per recommended package of practices. The observations on growth attributes, green turmeric yield and monetory returns which were statistically analysed and pooled data is presented in Tables 27.1, 27.2 and 27.3.

Table 27.1: Mean Turmeric Yield Influenced by Different Layouts and Mulches

Sl.No.	Treatment	Mean Yield (q/ha)	Yield (kg/plot)	Recovery	Plant Height	Av. No. of Tillers
1.	Ridges and furrow + polythene mulch	67.13	372.92	20.00	115.05	1.5
2.	Ridges and furrow method + soybean straw mulch	68.44	280.24	21.00	118.4	1.8
3.	Broad bed furrow method + polythene mulch	43.57	242.05	19.30	110.00	1.5
4.	Broad had furrow method + soybean straw mulch	41.61	231.17	19.50	116.00	1.7
5.	Ridges and furrow method	49.84	276.88	20.10	113.02	1.0
6.	Broad bed furrow method	46.29	257.19	19.6	114.00	1.5
	SE±	3.67	20.43			
	CD at 5 per cent	11.03	61.59			
	CY per cent	13.93	13.93			

Table 27.2: Green Yield of Turmeric as Influenced by Different Layouts and Mulches (Pooled)

Sl.No.	Treatment	Green Yield (kg/plot)			Pooled (mean)	Yield (qt/ha)
		1999–2000	2000–2001	2001–2002		
1.	Ridges and furrow + polythene mulch	35.59	58.37	67.12	45.29	251.60
2.	Ridges and furrow method + soybean straw mulch	40.08	59~42	68.44	48.32	268.44
3.	Broad bed furrow method + polythene mulch	31.85	51.06	43.57	39.99	222.16
4.	Broad had furrow method + soybean straw mulch	35.37	59.03	41.61	44.11	245.05
5.	Ridges and furrow method	33.79	54.46	49.84	42.56	236.44
6.	Broad bed furrow method	32.04	54.08	46.29	41.38	229.88
	SE±	1.287	1.502	20.43	0.597	3.32
	CD at 5 per cent	3.87	4.525	61.59	1.734	9.63

Table 27.3: Mean Monetary Returns from Green Turmeric (Rs/ha) as Influenced by Different Layouts and Mulches

Sl.No.	Treatment	Gross Returns	Cost of Cultivation	Net Returns (Rs/ha)	B : C Ratio
1.	Ridges and furrow + polythene mulch	100640	75000	25640	0.34
2.	Ridges and furrow method + soybean straw mulch	107376	68100	39576	0.58
3.	Broad bed furrow method + polythene mulch	88864	75000	13864	0.18
4.	Broad had furrow method + soybean straw mulch	98020	68100	29920	0.44
5.	Ridges and furrow method	94576	65000	29576	0.46
6.	Broad bed furrow method	91952	65000	26952	0.42

1. Cost of polythene film Rs. 10000/ha.
2. Cost of soybean straw @ Rs. 1.20/ton.
3. Selling rate of green turmeric @ Rs. 2000/qt.

Results and Discussion

The data on growth characters (Table 27.1) revealed that the interaction effect of different layouts and mulches on per cent recovery, plant height and overage number of tillers were differ as statistically significant. These parameters were highest (21,118.4 and 1.8) respectively, in the treatment of ridges and furrow method of planting with soybean straw mulch @ 20 tonnes/ha (T_2). These results are in agreement with Hussain *et al.* (1969) who reported the mulching of turmeric with sugarcane trash as useful in Punjab for increasing the germination rate, plant height as well as the fresh rhizome yield. Mulching of turmeric with daincha leaves (*Sesbania aculeata*) and sunhemp leaves (*Crotolaria juncea*) leaves were reported beneficial by Aiyadyrai (1966).

The pooled data in Table 27.2 revealed that the difference in green yield of turmeric due to various treatments was statistically significant. Among the treatments, ridges and furrow method of planting and soybean straw mulch resulted in highest yield (268.44 qt/ha) as against same layout but not mulched with soybean straw (236.44 qt/ha) and mulched with polythene (251.60 qt/ha). The ridges and furrow layout has resulted in higher yield as compare to Broad bed furrow irrespective of mulching. The results are again in agreement with Hussain *et al.* (1969) and Aiyadurai (1966). Suryanarayana and Venkatesulu (1988) found the maximum photosynthetic rate of tomato after 55 and 75 days after planting due to mulching with paddy straw which could be attributed to the effective conservation of moisture in plant base and supported the results. These results indicates the additional effects of organic or synthetic mulches on yield of turmeric of increased yields in present study. The similar views about mulching were kept by Rethinam *et al.* (1994) as mulching treatments hasten germination by a week and allow less number of weeds thereby resulting in higher rhizome yield than non-mulched treatments in turmeric. Jha, *et al.* (1984) reported that mulching turmeric with *Dalbergia* leaves increased the yield in Bihar. However, Mohanty *et al.*, in Orissa reported any of the locally available leaves or straw can be profitably used as mulch for turmeric @ 12.5 tonnes/ha/mulching for 3 times resulted in maximum yield.

The increase in green turmeric yield in present investigation is not only the effect of soybean straw mulch, however it could be attributed to the effective planting method *i.e.* ridges and furrows. Because these results are on the line of Shaikh and Mungse (1998) as they have reported the significantly higher leaf area index and leaf area duration (days) and total dry matter/ plant in ridges and furrow method of planting than flat bed system at all the growth states of chickpea. They further elaborated that the plants under ridges and furrow layout may be attributed to fact that this layout must have provided loose and friable soil at the base of plant and may have stored higher amount of soil moisture in addition to proper root aeration resulting in luxuriant growth.

The data on mean monetary returns (Table 27.3) revealed that ridges and furrows layout and soybean straw mulch was found to produce maximum net monetory returns of Rs. 39276/ha than other layouts and mulch material *i.e.* polythene in ridges and furrow (T_1) (Rs. 25640/ha) while in Broad bed furrow (Rs. 13864/ha). As regards the B: C ratio also the T_2 (1 : 0.58) was found beneficial than T_1 (1 : 0.34) and T_3 (1 : 0.18). The results on monetory returns were in agreement with those of Venkatesha *et al.* (1998) as he reported the linear increase in cured yield of turmeric and B : C ratio as found in present study as interaction effect of different layouts and mulches, Shinde and Firake (1998) have also reported the highest B : C ratio of 3.80 : 1 with net extra income of Rs. 30114/ha over control as an effect of sugarcane trash (organic) mulch in production of summer chilli and supported the results of present investigation.

The results thus concluded that for obtaining the maximum yield of turmeric, ridges and furrow method of planting followed by soybean straw as mulch @ 20 tonnes/ha is best and economical.

References

Hussain, A., Washeeda, A. and Zafar, M.A., 1969. Effect of mulches on germination, growth and yield of turmeric. *W. Pakistan J. Agric. Res.*, 7: 153–157.

Jha, R.C., Sharma, N.N. and Maurya, K.R., 1984. Effect of sowing dates and mulching materials on the yield of turmeric. *Proc. PLACROSYM-V*, pp. 495–498.

Kotoky, U. and Bhattacharya, R.K., 1991. Bunch weight and yield of Banana as influenced by organic mulches. *Indian J. Hort.*, 48(2): 121–123.

Aiyadurai, S.G., 1966. A review of research on spices and cashewnut in India. Regional office (Spices and cashew). Indian Council of Agricultural Research, Ernakulam, pp. 209.

Mohanty, D.C., Sharma, Y.N. and Panda, B.S., 1991. Effect of mulch materials and cover crops on the yields of turmeric Cv. Suvarna, under rainfed condition. *Indian Cocoa Arecanut and Spices J.*, 15: 8–11.

Suryanarayana, V. and Venkatesulu, A., 1988. Effect of anti-transpirants and mulching at different irrigation frequencies on cell sap concentration and photosynthesis in relation to yield of tomato. *The Andhra Agric. J.*, 33(2): 132–136.

Shaikh, R.S. and Mungse, H.B., 1998. Influence of planting system; plant population and fertilizer levels on physiological parameters of late sown chickpea. *J. Maharashtra Agric. Univ.*, 23(1): 73–74.

Mishra, M., Mishra, S.N. and Patra, G.J., 2000. Effect of depth of planting and mulching on rainfed turmeric (*Curcuma longa*). *Indian J. of Agron.*, 45(1): 210–213.

Rethinam, P., Srinivasan, K. and Sushama, P.K. 1994. Nutrition of turmeric. In: *Advances in Horticulture, Vol. 9: Plantation and Spice Crops*, Part 1, (Eds.) K.L. Chadha and P. Rethinam. Malhotra Publishing House, New Delhi India.

Shinde, U.R. and Firake, N.N., 1998. Economics of summer chilli production with mulching and micro irrigation. *J. Maharashtra Agric. Univ.*, 23 (1): 14–16.

Venkatesha, J., Khan, M.M. and Chandrappa, H., 1998. Studies on uptake on NPK nutrients by turmeric cultivars. *J. Maharashtra Agric. Univ.*, 23(1): 12–14.

Chapter 28

Genetic Diversity Studies in Cowpea (*Vigna unguiculata* L. Wap)

H.N. Shinde, K.V. Sushir, S.B. Kharbade and M.D. Tandale
Mahatma Phule Krishi Vidyapeeth, Rahuri – 413 722 (M.S.)

ABSTRACT

Genetic divergence among 45 genotypes of cowpea using Mahalnobis D^2 statistics showed considerable variability among these genotypes for thirteen characters. These genotypes were grouped into five clusters. Genetic diversity was found to be geographical diversity. On the basis of analysis, genetically diverse genotypes with desirable level of particular characters have been identified and crossing program has been suggested.

***Keywords**: Cowpea, Genetic Diversity, Diversity, Clustering behavior.*

Introduction

Cowpea is grown throughout India for it's long pods as vegetable seeds as a pulse. Foliage as a fodder and for green manuring as well. The most and important difficult task in initiation of hybridization program is selected of genotypes with high per se performance for yield and yield contributing components with suitable genetic divergence among them. D^2 statistics measures the forces of differentiation at intra cluster and intra cluster level and thus help in the selection of genetically divergent parents for their exploitation in hybridization program. The D^2 statistics also measures the degree of each component character to the total divergence. Looking to the immediate need of undertaking improvement of the crop it is felt necessary to measures the genetic divergence between different genotypes of cowpea.

Materials and Methods

The material comprised of 45 indigenous and exotic genotypes of cowpea and grown in randomized block design with three replication at College of Agriculture, Pune during *kharif* 2000. Each entry was represented by single row of 4.5 m length with spacing 45 × 20 cm. Five plants were selected randomly from each plot and observations were recorded on days to 50 per cent flowering, days to maturity, plant height (cm), plant spread (cm), number of pods per plant, number of seeds per pod, pod length (cm), number of branches per plants biomass (dry weight) at flowering (g), biomass (dry weight) at harvesting (g), harvest index (per cent), test weight (g) and grain yield per plant (g). Mahalnobis D^2 analysis (1936) was used to estimate genetic divergence among 45 genotypes of cowpea. The genotypes were grouped into different clusters according to Tocher's method as described by Rao (1952). The contribution of different characters towards divergence was also estimated.

Results and Discussion

The analysis of variance revealed highly significant differences among the 45 genotypes. This indicated the existence of substantial genetic variability for all the traits under study. The D^2 value corresponding to the pair of comparison between these genotypes was the lowest (11.32) between the genotypes IC-201095 and IC-201082, whereas, high D^2 value (254441.3) was recorded between Konkan Sadabahar and EC-343002. The 45 genotypes grouped into five clusters (Table 28.1). Cluster I containing 28 genotypes was the biggest cluster. Cluster IV and V were monogenotypic indicating wide diversity from the rest and also from each other. Cluster III, contain four genotypes and cluster II contains 11 genotypes. The seven exotic lines included in the evaluation have been distributed in cluster I (5), cluster II (1), and cluster III (1), likewise the local, Konkan Sadabahar, Sangvi local, Pune local etc. were also distributed in different clusters, indicating that the geographic diversity is not reflected in genetic diversity. The earlier findings of Rewale *et al.* (1996) and Tyagi *et al.* (1999) were similar to the present results.

Table 28.1: Distribution of 45 Genotypes into Different Clusters

Cluster Number	Number of Genotypes Included	Genotypes
I	28	No. 889, B-2, Sangvi local, GC-3, RBC-2, VWB-4, V-265, Pusa-do-phasali, Dapoli safed, V-5, VWB-5, Arangaon local, Pune local, NSP-27, Cowpea-18, V-322, V-262, EC-367678, EC-343057, EC-309498, EC-343036, IC-201098, IC-201092, IC-201095, IC-201082, IC-201097, IC-201081 and IC-201080
II	11	V-8981, OSP-27, 23/2, V-130, FTC-5, Chirodi(P) GC-10, G-1, JTC-22, EC-367684 and EC-367690
III	4	VWB-7, VWB-3, EC-343002 and TC-201100
IV	1	Manjarkheda local
V	1	Konkan sadabahar

Average intra and inter cluster D values represented in Table 28.2. The maximum intra-cluster distance was observed for cluster III (CD = 38.880) and followed by cluster I (CD = 32.880) and cluster II (CD = 27.045) suggesting that genotypes included in these clusters may have different genetic architecture. The cluster IV and V being monogenotypic showed no intra-cluster distance. The cluster II and V recorded the lowest inter cluster distance (CD = 49.489) suggesting that genotypes in these cluster were genetically close to each other. The cluster III and V exhibited the maximum intercluster distance (CD = 148.90) indicating that genetic makeup of genotypes falls in these clusters may be entirely different from one another. The cluster IV and V were monogenotypic. The monogenotypic clusters were also reported by Jindal and Gupta (1985) and Rewale *et al.* (1996) in cowpea.

Table 28.2: Average Intra and Inter-cluster D-values of 45 Genotypes of Cowpea

Cluster	*Cluster-I*	*Cluster-II*	*Cluster-III*	*Cluster-IV*	*Cluster-V*
Cluster-I	32.280	66.868	54.829	60.065	107.070
Cluster-II		27.043	107.652	77.122	49.489
Cluster-III			38.880	76.673	148.901
Cluster-IV				0.0	107.856
Cluster-V					0.0

The cluster means for each character are presented in Table 28.3. The highest cluster mean for grain yield per plant (15.367) recorded by cluster III appeared blue to contribution of component characters *viz.*, plant height (131.667), plant spread (40.658), number of branches per plant (4.50), biomass (dry weight) at flowering (20.733) and at harvesting (43.525), harvest index (35.157) and test weight (13.008). Like wise, cluster V (Kokan Sadabahar) was earlier for days to 50 per cent flowering (41.00) and days to maturity (60.33) and cluster IV (Manjarkheda local) recorded maximum pod length (34.933) and test weight (13.633). In a breeding programme aimed to crop improvement, the choice of yield should be taken into account for selecting genetically divergent parents. The hybridization among genetically diverse parental genotypes for specific traits may be helpful in bringing the new gene pool in population and expanding the range of adaptation. For example the genotypes of cluster V for earliness and cluster IV for pod length and test weight can be used as doner parents for enhancing the yield of other genotypes grouped in a cluster in F_1's and can be fixed by selecting transgressive segregants followed by continued selection in advanced generations which may lead to development of a high yielding varieties with desired component characters.

Table 28.3: The Cluster Means for 13 Characters of Cowpea

Sl.No.	*Characters*	*Clusters*					*Cluster Average*
		I	*II*	*III*	*IV*	*V*	
1.	Days to 50 per cent flowering	52.024	48.061	60.750	53.333	41.000	51.615
2.	Days to maturity	87.548	79.242	96.500	87.000	60.333	87.548
3.	Plant height (cm)	76.841	46.933	131.667	97.833	24.133	76.841
4.	Plant spread (cm)	29.196	19.103	40.658	28.567	9.533	29.196
5.	Number of pods per plant	15.164	15.255	14.617	8.467	5.867	15.164
6.	Number of seeds per pod	11.000	10.479	9.800	9.767	8.347	11.000
7.	Pod length (cm)	13.972	11.579	16.033	34.933	10.100	13.972
8.	Number of branches per plant	3.730	3.315	4.500	2.800	2.800	3.730
9.	Biomass (dry wt.) at flowering (g)	17.910	14.424	20.733	19.833	7.567	17.910
10.	Biomass (dry wt.) at harvesting (g)	37.456	31.148	43.525	35.100	11.967	37.456
11.	Harvest index (per cent)	34.514	30.759	35.157	22.633	25.300	34.514
12.	Test weight (100 seed weight) (g)	10.632	7.385	13.008	13.633	7.000	10.080
13.	Grain yield per plant (g)	13.225	9.697	15.367	7.933	3.040	13.225

References

Jindal, S.K. and Gupta, D.S., 1988. Genetic divergence in cowpea [*Vigna unguiculata* (L.) Walp.]. *Indian J. Agric. Sci.*, 55(1): 1–3.

Mahalnobis, P.C., 1936. On the generalised distance in statistics. In: *Proc. Nat. Ins. Sci., India*, 2: 49–55.

Rao, C.R., 1952. *Advanced Statistical Methods in Biometrical Research.* I. John Wiley and Sons Inc., New York.

Rewale, A.P., Birari, S.P. and Apte, U.B., 1996. Genetic divergence in cowpea. *Indian J. Agric. Res.*, 30(2): 73–79.

Tyagi, P.C., Kumar, N. and Agrawal, M.C., 1999. Genetic divergence in early maturity cowpea [*Vigna unguiculata* (L.) Walp]. *Agric. Sci. Dig., Karnal*, 19(3): 162–166.

Chapter 29

The Incidence of Helminth Parasites in Amphibian and Reptilian Population from Rohilkhand Region, India

Pooja Chandra, Neelima Gupta* and Manju Bhaskar
Department of Animal Science, MJP Rohilkhand University, Bareilly, U.P., India

ABSTRACT

A total of 300 amphibians (250 *Bufo melanostictus* and 50 *Rana tigrina*) and 150 reptiles (100 *Hemidactylus flaviviridis* and 50 *Calotes versicolor*) were examined for parasitic helminth infection from Rohilkhand region, India. Parasites from GI tract of freshly killed animals were isolated, washed in normal saline, processed as per helminthological techniques and examined under the microscope for identification. Five helminth species were discovered: 3 nematodes (*Oxysomatium, Oswaldocruzia* and *Thelandros*), 1 Cestode (*Proteocephalus*) and 1 unidentified trematode. 100 per cent infectivity of nematode *Oxysomatium* occurred in *Bufo melanostictus,* and the next prevalent nematode was *Thelandros* (92.5 per cent infectivity in *Hemidactylus flaviviridis*). On the other hand, *Rana tigrina* showed only 20 per cent infectivity of *Oxysomatium* and the remaining species were absent. *Calotes versicolor* was totally free from any helminth infection. It was further observed that parasitaemia in *Bufo* was maximum (*Oxysomatium:* 32 par/host) in lowest weight group (Group A: 1–20 gms) whereas in *Hemidactylus* it was maximum (*Thelandros:* 9 par/host) in the highest weight group (Group C: 10–20 gms).

***Keywords**: Incidence, Helminths, Amphibians and Reptiles.*

* E-mail: guptagrawal@rediffmail.com.

Introduction

The parasites of vertebrates and their community ecology have attracted interest of many workers in the last few decades, thus developing a conceptual framework for the hierarchical structure of parasites community (Poulin, 1998). There are several reports on ecological studies on helminth infra-communities of marine mammals, birds, fish etc. (Goater *et al.*, 1987; Goater and Bush, 1988; Edwards and Bush, 1989; Kennedy and Williams, 1989; Balbuena and Raga, 1993). However relatively limited number of investigations have been made on the helminth communities of amphibians and reptiles.

Although there are numerous reports and general ecological studies on the helminths of amphibians and reptiles, a very less amount of work has been done on specific studies utilizing community measures (species richness, helminth intensity, diversity indices and evenness). Aho (1990) opined, Helminth communities of amphibians and reptiles are highly variable, depauperate, and have traits characteristics of non interactive community structure.

Few reports exist on the helminth fauna of amphibians and reptiles from various parts of the world. Lees (1962) studied the incidence of helminth parasites in amphibians and reptiles. Frandsen (1974) investigated the parasite fauna of Danish amphibians. Baker (1987) provided a synopsis on the helminth fauna of amphibians and reptiles.

A report of parasitic nematodes from amphibians and reptiles was given by Moravec *et al.*, (1987). Goldberg and Bursey (1991) studied helminths of toads, *Bufo alvarius, Bufo cognatus* (Bufonidae) from Southern Arizona. Goldberg *et al.* (1998) reported gastrointestinal helminth of four gekkonid lizards from Marina Island, Micronesia. Goldberg *et al.* (2000) studied helminth fauna of plain leopard frog, *Rana blairi.* Goldberg and Bursey (2002a) reported helminths of 10 species of Anurans from Honshu Island, Japan. Recently, Goldberg *et al.* (2004) made an exhaustive survey on the helminths of six species of snakes from Honshu Island, Japan.

A review of literature on parasites of Indian amphiabian and reptilian hosts revealed a paucity of information. The major ones are those of Saxena and Nama (1978) who reported reptilian nematodes from Rajasthan. Subba Laxmi Devi and Subba Rao (1990) studied helminths of *Rana cynophylactis* and Tandon *et al.* (2001) reported helminth parasites of anuran host species from different regions of Nagaland. In context with Rohilkhand region India, the present communication is the first attempt to investigate helminth infracommunities of amphibians and reptilian hosts carried out by the authors from the year 2002 to 2005.

Materials and Methods

The experimental hosts, amphibians (*Bufo melanostictus* and *Rana tigrina*) and reptiles (*Hemidactylus flaviviridis* and *Calotes versicolor*) were collected regularly for 3 years (June 2002 to June 2005) from Rohilkhand region, India. Amphibian hosts were collected from different parks, ponds, marshy places, villages etc. whereas reptilian hosts were collected from houses, buildings, godowns, gardens etc. They were brought from wild conditions and maintained in the animal house under normal conditions of food, water and aeration. The hosts were pithed, dissected and the gastrointestinal tract and urinary bladder were scanned for any parasitic infection. The parasites were taken in normal saline wherein 2–3 washings were done. Identification and counting of parasites were done of live helminths. They were further processed and permanent slides made according to the techniques of Gibson (1984). Drawings of the parasites were made with the help of drawing tube fitted on a Nikon microscope or with camera lucida. Measurements were done with ocular and stage micrometer. Photographs of parasites were taken under LEICA DMLB Photoautomat.

Results and Discussion

A total of 5 species of helminth parasites were collected from the gastrointestinal tract of amphibian and reptilian hosts.

The list of parasites found in these hosts in all localities investigated are given in Tables 29.1 and 29.2. The list comprises of 5 species of the parasites including 3 nematodes (*Oxysomatium, Oswaldocruzia* and *Thelandros*), 1 cestode (*Proteocephalus*) and 1 unidentified trematode.

Out of the four hosts *i.e. Bufo melanostictus, Rana tigrina, Hemidactylus flaviviridis* and *Calotes versicolor; Bufo* and *Hemidactylus* were found to be heavily infected with helminth parasites while *Rana* and *Calotes* were comparatively less infected.

Table 29.1: Helminth Parasites and Per cent Infectivity in *Bufo melanostictus* According to Different Weight Groups

Helminth Parasite	*Site of Infection*	*Prevalence*	*Weight Groups (gms)**	*Intensity (Par/Host)*
Oxysomatium	Rectum	100 per cent	Group A	32
			Group B	20
			Group C	2–5
Oswaldocruzia	Stomach/small Intestine	40 per cent	Group A	2
			Group B	20
			Group C	6
Proteocephalus	Small Intestine	28 per cent	Group A	6
			Group B	2
			Group C	Nil
Unidentified Trematode	Small Intestine	0.8 per cent	Group A	Nil
			Group B	1
			Group C	Nil

Weight Groups (gms)*

Group A: 1–20 gms; Group B: 21–50 gms; Group C: 51–75 gms

Oxysomatium

The parasite showed absolute infectivity (100 per cent) in *Bufo melanostictus* and 20 per cent infectivity in *Rana tigrina.* On the other hand, its infectivity was comparatively low in *Hemidactylus flaviviridis* (4.62 per cent) and nil in *Calotes versicolor.* The worms occur in the rectum, but few worms were also seen in the posterior part of the intestine. These worms do not appear to be pathogenic to the host. They are mainly found in the food content of the rectum and have less contact with the epithelial lining.

Oswaldocruzia

The parasite occurs in an average of 40 per cent in *Bufo,* but surprisingly it showed a 100 per cent infectivity in the year 2005. 3.7 per cent infection was observed in *Hemidactylus* and *Rana* and *Calotes* were free from infection. It occurs in the intestine, less frequently in the terminal part of the stomach. Rarely it is also found in the anterior part of the rectum, and is thus less site specific.

Table 29.2: Helminth Parasites and Per cent Infectivity in *Hemidactylus flaviviridis* According to Different Weight Groups

Helminth Parasite	*Site of Infection*	*Prevalence*	*Weight Groups (gms)**	*Intensity (Par/Host)*
Thelandros	Rectum	92.59 per cent	Group A	3
			Group B	6
			Group C	9
Oxysomatium	Rectum	4.62 per cent	Group A	4
			Group B	3
			Group C	Nil
Oswaldocruzia	Small Intestine	3.70 per cent	Group A	3
			Group B	2
			Group C	Nil
Proteocephalus	Small Intestine	9.25 per cent	Group A	2
			Group B	5
			Group C	5

Weight Groups (gms)*

Group A: 1–20 gms; Group B: 21–50 gms; Group C: 51–75 gms

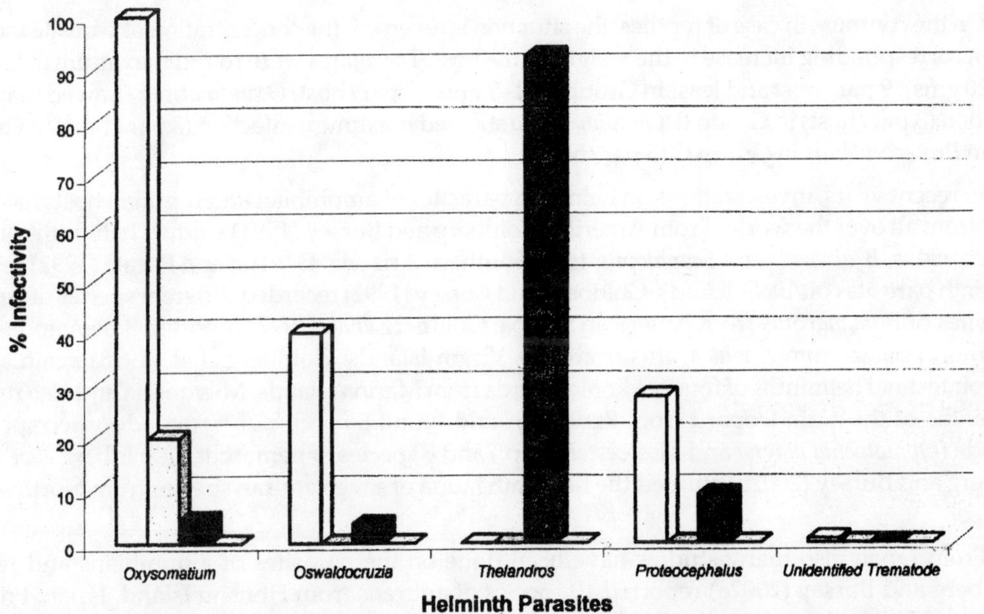

Figure 29.1: Percentage Infectivity of Parasites in Amphibian and Reptilian Hosts

Thelandros

This parasite was more prevalent in reptiles in contrast to *Oxysomatium and Oswaldocruzia.* 92.5 per cent infection of this worm was found in *Hemidactylus* and the remaining host species were devoid of infection. This worm occurs in the terminal part of the rectum entangled with the food particles.

Proteocephalus

The infectivity of this cestode was 28 per cent in *Bufo* and 9.2 per cent in *Hemidactylus.* On the other hand, *Rana* and *Calotes* were free from this infection. This worm occurs in the whole of the intestine, and was rarely encountered in the rectum.

Unidentified Trematode

A single trematode was found in *Bufo melanostictus* and showed 0.8 per cent infectivity. Trematode infection was not found in the remaining three host species. This trematode is not yet identified and is being further processed for identification.

Investigations on the concentration index of the parasite indicated that in case of amphibians, the values showed a decline with the corresponding increase in the weight of the host. *Oxysomatium* occurred in maximum numbers in Group A (1–20 gms.): 32 par/host, and minimum in Group C (51–75 gms.): 2–5 par/host. Following a similar trend, *Proteocephalus* were present at a concentration index of 6 par/host (Group A); 2 par/host (Group B) and completely disappeared in Group C. On the other hand *Oswaldocruzia* were maximum in host of Grøup B (20 par/host), minimum in Group A (2 par/host) and moderate in Group C (6 par/host).

On the contrary, in case of reptiles, the situation is reversed: the concentration of parasite increases with a corresponding increase in the weight of the host. *Thelandros* were found maximum in Gróup C (10–20 gms): 9 par/post and least in Group A (1–5 gms): 3 par/host. *Oxysomatium* showed maximum infection (3 par/host) in Group B. *Oswaldocruzia* showed maximum infection (3 par *I* host) in Group A and in *Proteocephalus* in Group C (5 par/host).

In recent years, investigations on helminth parasites of amphibian and reptilian hosts have been done from all over the world. From America, Goldberg and Bursey (1991) studied helminths of toads, *Bufo cognatus, B. alvarius* and *Scaphiopus,* from Southern Arizona. Dobson and Pacala (1992) reported helminth parasites of *Anolis* lizards. Goldberg and Bursey (1992) recorded different species of helminth parasites of *Bufo marinus* from American Samoa. Goldberg *et al.* (1997) reported 3 nematodes and 1 acanthocephalan from *Anolis acutus* from U.S. Virgin Islands. Goldberg *et al.* (1998) again studied gastrointestinal helminths of four gekkonid lizards from Marina Islands, Micronesia and also reported helminths of the plain leopard frog, *Rana blairi* and found 5 species of trematodes, two species of cestode (*Ophiotaenia magna* and *Mesocestoides* sp.) and 6 species of nematodes (Goldberg *et al.*, 2000). Golberg and Bursey (2002b) studied the helminth fauna of seven anuran species from Northwestern Mexico.

From Japan also many studies have been done on the parasites of amphibians and reptiles. Goldberg and Bursey (2002a) reported 10 species of anurans from Honshu Island, Japan. From the snakes of Honshu Island, Japan, Goldberg *et al.* (2004) found 6 species of helminths.

Frandsen, (1974) studied parasite fauna of Danish amphibians which yielded 25 trematodes, 11 nematodes and 1 acanthocephalan. Different studies on helminth fauna were done from Poland also. Kuc and Sulgostowska (1998) reported 17 digeneans, 3 nematodes, and 1 acanthocephalan from frogs in the forest of Kampinos near Warnszawa.

Mc Alpine (1997) found 23 helminth species from *Rana catesbeiana, R. clamitans* and *R. pipiens* from New Brunswick, Canada. Bursey and Goldberg (1998) found four species of helminths, *Gordgoderina, Cosmocercoides, Oswaldocruzia* and *Rhabdias* from Alberta Canada. Goldberg *et al.* (2002) reported 2 species of trematodes (*Apharygostrigea* and *Choledocystus*) one species of cestode (*Cylindrotaenia*) and 2 species of nematodes (*Oswaldocruzia* and *Rhabdias*) and a new host record for *A. pipientis* was put on record from Alberta, Canada.

Although surveys, reports and investigations on helminth fauna of amphibians and reptiles are numerous but unfortunately there are limited numbers of investigations on parasite fauna from India. Saxena and Nama (1978) gave a report of reptilian nematodes from Rajasthan. Subba Laxmi Devi and Subba Rao (1990) recorded 5 helminth species from *Rana cynophylictis*. Tandon *et al.* (2001) reported the species richness of helminth parasites in anuran host species from climatically varying regions of Nagaland.

The present findings indicate variability in infectivity of the helminth fauna in amphibian (*Bufo melanostictus* and *Rana tigrina*) and reptilian (*Hemidactylus flaviviridis* and *Calotes versicolor*) hosts with particular reference to different weight groups. Specificity of the parasites, micro and macro environmental factors, parasite biology and physiology and factors related with co-existence, dominance and elimination of parasite species may serve as viable agents for the differential infectivity.

Acknowledgement

One of us (Manju Bhaskar) expresses her thanks to the Council of Scientific and Industrial Research, New Delhi for financial help rendered by awarding Junior Research Fellowship.

References

Aho, J.M., 1990. Helminth communities of amphibians and reptiles: comparative approaches to understanding patterns and processes. In: *Parasite Communities: Patterns and Processes*, (Eds.) G. Esch, A. Bush, and J. Aho. Chapman and Hall, London, pp. 157–195.

Baker, M.R., 1987. Synopsis of the nematoda parasitic in amphibians and reptiles. *Occ. Papers Biol.*, 11: 325.

Bursey, C.R. and Goldberg, S.R., 1998. Helminths of Canadian toad, *Bufo hemiophrys* (Amphibia : Anura) from Alberta. *Canada. J. Parasitol.*, 84(3): 617–618.

Dobson, A.P. and Pacala, S.W., 1992. The parasites of *Anolis* lizards the northern lesser Antilles. *Oecologia*, 92: 118–125.

Edwards, D.D. and Bush, A.O., 1989. Helminth communities in avocets: Importance of compound community. *J. Parasitol.*, 98: 439–445.

Falbuena, J.A. and Raga, J.A., 1993. Intestinal helminth communities of long-firmed pilot whale (*Globicephala melas*) off the Faroe Islands. *Parasitology*, 106: 327–333.

Frandsen, F., 1974. A study of Danish amphibian parasite fauna. *Acta Parasitologica Polonica*, 22(6): 49–66.

Gibson, D.I., 1984. Technology as applied to museum collection: The collection, fixation and conservation of helminths. *Syst. Parasitol.*, 6: 241–255.

Goater, C.P. and Bush, A.O., 1988. Intestinal helminth communities in long-billed curlews: The importance of congeneric host-specialist. *Holarct. Ecol.*, 11: 140–145.

Goater, T.M., Esch, G.W. and Bush, A.O., 1987. Helminth parasites of sympatric salamanders: Ecological concepts at infracommunity component and compound community levels. *Am. Midl. Natl.*, 118: 289–300.

Goldberg, S.R. and Bursey, C.R., 1991. Helminths of three toads, *Bufo alvarius, Bufo cognatus* (Bufonidae), and *Scaphiopus couchil* (Pelobatidae), from Southern Arizona. *J. Helminthol Soc., Wash.*, 58(1): 142–146.

Goldberg, S.R. and Bursey, C.R., 1992. Helminths of the marine toad, *Bufo marinus* (Anura : Bufonidae), from American Samoa. *J. Helminthol. Soc., Wash.*, 59(1): 131–133.

Goldberg, S.R. and Bursey, C.R., 2002a. Helminths of 10 species of anurans from Honshu Island, Japan. *Comp. Parasitol.*, 69(2): 162–176.

Goldberg, S.R. and Bursey, C.R., 2002b. Helminth parasites of seven anuran species from Northwestern Mexico. *Western North American Naturalist*, 62(2): 160–169.

Goldberg, S.R., Bursey, C.R. and Cheam, H., 1997. Helminths of *Anolis acutus* (Sauria: Polychrotidae) from St. Croix, U.S. Virgin Islands. *J. Parasitol.*, 83(3): 530–531.

Goldberg, S.R., Bursey, C.R. and Cheam, H., 1998. Gastrointestinal helminths of four Gekkonid lizards, *Gehya mutilata, Gehyra oceanica, Hemidactylus frenatus* and *Lepidodactylus lugubris* from the Marina Islands, Micronesia. *J. Parasitol.*, 84(6): 1295–1298.

Goldberg, S.R., Bursey, C.R. and Platz, J.E., 2000. Helminths of the plains Leopard frog, *Rana blairi* (Ranidae). *The Southwestern Naturalist*, 45(3).362–366.

Goldberg, S.R., Bursey, C.R. and Telford, S.R., 2004. Helminths of six species of snakes from Honshu Island, Japan. *Comp. Parasitol.*, 71(1): 49–60.

Goldberg, S.R., Bursey, C.R. and Wong, C., 2002. Helminths of the Western Chorus frog from Eastern Alberta, Canada. *Northwest Science*, 76(1): 77–79.

Kennedy, C.R. and Williams, H.H., 1989. Helminth parasite community diversity in marine fish, *Raja batis* L. *J. Fish Biol.*, 34: 971–972.

Kuc, I. and Sulgostowska, T., 1988. Helminth fauna of frogs in the Forest of Kampinos near Warzawa. *Acta Parasisologica Polonica*, 33(4): 267–272.

Lees, E., 1962. The incidence of helminth parasite in a particular frog population. *Parasitology*, 52: 95–102.

McAlpine, D.F., 1997. Helminth communities in Bullfrog (*Rana catesbeiana*), green frog (*Rana clamitans*) and leopard frog (*Rana pipiens*) from New Brunswick, Canada. *Can. J. Zool.*, 75: 1883–1890.

Moravec, F., Barus, V. and Rysavy, B., 1987. Some parasitic nematodes, excluding Heterakidae and Pharyngodonidae from amphibians and reptiles in Egypt. *Folia Parasitologica*, 34: 255–267.

Poulin, R., 1998. *Evolutionary Ecology of Parasites: From Individuals to Communities*. Chapman and Hall, London, 212 pp.

Saxena, A. and Nama, H.S., 1978. Some reptilian nematodes from Rajasthan. *Geobios.* 5(1): 25–26.

Subba Lakshmi Devi, S. and Subba Rao, M.V., 1990. An ecological study of helminth infestation in *Rana cyanophlyctis. Comp. Physiol. Ecol.*, 15(2): 70–74.

Tandon, V., Imkongwapang, R. and Kar, P.K., 2001. Helminth infracommunities in anuran amphibia of Nagaland India. *J. Par. Dis.*, 25(1): 8–20.

Chapter 30

Studies on the Effect of Graded Levels of NaCl of *in vitro* IAA Production, Wheat Germination and Salinity Associated Soil Parameters

N.S. Kulkarni[1], M.G. Bodhankar[2], R.B. Somani[3] and J.V. Jaiswal[1]

[1]Department of Microbiology, R.A. College, Washim

[2]Department of Microbiology, Bharti Vidyapeeth's M. B. S. K. College, Kadegaon, Sangli

[3]Department of Plant Pathology, Dr. P.D.K.V., Akola

ABSTRACT

Effect of NaCl concentration on *in vitro* IAA production by *Azotobacter* isolates and their interaction with different salt concentration on seed germination in wheat were investigated, simultaneously, the soil salinity associated parameters were studied. Improvement in *in vitro* IAA production by *Azotobacter* isolates was noticed only at low levels *i.e.* up to 1.5 per cent of NaCl concentration. However, above 2 per cent NaCl, decreased IAA production was recorded. *Azotobacter* species at salt concentration at less than 1.5 per cent, showed significant improvement in seed germination. However, reduction in germination at 1.5 per cent and above salt concentration was noticed. It was also observed that as salt concentration in soil increases the salinity associated properties also fluctuates which may retard the germination in seed.

Keywords: *Azotobacter species, NaCl per cent, IAA production, Seed germination, Soil characteristics.*

Introduction

Indian farmers since long back had adopted physical and chemical methods for soil management. However, due to several constraints *viz.*, loss of soil fertility, high cost and increasing pollution, chemical methods proved set back for their use (Malik *et al.*, 1998) in soil management.

Recent developments in agricultural research are concerned with enhanced crop yield leaned on the use of microbial products, which not only acts as a source and sink of plant nutrients but synthesizes growth promoting substances lates of *Azotobacter* were further tentatively identified as *A. vinclandii, A. chroococcum, A. nigricans, A. beijerinckii* etc. on the basis of conventional morphological, biochemical and cultural studies (Bergey's 1989). IAA production was done by inoculating all the isolates separately, in 0.2 per cent tryptophan enriched Asbhy's media (Castillo, 2000) containing graded concentration of NaCl *viz.* (0.5 per cent, 1.0 per cent to 3.5 per cent). All the isolates were qualitatively analysed for IAA production using TCL (Sadashivam *et al.*, 1982); 2,6-dichlorphenol indophenol was used as a locating agent. The culture supernatant was separated using cooling centrifuge and quantitative analysis was done by adopting spectrophotometric method (Dubey, 2002). In 2 ml of supernatant, 2–3 drops of orthophosphoric acid, and 4 ml of Salper's reagent was added and kept for incubation at 37°C for 25 min. The absorbance was recorded at 560 nm. IAA concentration was estimated by comparing O.D. values with standard caliberation curve.

The pot culture experiments were conducted to assess the expression of Azotobacter species at different salt concentration. The vertisol soil was collected from local wheat farm and prepared by fortifying the calculated dose of NaCl *viz.* (0.5 per cent, 1.0 per cent to 3.5 per cent) in the soil. Pot without NaCl was maintained as control. Seed treatments were done separately on wheat (cv. Lok-1) as test crop using *Azotobacter* isolates. Wheat seeds (100 gm) were first surface sterilized with acidic alcohol followed by bacterization using charcoal based bio inoculant and jaggery solution (10 per cent) as an adhesive for each isolate separately so as to have 10^4 bacterial cells per seed. Sowing was done at the rate of 10 seeds per plot and all plots were irrigated as and when required and germination count as plant stand was noted on 20th DAS.

The soil sample used in pot cultures was analyzed for salinity associated parameters. The water holding capacity (WHC) of soil samples was analyzed adopting KR Box method suggested by Chopra (1991). Electrical conductivity was determined at 25°C (Black, 1965). pH, extractable chloride and soluble Na per cent was estimated as described by Jackson (1973).

Results and Discussion

The soil salinity had induced the major problem in the productivity of agriculture. Hence efforts have been made to investigate the effect of different salt concentration on *in vitro* Indole-3-acetic acid (IAA) production, germination percentage and some salinity associated parameters of soil.

In vitro IAA Production (Table 30.1)

It was observed that, the IAA production increases along with increase in salt concentration upto 1.5 per cent w/w NaCl. However, above 2.0 per cent level of salt concentration the IAA (37.76 µg/ml) was recorded in the treatment having 1.5 per cent NaCl and found statistically significant over control at which the IAA recorded was (35.12 µg/ml) where as the minimum *i.e.* (1.62 µg/ml) IAA was recorded at 3.5 per cent NaCl concentration. The result reveals that salt concentration at low levels *i.e.* upto 1.5 per cent w/w may positively modulates the IAA production. However, at higher levels *i.e.* above 2 per cent w/w, the negative modulation in IAA synthesis may be due to the salt stress on cell physiological activities.

Table 30.1: IAA Production by *Azotobacter* Isolates at Different Salt Concentration

Salt Conc. in Soil w/w (%)	Conc. of IAA (µg/ml) Produced				Mean
	A. vinelandii	A. chroococcum	A. nigricans	A. beigernickii	
Control	45.5	34.0	30.0	31.0	35.12
0.5 per cent	46.0	32.0	34.0	39.0	37.75
1.0 per cent	41.0	46.0	30.0	25.0	35.5
1.5 per cent	13.0	22.0	29.0	22.0	21.5
2.0 per cent	9.5	12.0	11.0	9.0	10.37
2.5 per cent	7.0	8.0	5.0	3.0	5.75
3.0 per cent	2.5	5.0	2.0	1.0	2.62
3.5 per cent	1.0	2.0	2.5	1.0	1.62

For comparing mean IAA (µg/ml) S.E.: 0.40, CD at 5 per cent: 1.15.

Germination Studies (Table 30.2)

The germination study on wheat cv. Lok-1 was carried out in pots. It was observed that, with the increase in salt concentration up to certain level there was increase in germination. However, at salt concentration of 1.5 per cent w/w and above germination per cent decreases. The maximum germination (87.7 per cent) followed by (84 per cent) was recorded in the treatment with 1.0 per cent and 0.5 per cent salt concentration respectively and proved to be statistically significant over control. The treatment with NaCl above 1.5 per cent proved to be statistically insignificant for wheat germination over control. Our results are in accordance with experimental-findings of Docuyucu *et al.* (2005). They reported the decrease in callus weight due to increasing residual effects of NaCl concentration. The reduction in germination per cent may be due to reduction in growth hormone secreting microflora or it may be due to alteration in soil characteristics that interfere with the physiological activities in germination seeds. Soil salinity adversely affects the soil microbial activities and microbial transformation of nutrients was reported by Patra *et al.* (1987).

Table 30.2: Effect or *Azotobacter* Inoculation on Germination (per cent) in Wheat at Different Salt Concentration

Salt Conc. in Soil w/w (%)	Germination % in Wheat at 20th DAS				Mean
	A. vinelandii	A. chroococcum	A. nigricans	A. beigernickii	
Control	87.0	94.0	82.0	69.0	83.0
0.5 per cent	87.0	95.0	84.0	70.0	84.0
1.0 per cent	89.0	98.0	86.0	78.0	87.7
1.5 per cent	78.0	83.0	73.0	72.0	76.5
2.0 per cent	64.0	70.0	59.0	35.0	57.0
2.5 per cent	40.0	29.0	18.0	21.0	27.0
3.0 per cent	23.0	27.0	14.0	16.0	20.0
3.5 per cent	20.0	20.0	10.0	16.0	16.5

For comparing mean Germination (per cent) S.E.: 1.22, CD at 5 per cent: 3.51.

Salinity Associated Parameters of Soil (Table 30.3)

It was observed that the parameters *viz.*, water holding capacity (WHC), electrical conductivity (EC), pH, Cl_2, soluble Na per cent increased with the increase in salt concentration as compared to control. The average WHC per cent of salt fortified soil was in the range of 40 to 48.57 per cent, where as 33.85 per cent WHC was detected in control. This might be due to higher accumulation of soluble salt in salt affected soils than normal soil. The pH of salt affected soils was detected in the range of 7.84 to 8.4. However, the treatment with 1.5 per cent NaCl and above showed increase in soil alkalinity over control. The electrical conductivity of salt treated soil varied from 2.17 to 11.55 dsm^{-1} where as it was only 1.26 dsm^{-1} in control treatment, which indicates the fluctuation in soil electrical conductivity due to the salt accumulation. The anions *i.e.* chlorides increases with increase in salt concentration. The average soluble–Na per cent value in salt treated soil was 11.12 to 44.42 per cent, which was more as compared to normal soil (8.70 per cent). This indicates that NaCl induces the changes in soil salinity associated parameters and which may ultimately disequilibrates the microbial activity in soil and hence, retardation in plant growth. Fluctuation in the values of pH, EC, Na per cent and Cl_2 due to saline water-irrigation significantly reduces the growth of the crop at growth, shooting and harvesting stages respectively, was reported by Palaniappan *et al.* (1996). Hence, there is an urgent need to screen the specific microbial consortia for each plant with special interest to their salt tolerance.

Table 30.3: Effect of Graded Doses of NaCl Per cent on Salinity Associated Parameters of Soil at 20th DAS

Salt Conc. in Soil w/w (%)	*WHC%*	*pH*	*E.C. dsm^{-1}*	*Cl_2 m'eq*	*Na%*
Control	33.85	7.86	1.26	7.89	8.70
0.5 per cent	40.0	7.86	2.70	10.16	11.12
1.0 per cent	43.07	7.86	3.14	11.27	12.33
1.5 per cent	43.15	8.03	4.2	32.2	39.81
2.0 per cent	44.81	8.2	6.62	33.93	40.12
2.5 per cent	44.98	8.5	7.75	34.85	40.18
3.0 per cent	46.92	8.3	9.45	35.4	42.59
3.5 per cent	48.57	8.4	11.15	35.76	44.42

References

Agrawal, N. and Singh, H.P., 2002. Antibiotic resistance and inhibitory effect of Azotobacter on soil borne plant pathogens. *Ind. J. Microbiol.*, 42: 245–246.

Black, C.A., Evan, D.D., Ensminger, L.E., White, I.L., Clark, F.C. and Dinarer, R.C. (Eds.), 1965. *Methods of Soil Analysis*, Part-II. Amer. Soc. Agron. Inc., Publisher, Madison, Wisconsin, USA.

Chopra, S.L. and Kanwar, I.S., 1991. *Analytical Agriculture Chemistry*, 4th Edition.

Docuyutu, T., Akkececi, S., Akkaya, A. and Kara, R., 2005. Investigation of the response of bread wheat cultivars to salinity by using callus cultures. *J. Environ. Biol.*, 26(2): 251–255.

Dubey, R.C. and Maheshwary, D.K., 2000. Cultivation and isolation of microorganism soil microbiology. In: *Practical Microbiology*. S. Chand and Company Limited, pp. 90–94, 197–198.

Jackson, M.L., 1973. *Soil Chemical Analysis*. Prentice Hall, New Delhi.

Loper, J. and Schroth, M., 1986. Influence of bacterial sources of Indole-3-acetic acid on root elongation of sugar beet. *Physiol. Biochem.*, 76: 386–398.

Martinez-Toledo, M.V., Sairneron, V. and Gonzalex-Lopez, J., 1988. Root exudate of Zea Mays and production of auxins, gibberelins and cytokinis by *Azotobacter chroococcum. Plant Soil,* 110: 149–152.

Narula, N., Lakshminarayana, K. and Tauro, P., 1981. Ammonia excretion by *Azotobacter chroococcum. Biotech. Bioengg.*, 23: 467–470.

Nileema, 1995. Inhibitory effect of *Azotobacter chroococcum* on a range of rhizosphoric fungi. *Ind. J. Expl. Biol.*, 33: 52–54.

Palaniappan, R. and Yerriswamy, R.M., 1996. Effect of saline water irrigation on growth, yield, quality and nutrition of Robusta Banana. *J. Ind. Soc. Soil. Sci.*, 44(1): 143–146.

Patra, D., Prasad, A., Nirpendra, K. and Singh, D.V., 1996. Nitrogen transformation in salt affected soil. *J. Ind. Soc. Soil Sci.*, 44(1): 151–153.

Sadashivam and Manikam, 1987. *Methods in Biochemistry*, pp. 232.

Torres-Rubio, M.G., Valencia-Plata, S.A., Bernal-Castitlo, J. and Martinex-Neito, P., 2000. Isolation of enterobacteria, *Azotobacter* spp. and *Pseudomona* spp. producers of indole-3-acetic acid and siderphores, from colombian rice rhizosphere. *Ravisto Latinoamerica de Microbiologia.* 42: 171–176.

Verma, A., Kukerja, K., Pathak, D.V., Suneja, S. and Narula, N., 2001. *In vitro* production of plant growth regulators by *Azotobacter chroococcum. Ind. J. Microbiol.*, 41: 305–307.

Yao-Yseng Tachan, 1933. *Azotobacteriacae in Burgey's Manual of Systematic Bacteriology*. William and Wilkins, Baltimore, pp. 218–229.

Chapter 31

Influence of Hydrographic Features on Pelagic Fisheries Along Dakshina Kannada Coast

T.V. Ramana and M.P.M. Reddy
College of Fishery Science, ANGRAU, Muthukur – 524 344, Andhra Pradesh, India

ABSTRACT

This article deals with a study carried out to find out the effect of different hydrographical features on pelagic fish landings along Dakshina Kannada coast. The total pelagic fish landings were found to be high during the period when the columnar water temperature ranged from 28.70 to 29.76°C and salinity from 33.46 to 34.13 per cent. Similar optimum temperature and salinity ranges were found for better catches of oil sardine und mackerel along Dakshina Kannada coast. Dissolved oxygen, transparency, currents and standing crop of plankton were found to have some effect on the pelagic fisheries.

Keywords: *Hydrography, Pelagic, Plankton, Transparency, Currents.*

Introduction

The conditions in the oceanic environment and their changes have great effect on the periodic and seasonal migrations and occurrence of marine fish population. The physical, chemical and biological features of the sea waters are subject to various changes monthly, seasonally and regionally. In the marine fisheries map of India, pelagic fishes dominated by clupeids undoubtedly occupy an important place by virtue of relative magnitude of their fishery. Clupeids, in general, constitute about a third of the marine fish caught in India and within this group, oil sardine (*Sardinella longiceps*) ranks first. Indian mackerel (*Rastrelliger kanagurta*) contributes other major supplement of this pelagic fishery. The Dakshina Kannada area on the west coast of India supports a rich pelagic fishery constituted by oil sardine and Indian mackerel. It is well known that the pelagic fisheries, particularly the oil sardine and Indian mackerel are varying from month to month, season to season and year to year.

In the present investigation, data pertaining to several hydrographic parameters obtained from intensive studies in the near shore waters of Dakshina Kannada coast extending for about 100 km in length have been used to find out the influence of these parameters on pelagic fish catches from this area. An understanding of the influence of the various hydrographical parameters on pelagic fishery would help not only in predicting the fishery but also in better management and exploitation.

Materials and Methods

Data on sea water temperature, salinity, dissolved oxygen, extinction coefficient, currents, phytoplankton and zooplankton were collected using standard techniques in the Arabian Sea at twelve stations along three sections. one off Mangalore, one off Maple and one off Gangolli (Figure 31.1). These oceanographic collections were made at monthly intervals for one year from April 1987 to May 1988 except in the southwest monsoon months of June, July and August during the cruises of fishing vessel M.F.V. Dolphin of the College of Fisheries, Mangalore. In addition, oceanographic data (Unpublished) collected by the Department of Fishery Oceanography, College of Fisheries, Mangalore, during the period from May 1979 to May 1980 off Mangalore, Malpe, and Gangolli were used in the present study to investigate the possible influence of oceanographic parameters on pelagic fish catches. The data on monthly landings of oil sardine, mackerel and other pelagic fishes landed at Mangalore, Udupi (Maple) and Coondapur (Gangolli) for the periods from April 1987 to May 1988 and from May 1979 to May 1980 were obtained from the State Department of Fisheries, Government of Karnataka, Mangalore.

Results and Discussion

Pelagic Fishery in Relation to Hydrography

Seawater Temperature and Salinity

Oil sardine and mackerel mainly contribute to pelagic fishery in the area of investigation, though other groups such as lesser sardines and anchovies occurred in the landings with comparatively low abundance. The major groups of pelagic fishes are well known to occur in the near shore waters along the west coast of India up to 20 m depth. It is also known that purse seine boats operating with in 20 m depth mostly obtain the pelagic fishes along the Dakshina Kannada coast. Hence, the average values of temperature and salinity for four stations in each of the three sections for columnar waters from 0 to 20 meter depth were calculated separately for each month and presented in the Tables 31.1 and 13.2, with average biomass values of phytoplankton and zooplankton.

From the data it is clear that the pelagic fish landings along the three sections were relatively high during September, October, November and December of 1979 and 1987 when the average columnar water temperature ranged between 28.70 and 29.70°C and salinity fluctuated from 33.46 to 34.13 per cent. The oil sardine landings along the three sections during the study period were comparatively high during September, October, November and December in 1979 and in 1987. The average columnar water temperature and salinity during peak landings ranged between 28.74 and 29.71 °C and between 33.21 and 34.05 per cent respectively. Chidambaram and Menon (1945) observed comparatively low catches during March to May and attributed thus to temporary migration of fish to off shore regions

Table 31.1: Average Values of Column Water Temperature (°C), Salinity (%), Phytoplankton (mg/m³), Zooplankton (mg/m³), Oil Sardine (MT), Mackerel (MT) and Total Pelagic Fish Findings (MT) During 1979–80

Month	*Temperature (°C)*	*Salinity (%)*	*Phytoplankton (mg/m³)*	*Zooplankton (mg/m³)*	*Oil Sardine (MT)*	*Mackerel (MT)*	*Total Pelagic Landings (MT)*
				Mangalore			
Apr 1979	30.06	34.81	0.59	22.20	1359	228	1722
May	29.63	35.28	2.16	30.40	39	75	435
Aug	28.53	32.83	5.44	32.22	Nil	Nil	Nil
Sep	28.75	33.67	4.21	129.87	4440	5792	10232
Oct	29.38	33.96	0.97	31.83	7261	3714	10975
Nov	29.22	33.80	3.18	20.83	6106	1794	7900
Dec	28.85	33.21	0.69	38.41	3009	207	3216
Jan 1980	28.00	33.25	1.10	14.48	1769	97	1866
Feb	29.86	33.44	3.73	41.52	921	159	10
Mar	30.31	34.43	1.02	16.24	466	49	515
Apr	30.57	34.77	0.84	21.23	643	92	735
May	29.83	35.25	2.10	20.43	873	87	960
				Malpe			
Apr 1979	29.97	34.71	0.96	12.65	8	16	316
May	29.62	35.72	2:26	18.47	Nil	18	143
Aug	28.15	32.96	12.48	51.12	Nil	Nil	Nil
Sep	28.76	33.94	7.36	66.99	2890	2713	5603
Oct	29.35	34.26	110	2127	1162	1259	2421
Nov	29.25	34.16	4.42	32.72	1548	483	2031
Dec	28.21	33.33	0.67	33.73	334	106	440
Jan 1980	27.78	33.27	1.32	28.43	297	115	412
Feb	29.91	33.41	6.26	27.52	506	125	631
Mar	30.07	34.48	1.38	16.76	522	95	617
Apr	30.86	34.88	0.95	15.69	165	45	212
May	29.76	35.55	1.70	33.82	41	17	58
				Gangolli			
Apr 1979	29.97	34.87	0.32	11.127	Nil	1566	2317
May	29.68	35.50	1.76	12.50	Nil	77	260
Aug	28.25	33.28	17.61	28.94	Nil	Nil	Nil
Sep	28.70	33.60	7.22	62.42	1651	2554	4205
Oct	29.35	34.01	1.61	31.93	3460	4260	7720
Nov	29.39	33.87	5.77	27.40	4265	835	5100
Dec	28.21	33.19	0.99	19.60	562	424	986
Jan 1980	27.86	33.23	0.63	2.76	1153	244	1397
Feb	29.81	33.37	2.58	31.73	466	202	668
Mar	30.40	34.34	0.47	12.86	494	81	575
Apr	31.36	34.95	0.15	3.54	84	Nil	84
May	39.72	35.47	1.40	14.82	Nil	2	2

Table 31.2: Average Values of Column Water Temperature (°C), Salinity (%), Phytoplankton (mg/m³), Zooplankton (mg/m³), Oil Sardine (MT), Mackerel (MT) and Total Pelagic Landings During 1987–88

Month	*Temperature (°C)*	*Salinity (%)*	*Phytoplankton (mg/m³)*	*Zooplankton (mg/m³)*	*Oil Sardine (MT)*	*Mackerel (MT)*	*Total Pelagic Landings (MT)*
				Mangalore			
Apr 1987	30.75	34.84	0.60	7.17	1.0	0.8	115.70
May	30.15	35.20	3.60	13.57	0.8	13.0	14.00
Sep	28.74	33.46	19.46	48.04	4753.0	1210.0	5965.00
Oct	29.66	34.15	4.58	17.21	971.0	176.5	1626.20
Nov	30.21	34.26	44.85	17.79	28.5	1.5	1076.70
Dec	28.66	33.91	5.18	12.68	544.8	147.0	1792.00
Jan 1988	28.61	33.45	0.65	14.05	312.0	22.1	1950.70
Feb	29.92	34.02	0.28	24.06	176.0	Nil	949.00
Mar	30.33	34.36	0.35	13.06	90.0	12.0	459.00
Apr	30.70	34.67	0.48	9.23	102.5	9.0	375.50
May	30.25	35.10	4.69	16.71	31.0	7.5	129.00
				Malpe			
Apr 1987	30.84	34.83	0.38	5.51	1.5	1.0	45.50
May	30.24	35.23	3.14	11.91	551.0	13.3	571.20
Sep	28.73	33.46	18.36	39.72	1005.0	2170.0	3625.50
Oct	29.71	34.05	7.04	28.26	3508.5	36.4	4852.50
Nov	30.25	34.21	47.69	20.29	185.5	13.9	235.40
Dec	28.64	33.94	3.96	16.01	17.0	38.0	930.90
Jan 1988	28.44	33.59	0.36	4.35	841.7	26.3	879.70
Feb	29.86	34.09	0.26	17.32	731.3	Nil	757.60
Mar	30.2	34.39	0.22	15.53	355.4	43	355.40
Apr	30.73	34.63	0.17	15.53	533.0	29.6	629.00
May	30.37	35.12	4.84	10.92	338.5	Nil	368.10
				Gangolli			
Apr 1987	30.92	34.87	0.48	5.35	Nil	2.0	99.50
May	30.41	35.27	3.12	10.20	357.6	Nil	394.30
Sep	28.73	33.47	20.24	45.90	613.4	2208.0	3360.80
Oct	29.76	34.13	7.95	19.45	804.7	769.9	3267.50
Nov	30.26	34.21	54.08	18.14	907.2	40.4	1805.90
Dec	28.67	34.00	5.50	13.93	1041.3	31.3	2498.10
Jan 1986	28.27	33.72	0.69	3.68	1218.1	1.2	2599.70
Feb	29.76	34.18	2.61	12.64	1191.0	Nil	1545.90
Mar	30.29	34.49	0.50	16.34	931.4	Nil	1179.90
Apr	30.80	34.66	0.13	16.37	569.9	3.8	745.80
May	30.53	35.21	4.82	7.32	351.0	Nil	402.20

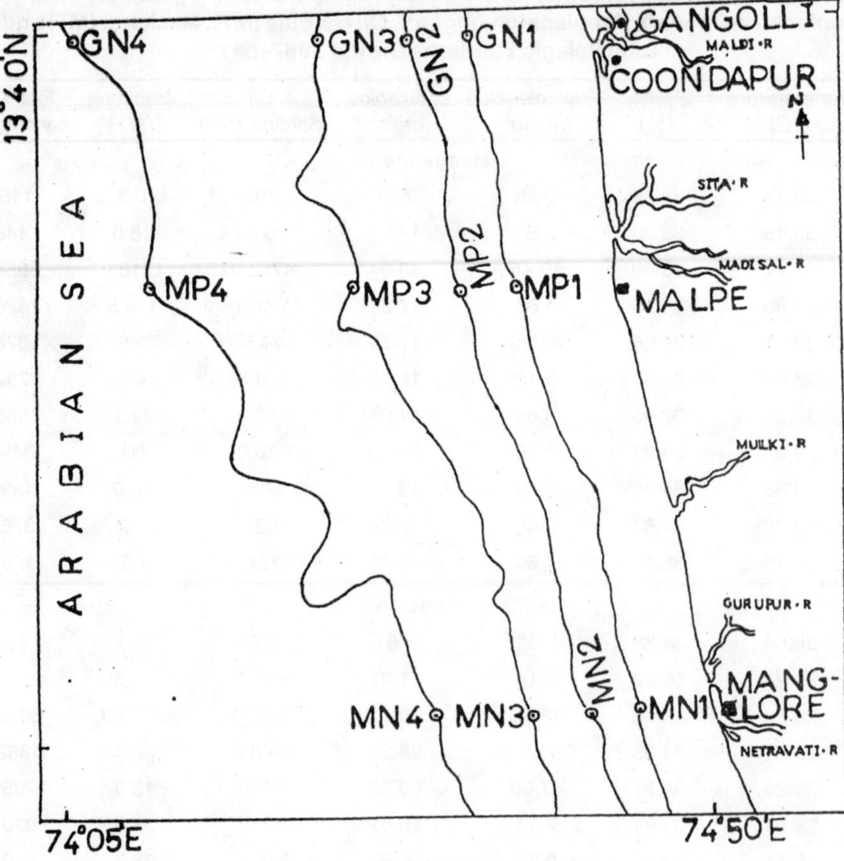

Figure 31.1

due to increase in temperature. Sekharan and Dhulkhad (1963) while dealing with the oil sardine fishery of Mangalore zone stated that the length of fishery season apparently depends on the duration of the period of intermediate values of temperature and salinity. Good landing of sardine at Calicut were observed during 1957–58, when the average values of temperature and salinity ranged from 28.5 to 29.9°C and 33.08 to 35.04 per cent respectively and poor fishing was noticed when these values exceeded (Sekharan, 1962). According to Prabhu *et al.* (1967) during the peak abundance of oil sardine, the temperature and salinity varied from 27.0 to 29.0 °C and from 33.0 to 35.0 per cent respectively. Bensam (1970) reported that juvenile sardines enter into fishery quite suddenly when temperature and salinity approach the optimum of 27–28 °C and 34–35 per cent respectively during September/ October. Annegiri (1972) observed good catches of oil sardine when the temperature ranged from 27.0 to 29.5 °C white phosphate and salinity contents in the waters were very low. In the present study, the ranges of temperature and salinity were narrow compared to the studies done elsewhere along the west coast of India. The present investigation also showed intermediate values of temperature and salinity to be favorable for higher landings of oil sardine and values exceeding these optimum ranges resulted in poor catches.

During the peak landings of mackerel in both the periods, the average columnar temperature fluctuated from 28.70 to 29.38°C, while salinity from 33.47 to 34.01 per cent. These optimum ranges

Table 31.3: Vertical Distribution of Dissolved Oxygen (ml/l) at Different Stations Located Along Dakshina Kannada Coast

Section	Station	Level	Period											
			1979							1980				
			Apr	May	Aug	Sep	Oct	Nov	Dec	Jan	Feb	Mar	Apr	May
I	MN1	S	4.30	3.87	4.98	4.95	4.85	4.88	4.96	4.88	4.76	4.43	4.31	3.78
		B	4.27	3.11	2.84	4.76	4.51	4.61	4.88	4.76	4.59	4.21	4.06	2.69
	MN2	S	4.33	3.96	5.00	4.92	4.83	4.88	4.99	4.88	4.79	4.51	4.37	4.00
		M	4.21	3.68	4.18	4.86	4.74	4.61	4.98	4.60	4.65	4.35	4.10	3.52
		B	4.05	3.00	2.95	4.81	4.57	4.54	4.71	4.48	4.52	4.18	4.03	2.88
	MN3	S	4.31	3.91	5.00	4.85	4.89	4.86	4.95	4.80	4.70	4.41	4.36	3.89
		M	4.21	3.54	3.53	4.85	4.67	4.61	4.77	4.54	4.62	4.37	4.12	3.43
		B	4.12	3.69	2.88	4.60	4.57	4.47	4.58	4.46	4.53	4.10	4.00	2.72
	MN4	S	4.23	4.06	4.67	4.83	4.82	4.84	4.92	4.88	4.72	4.52	4.32	3.76
		M	4.07	3.44	3.68	4.85	4.68	4.62	4.71	4.46	4.69	4.42	4.16	3.39
		B	4.08	2.96	2.00	4.22	4.31	4.36	4.63	4.36	4.54	4.21	4.08	2.68
II	MP1	S	4.79	3.60	4.31	4.72	4.93	4.96	4.53	5.16	4.94	4.64	4.24	3.62
		B	4.43	2.87	2.63	3.98	4.42	4.55	4.44	5.03	4.74	4.56	4.09	2.56
	MP2	S	4.44	4.11	4.05	4.78	4.89	5.04	5.19	5.15	4.93	4.67	4.28	3.79
		M	4.17	3.26	2.07	4.48	4.51	4.69	4.90	4.87	4.74	4.59	4.16	3.08
		B	4.01	2.48	1.45	4.02	4.50	4.52	4.99	4.79	4.67	4.47	4.07	2.25
	MP3	S	4.39	3.91	4.12	4.91	5.00	5.11	5.13	4.96	4.21	4.57	4.34	3.87
		M	4.32	3.65	3.01	4.09	4.78	4.53	4.91	4.89	4.77	4.52	4.17	3.32
		B	3.63	2.45	1.86	4.01	4.13	4.28	4.65	4.83	4.47	4.47	4.05	2.13
	MP4	S	4.29	4.27	4.18	4.82	4.92	4.98	4.99	5.01	4.88	4.51	4.42	4.04
		M	4.09	3.41	3.92	4.62	4.63	4.63	4.90	4.88	4.72	4.44	4.21	3.62
		B	4.00	2.72	2.09	4.13	4.36	4.31	4.68	4.73	4.52	4.39	4.02	2.34
III	GN1	S	4.60	3.75	4.16	4.68	4.86	4.89	5.05	5.27	5.12	4.85	4.27	3.62
		B	4.52	3.37	2.63	4.43	4.72	4.61	4.79	4.96	4.89	4.65	4.15	2.35
	GN2	S	4.60	3.76	4.02	4.80	4.89	4.91	4.98	5.12	5.05	4.83	4.30	3.78
		M	4.60	3.42	2.76	4.36	4.54	4.74	4.86	5.03	4.98	4.79	4.13	2.96
		B	4.50	2.85	1.82	4.12	4.03	4.23	4.62	4.87	4.75	4.62	4.09	2.41
	GN3	S	4.53	4.12	4.00	4.61	4.83	4.87	4.83	5.12	4.99	4.79	4.15	3.89
		M	4.37	3.86	2.84	4.24	4.42	4.59	4.75	5.08	4.96	4.71	3.98	3.23
		B	4.37	2.17	2.03	3.92	4.18	4.23	4.43	4.97	4.73	4.61	3.86	2.45
	GN4	S	4.40	4.00	4.35	4.62	4.88	4.89	4.87	5.11	5.03	4.82	4.19	4.01
		M	4.21	3.46	3.06	4.19	4.34	4.61	4.79	4.93	4.91	4.68	3.82	3.42
		B	4.00	2.25	2.42	3.60	4.22	4.21	4.62	4.82	4.79	4.67	3.69	2.69

S: Surface; M: Mid; B: Bottom.

Table 31.4: Vertical Distribution of Dissolved Oxygen (ml/l) at Different Stations Located Along Dakshina Kannada Coast

Section	*Station*	*Level*	*Period*										
			1987						*1988*				
			Apr	*May*	*Sep*	*Oct*	*Nov*	*Dec*	*Jan*	*Feb*	*Mar*	*Apr*	*May*
I	MN1	S	4.40	4.04	3.35	4.02	3.19	4.66	4.94	4.80	4.89	4.48	4.45
		M	4.31	3.96	3.26	3.85	4.19	4.51	4.83	4.71	4.61	4.36	3.97
		B	4.21	3.87	3.09	3.78	4.08	4.47	4.36	4.69	4.54	4.20	3.80
	MN2	S	4.52	4.11	3.27	4.04	4.26	4.69	5.19	4.72	4.78	4.56	4.19
		M	4.38	4.05	3.19	3.36	4.08	4.65	4.83	4.60	4.64	4.42	4.06
		B	4.02	3.91	3.01	3.76	3.98	4.60	4.74	4.54	4.48	3.98	3.32
	MN3	S	4.48	3.98	3.27	4.32	4.34	4.68	4.79	4.81	4.61	4.49	4.32
		M	4.35	3.82	3.20	3.76	4.26	4.53	4.68	4.76	4.57	4.36	3.91
		B	4.21	3.71	2.64	3.65	4.08	4.45	4.44	4.39	4.48	3.81	3.65
	MN4	S	4.41	4.02	3.75	4.42	4.28	4.59	5.03	4.69	4.71	4.42	4.13
		M	4.32	3.89	3.27	3.96	4.22	4.48	4.45	4.75	4.59	4.35	3.72
		B	4.19	3.56	253	3.54	4.16	4.37	4.40	4.62	4.39	3.63	3.40
II	MP1	S	4.36	4.01	3.38	4.28	4.28	5.24	4.82	4.81	4.65	4.43	4.09
		M	4.31	3.82	3.21	4.05	4.22	5.12	4.61	4.63	4.50	4.20	3.79
		B	4.22	3.51	3.12	4.00	3.98	4.62	4.48	4.40	4.35	3.94	3.41
	MP2	S	4.41	3.96	3.25	4.38	4.15	5.15	4.83	4.63	4.75	4.56	3.97
		M	4.39	3.64	3.25	4.38	4.15	5.15	4.83	4.63	4.75	4.56	3.97
		B	4.23	3.46	3.02	4.12	3.40	4.87	4.42	4.28	4.39	3.88	3.82
	MP3	S	4.38	3.91	3.22	4.11	4.33	5.08	4.88	4.69	4.63	4.37	4.02
		M	4.22	3.79	3.18	4.00	4.22	4.89	4.65	4.53	4.59	4.25	3.96
		B	4.19	3.64	2.72	3.97	3.91	4.07	4.43	4.33	4.42	3.86	3.65
	MP4	S	4.42	3.98	3.63	4.21	4.71	4.91	4.65	4.68	4.72	4.45	4.15
		M	4.21	3.87	3.21	4.08	4.60	4.75	4.51	4.48	4.51	4.26	3.90
		B	3.88	3.18	2.52	3.96	4.44	4.55	4.40	4.40	4.00	3.79	3.12
III	GN1	S	4.39	4.27	3.48	4.30	4.59	4.87	4.82	4.66	4.76	4.38	4.42
		M	4.27	3.82	3.32	4.00	4.27	4.46	4.70	4.57	4.63	4.26	3.70
		B	4.11	3.21	3.16	3.82	3.99	4.21	4.59	4.40	4.52	4.12	3.30
	GN2	S	4.37	4.16	3.32	4.47	4.64	4.90	4.73	4.90	4.73	4.40	4.38
		M	4.25	3.72	3.22	4.08	4.35	4.34	4.66	4.60	4.63	4.28	3.92
		B	4.08	3.19	3.11	3.98	4.16	4.19	4.43	4.45	4.46	3.95	3.15
	GN3	S	4.32	4.08	3.28	4.32	4.56	4.99	4.91	4.64	4.71	4.31	4.20
		M	4.21	3.69	3.21	4.04	4.63	4.73	4.70	4.52	4.59	4.26	3.86
		B	4.16	3.08	2.63	3.07	4.34	4.58	4.54	4.20	4.30	3.88	3.09
	GN4	S	4.41	3.97	3.22	4.39	4.00	5.01	4.86	4.65	4.79	4.58	4.08
		M	4.22	3.69	3.11	4.12	4.65	4.83	4.69	4.39	4.51	4.26	3.72
		B	4.15	3.01	2.61	3.92	4.39	4.69	4.50	4.29	4.30	3.69	3.12

S: Surface; M: Mid; B: Bottom.

were comparable with the optimum values recorded by earlier workers. Pradhan and Reddy (1962) while comparing the mackerel landings and monthly average temperature and salinity values for three years at Calicut considered 29.1°C and 33.27 per cent as the optimum values of temperature and salinity for good catches of mackerel. Venkataraman and Rao (1973) observed high abundance of mackerel at Calicut in the fishing season when temperature and salinity ranged from 27.1 to 29.9 °C and 30.4 to 33.9 per cent respectively. In general, the abundance of mackerel is reported to be more in the region north of 13°N than south of it due to relatively high salinity (Rao *et al.*, 1973).

Dissolved Oxygen

In the present study, surface waters were well aerated with dissolved oxygen values of 4.0 ml/l and above when there were good pelagic fish landings (Table 31.3 and 31.4). It is felt that fish distribution could be limited by dissolved oxygen when its concentration is too low in the vertical column.

Transparency of Water

During the period of good pelagic fish landings along Dakshina Kannada coast, transparency of water was considerably lower as indicated by the relatively high values of extinction coefficient during that period (Table 31.5).

Table 31.5: Extinction Coefficient Value at Different Stations Located Along Dakshina Kannada Coast

Section	*Station*	*Period*										
		1987						*1988*				
		Apr	*May*	*Sep*	*Oct*	*Nov*	*Dec*	*Jan*	*Feb*	*Mar*	*Apr*	*May*
I	MN1	0.71	0.60	0.95	0.79	0.66	0.62	0.60	0.59	0.70	0.41	0.40
	MN2	0.58	0.46	0.87	0.63	0.49	0.44	0.42	0.40	0.47	0.38	0.29
	MN3	0.48	0.21	0.75	0.56	0.43	0.39	0.38	0.38	0.41	0.32	0.18
	MN4	0.42	0.18	0.52	0.41	0.36	0.35	0.36	10.32	0.38	0.22	0.16
II	MP1	0.75	0.64	1.03	0.88	0.72	0.66	0.59	0.68	0.76	0.56	0.54
	MP2	0.63	0.38	0.94	0.66	0.61	0.53	0.50	0.55	0.59	0.28	0.25
	MP3	0.51	0.15	0.62	0.53	0.49	0.41	0.44	0.49	0.51	0.22	0.20
	MP4	0.32	0.12	0.48	0.36	0.34	0.29	0.35	0.40	0.43	0.19	0.14
III	GN1	0.63	0.58	1.04	0.86	0.79	0.71	0.69	0.67	0.72	0.54	0.48
	GN2	0.48	0.40	0.93	0.69	0.63	0.57	0.58	0.56	0.60	0.51	0.39
	GN3	0.39	0.19	0.68	0.61	0.49	0.44	0.42	0.42	0.48	0.36	0.26
	GN4	0.28	0.11	0.65	0.52	0.35	0.34	0.37	0.37	0.41	0.24	0.11

Ocean Currents

The geographical distribution of pelagic fish shoals seems to coincide with the north equatorial current region of the Indian Ocean. As the current from the southern boundary of the clockwise cell in the Bay of Bengal during the northeast monsoon period joins with the north equatorial current, there could be a concentrated flow south of Sri Lanka as opined by Murty (1965). This strong current takes a northerly deviation from its general westward flow with which the pelagic fish shoals are likely to migrate along the west coast of India. Thus, this northerly flow seems to be responsible for the high

Table 31.6: Currents at Different Stations Located Along the Dakshina Kannada Coast

Station	Level	1987 April Di	1987 April Sp	1987 May Di	1987 May Sp	1987 September Di	1987 September Sp	1987 October Di	1987 October Sp	1987 November Di	1987 November Sp	1987 December Di	1987 December Sp	1988 January Di	1988 January Sp	1988 February Di	1988 February Sp	1988 March Di	1988 March Sp	1988 April Di	1988 April Sp	1988 May Di	1988 May Sp
MN1	S	186	0.196	226	0.175	160	0.320	205	0.240	88	0.185	360	0.080	302	0.045	46	0.122	92	0.108	248	0.196	216	0.216
	M	248	0.242	219	0.142	170	0.308	190	0.232	75	0.142	348	0.132	311	0.032	53	0.116	118	0.075	236	0.208	232	0.224
	B	236	0.218	221	0.219	165	0.264	128	0.225	135	0.115	46	0.164	334	0.080	64	0.110	54	0.105	224	0.212	228	0.236
MN2	S	250	0.206	235	0.232	180	0.453	225	0.145	85	0.136	72	0.157	28	0.145	356	0.168	355	0.184	256	0.342	221	0.288
	M	239	0.218	242	0.236	204	0.408	228	0.192	105	0.085	355	0.108	36	0.172	344	0.182	348	0.196	242	0.308	219	0.305
	B	242	0.239	219	0.246	186	0.395	170	0.118	145	0.068	345	0.113	64	0.226	338	0.175	343	0.179	235	0.243	226	0.265
MN3	S	218	0.246	196	0.188	174	0.S02	196	0.108	294	0.163	306	0.270	11	0.090	357	0.130	196	0.031	126	0.189	198	0.205
	M	234	0.252	207	0.205	192	0.463	168	0.163	165	0.172	310	0.220	44	0.110	348	0.164	354	0.116	218	0.173	203	0.222
	B	239	0.229	21	0.216	166	0.416	192	0.182	156	0.190	338	0.278	35	0.104	336	0.136	358	0.163	222	0.195	215	0.216
MN4	S	226	0.235	215	0.244	184	0.29.8	176	0.145	340	0.109	340	0.120	10	0.060	351	0.168	182	0.089	219	0.353	194	0.288
	M	218	0.238	220	0.219	220	0.263	169	0.167	154	0.065	338	0.163	12	0.078	352	0.179	198	0.104	208	0.364	206	0.264
	B	225	0.296	218	0.236	186	0.235	148	0.203	142	0.052	326	0.209	23	0.095	344	0.182	202	0.135	232	0.382	212	0.275
MP1	S	192	0.116	208	0.153	172	0.306	190	0.182	14	0.112	42	0.103	33	0.150	332	0.150	30	0.154	176	0.184	163	0.189
	M	160	0.132	211	0.192	180	0.252	174	0.135	26	0.131	356	0.095	14	0.106	350	0.106	272	0.162	152	0.216	220	0.231
	B	210	0.146	219	0.181	191	0.268	180	0.106	42	0.125	310	0.112	21	0.102	43	0.095	294	0.139	210	0.222	222	0.208
MP2	S	194	0.162	227	0.176	173	0.314	226	0.098	18	0.106	25	0.082	353	0.090	326	0.080	10	0.146	195	0.209	218	0.192
	M	206	0.171	234	0.192	169	0.322	220	0.139	32	0.122	349	0.116	324	0.086	12	0.124	280	0.158	196	0.231	236	0.236
	B	218	0.168	210	0.209	182	0.258	1.24	0.162	34	0.131	335	0.124	318	0.112	42	0.106	309	0.144	210	0.232	232	0.212

Contd...

Table 31.6–Contd...

Station	Level	1987 April Di	1987 April Sp	1987 May Di	1987 May Sp	1987 September Di	1987 September Sp	1987 October Di	1987 October Sp	1987 November Di	1987 November Sp	1987 December Di	1987 December Sp	1988 January Di	1988 January Sp	1988 February Di	1988 February Sp	1988 March Di	1988 March Sp	1988 April Di	1988 April Sp	1988 May Di	1988 May Sp
MP3	S	182	0.172	198	0.144	204	0.287	236	0.080	42	0.152	320	0.109	10	0.085	352	0.152	286	0.139	170	0.206	228	0.215
	M	180	0.194	208	0.168	209	0.306	240	0.103	26	0.153	314	0.150	345	0.096	354	0.104	292	0.164	184	0.214	244	0.224
	B	163	0.206	212	0.191	186	0.314	210	0.162	31	0.148	330	0.130	330	0.102	12	0.102	310	0.125	163	0.218	232	0.206
MP4	S	158	0.184	225	0.159	195	0.292	206	0.082	11	0.108	18	0.095	358	0.065	344	0.134	10	0.156	142	0.212	198	0.192
	M	191	0.206	224	0.162	196	0.218	233	0.094	16	0.134	27	0.106	330	0.084	29	0.156	312	0.134	190	0.209	206	0.208
	B	194	0.222	228	0.173	187	0.268	242	0.116	24	0.141	42	0.125	349	0.100	32	0.142	320	0.138	196	0.231	224	0.189
GN1	S	251	0.162	186	0.267	184	0.248	262	0.143	310	0.106	304	0.092	25	0.132	330	0.109	230	0.0S8	195	0.215	220	0.290
	M	206	0.183	163	0.232	203	0.192	280	0.167	290	0.132	296	0.106	40	0.125	310	0.131	20	0.106	176	0.192	230	0.206
	B	238	0.159	192	0.215	190	0.236	195	0.182	306	0.145	316	0.121	51	0.122	316	0.122	196	0.085	162	0.179	170	0.192
GN2	S	216	0.210	184	0.345	170	0.365	290	0.217	352	0.132	332	0.108	20	0.105	314	0.116	110	0.095	180	0.196	180	0.223
	M	236	0.196	172	0.294	192	0.124	288	0.203	305	0.162	318	0.122	46	0.130	308	0.129	226	0.125	205	0.182	165	0.212
	B	222	0.220	204	0.266	206	0.263	286	0.232	321	0.178	309	0.125	352	0.126	321	0.118	230	0.135	210	0.154	172	0.187
GN3	S	253	0.252	186	0.192	166	0.364	294	0.168	62	0.201	23	0.145	338	0.075	318	0.106	218	0.116	235	0.150	200	0.280
	M	215	0.231	212	0.209	178	0.245	282	0.204	38	0.139	28	0.090	346	0.125	322	0.132	210	0.120	220	0.162	210	0.218
	B	219	0.204	192	0.168	185	0.218	290	0.216	355	0.144	31	0.103	352	0.118	319	0.144	240	0.125	210	0.144	209	0.253
GN4	S	198	0.215	198	0.290	210	0.298	295	0.169	20	0.152	358	0.132	321	0.115	315	0.096	250	0.122	236	0.145	185	0.189
	M	175	0.169	202	0.236	206	0.235	303	0.152	31	0.138	346	0.105	343	0.109	321	0.106	206	0.117	219	0.110	176	0.196
	B	203	0.232	214	0.208	225	0.264	282	0.187	50	0.142	339	0.116	319	0.132	326	0.121	223	0.109	222	0.132	190	0.164

S: Surface; M: Mid water; B: Bottom; Di: Direction; Sp: Speed.

pelagic fish catches along Dakshina Kannada coast. With the onset of the southwest monsoon, north equatorial current is replaced by an eastward flowing southwest monsoon current south of Sri Lanka and southerly current flows along the west coast of India. This may cause the pelagic fish shoals to move away from the west coast of India into the southwest monsoon current that flows to east. Current observations made along the Dakshina Kannada coast in the present study revealed that the surface currents were generally towards south from April to September and towards North from November till February, transition months being October and March (Table 31.6). In the present investigation, total pelagic fish landings were observed to be high during September, October, November and December. Higher catches found during September and October in the present study could probably be due to the other suitable environmental conditions such as seawater temperature, salinity, pH, etc., prevailing along this region in those months.

Plankton

In the present study, an attempt was made to correlate the phytoplankton biomass (dry weight) with oil sardine landings. From the tables, it can be seen that high phytoplankton production in general lead to higher sardine landings in succeeding months. This direct relationship was observed only during first peak period *i.e.*, August/September, but such relationship could not be drawn for the second peak of phytoplankton that occurred during February/November. In the remaining months, no exact relationship could be drawn. Similarly, peak landings of oil sardine followed a good biomass of zooplankton only in some months. An attempt was made to find out the relation between mackerel landings with zooplankton. In general, an increase in the zooplankton biomass lead to an increase in the mackerel landings in the succeeding months, though not proportionally and during all the months. This indicates that even when the food availability in any region is below the normal requirements, mackerel may prefer an area with favorable environmental conditions than to the medium of fish food. In this case also, exact relationship was not found between phytoplankton/zooplankton biomass and total pelagic fish landings.

Acknowledgements

The first author is thankful to the authorities of University of Agricultural Sciences, Bangalore, for providing Merit Scholarship during the doctoral study period. Authors are grateful to the Director of Instruction, College of Fisheries, Mangalore, for providing the facilities to carry out the research work.

References

Annigeri, G.C., 1972. Hydrological conditions in the inshore region of Karwar during 1965–1967. *Indian J. Fish.*, 19(1&2): 156–162.

Bensam, P., 1970. On the fluctuations of the oil sardine fishery at Cannanore during 1961– 64. *Indian J. Fish.*, 17(1&2): 132–140.

Chidambararn, K. and Menon, M.D., 1945. The correlation of the (Malabar and South, Kanara) fisheries with plankton and certain oceanographical factors. *Proc. Indian Acad. Sci.*, 22: 355–357.

Longhurst, A.R., 1958. The pelagic phase of pleuroncodes planipes in the California current. *Rep. California Coop. Fish. Invest.*, 11: 142–154.

Murty, A.V.S., 1965. Studies on the surface mixed layer and its associated thermocline off the west coast of India and the inference thereby working a prediction system of the pelagic fisheries of India. *Indian J. Fish.*, 12(1): 118–134.

Prabhu, M.S., Ramamurthy, S., Kuthalingam, M.D.K. and Dhulkhed, M.H., 1967. On the experimental fishing off Ullal (Mangalore). *Indian J. Fish.*, 14 (1&2): 225–231.

Pradhan, L.B. and Reddy, C.V.G., 1962. Fluctuations in mackerel landings at Calicut in relation to hydrological factors. *Indian J. Fish.*, 9: 100–109.

Rao, D.S., Ramamirtham, C.P. and Krishnan, T.S., 1973. Oceanographic features and abundances of the pelagic fisheries along the west coast of India. *Proceedings of the Symposium on the Living Resources of the Seas Around India.* Special Publication, CMFRI, pp. 400–413.

Sekharan, K.V., 1962. On the oil sardine fishery of the Calicut area during the years 1955–56 to 1958–59. *Indian J. Fish.*, 9: 679–700.

Sekharan, K.V. and Dhulkhed, M.H., 1963. On the oil sardine fishery of the Mangalore zone during the years 1957– 63. *Indian J. Fish.*, 10A(1): 601–626.

Venkataraman, G. and Rao, K.V.N., 1973. On the mackerel fishing of Calicut during the years 1959–60 to 1967–68. *Indian J. Fish.*, 20(2): 448–475.

Chapter 32

Biodiversity of Aphids (Order : Hemiptera) from Satara District of Western Ghats

B.V. Jadhav and T.V. Sathe
Department of Zoology, Shivaji University, Kolhapur – 416 004, India

ABSTRACT

Aphids (Order : Hemiptera) cause damage to agricultural and forest flora by sucking the cell sap and affecting growth and yield of the crops. Therefore aphids have been selected for biodiversity study. In all 21 species of aphids have been recorded from the Satara District belonging to the following eleven important genera *viz., Aphis Dactynotus, Myzus, Lipaphis, Rhopalosiphum, Melanophis, Siphocronyne, Brevicoryne, Teratneura, Pentalonia, and Toxoptera.* Their seasonal abundance and distribution in plain and hilly region of Satara district have been studied.

***Keywords**: Aphids, Satara, Western Ghats, Distribution, Abundance.*

Introduction

Aphids are notorious in damaging plants by sucking the cell sap. They are obligatory parasites of plants, polyphagus in habit and cosmopolitan in distribution. About 400 aphid species were recorded from the oriental region (Sathe, 1992), out of which 108 aphids were recorded on economically important plants. Aphids also cause the severe damage to forest plants and affect the vitality.

The western part of Satara District include Koyana, Ghatmatha, Kas, Pachgani, Mahabaleshwar etc. is situated in Western Ghats. It covers thick, dense and deciduous forests. Its altitude ranges from 550 mm to 1500 mm and rain fall 1500 to 4000 mm.

The Western Ghats are a chain of highlands running along the western edge of the Indian subcontinent, from Mumbai (Bombay) south to the southern tip of the Peninsula. through the district of Kolhapur, Sangli, Satara, Pune and Mumbai of Maharashtra. The Western Ghats are an area of exceptional biological diversity and conservation interest, one of the major Tropical Evergreen Forest regions in India and one of the hotspots of the world in biodiversity. The protection and conservation of biodiversity of plants and insects is on international agenda. Keeping in view all above facts the present work was carried out.

Materials and Methods

Aphids were collected directly from the plants with camel hair brush and palate aphids were trapped in yellow pan water traps. Sometimes, infested plant parts together with the aphids were collected in polythene bag. The insects thus collected were put into 70 per cent ethyl alcohol in small specimen tubes (45 cm × 2.7 cm). Later, microscopic studies were carried out for their identification. The twigs of the host plants have been collected for identification.

Results and Discussion

The aphid species recorded on various plants (agricultural and forest) have been tabulated in Tables 32.1 and 32.2.

Table 32.1: Aphids Recorded on Forest Plants from Western Ghats (Satara district)

Sl.No.	Aphids	Plants
1.	*Aphis gossypii* Glover	Eucalyptus: *Eucalyptus* species
2.	*Myzus persicae* Sulzer *Lipaphis erysimi* Kattenbuch	Subabul: *Leucaena*
3.	*Aphis craccivora* Koch. *Aphis gossypii* Glover	Neem: *Azadirachta indica*
4.	*Aphis gossypii* Glover *Aphis craccivora* Koch.	Mango: *Mangifera indica*
5.	*Aphis craccivora* Koch.	Jack: *Artocarpus integri folia*
6.	*Myzus persicae* Sulzer *Lipaphis erysimi* Kattenbuch	Khair: *Acasia catechu*
7.	*Aphis gossypii* Glover *Myzus persicae* Sulzer	Ain: *Terminalia tomentasa*
8.	*Lipaphis erysimi* Kattenbuch	Kinja: *Terminalia paniculata*
9.	*Myzus persicae* Sulzer	Teak: *Tectona grandis*
10.	*Aphis craccivora* Koch.	Hirda: *Terminalia chibula*
11.	*Aphis gossypii* Glover *Aphis craccivora* Koch.	Jamun: *S. cumunii*
12.	*Myzus persicae* Sulzer	Tamarind: *Tamarindus indica*

Table 32.2: Aphids Recorded on Agricultural Plants (Plain Region) from Western Ghats (Satara District)

Sl.No.	Aphids	Plants
1.		Oil seed crops:
	Aphis craccivora Koch.	Ground nut: *Arachis hypogae* Linn.
	Myzus persicae Sulzer	Mustard: *Brassica campestris* Var.
	Lipaphis erysim Kaltenbuch	
	Siphocronyne brassicae Linn.	
	Aphis craccivora Koch.	Sunflower: *Helianthus annus* Linn.
2.		Pulse crops:
	Aphis craccivora Koch.	Pigeon pea: *Cajanus cajan* Mill.
	Aphis gossypii Glover	
	Aphis craccivora Koch.	Bean: *Dolichos lablab* Linn.
	Aphis gossypii Glover	
	Aphis craccivora Koch.	Cow pea: *Vigna catjang* Walp.
	Aphis craccivora Koch.	Pea: *Pisum sativum* Linn.
	Aphis craccivora Koch.	Mung: *Phaseolus radiatus* Linn.
	Aphis craccivora Koch.	Gram: *Cicer aeriatum* Linn.
3.		Cereal crops:
	Rhopalosiphum maidis Fitch	Jowar: *Sorghum vulgare* Pers.
	Melanophis sacchart Zehntner.	
	Rhopalosiphum maidis Fitch	Maize: *Zea mays* Linn.
	Rhopalosiphum maidis Fitch	Wheat: *Triticum vulgare* Linn.
	Rhopalosiphum padi Linn.	Paddy: *Oryza sativa* Linn.
	Rhopalosiphum rufiabdominalis	
4.		Summer vegetables:
	Aphis gossypii Glover	Bringal: *Solanum melongena* Linn.
	Myzus persicae Sulzer	
	Aphis malvae Koch.	Cucumber: *Cucumis sativum* Linn.
	Aphis gossypii Glover	Chillies: *Capsicum annum* Linn.
5.		Winter vegetables:
	Lipaphis erysimi Kattenbuch	Cabbage: *Brassica capitata* Linn.
	Myzus persicae Sulzer	
	Brevicoryne brassicae Linn.	Cauliflower: *Brassica botrytis* Linn.
	Lipaphis erysimi Kattenbuch	
6.		Spring vegetables:
	Aphis gossypii Glover	Tomato: *Lycopersicon esculeatum* Mill.
	Myzus persicae Sulzer	Carrot: *Dancus carrota* Linn.
7.	*Melanophis sacchari* Zehntner	Sugarcane: *Saccharum officinalis*
8.		Fruit plants:
	Aphis gossypii Glover	Mango: *Mangifera indica* Linn.
	Aphis fabae Glover	Citrus: *Citrus limetta* Linn.
	Toxofera auratntii Boy.	Guava: *Pisidium quaiava* Linn.
9.		Ornamental plants:
	Aphis gossypii Glover	Rose: *Rose multiform* Thumb
	Aphis gossypii Glover	China rose: *Hibiscus rosa-sinensis* Linn
	Myzuz persicae Sulzer	Bryophylum: *Bryophyllum pinnatum*

M. sacchari is widely distributed in plain region on sugarcane. *N. caganae* was abundant on *Cajanus cajan* in plain region of western Maharashtra in cold season. The most prevalent species of aphids found on vegetables in plain region were as *Aphis gossypii, M. persicae, L. erysimi* and *B. brassicae. Aphis gossypii* was abundant in cold season on ornamental plants like rose and china rose.

In present ecosystem *Aphis gossypii* was highly polyphagus on plants more than sixty families of plants kingdom have been attacked by this species. The second dominant species of aphids found in agro ecosystem was *Aphis craccivora*, this species attacked the members of thirteen families of plant kingdom. The above two species migrated from plain region to forest region after harvest of monsoon crops and again after onset of monsoon, they returned to plain region. The migratory behavior of many aphids species is unknown and needs attention.

References

Basu, R.C., Ghosh, M.R. and Raychaudhari, D.N. 1975. Studies on the aphids (Homoptera : Aphididae) from eastern India. VIII. New records of aphids from Assam. *Sci. Cult.*, 40: 41–423.

David, S.K., 1969. Some rare aphids in new regions in India. I. *Bombay Nat. Hist. Soc.*, 66: 323–326.

Eastop, V.F. and D. Hill Rislambers, 1976. *Survey of Worlds Aphids*, p. 1–573.

Ghosh, A.K., 1969. On a collection of aphid (Homoptera : Aphididae) from India. *Orient. Ins.*, 8: 425–432.

Ghosh, A.K. and Raychaudhari, D.N. 1968. Two new records of aphids (Homoptera : Aphididae) from India. *Bull. Ent.*, 9: 147–148.

Sathe, T.V., 1992. Fauna of Aphids on plants of economic importance found in Western Maharashtra. India. *J. Curr. Biosci.*, 9: 27–31.

Verma, K.D., 1965. Addition to the aphid fauna of Northern West India. *Sci. Cult.*, 31: 389.

Chapter 33

Behavioural Surveillance of Indian Major Carp *Catla catla* (Hamilton) Exposed to Free Cyanide

M.S. Prashanth* and Yogesh B. Patil

P.G. Department of Environmental Management,
Chh. Shahu Central Institute of Business Education and Research (SIBER),
University Road, Kolhapur – 416 004, Maharashtra, India

ABSTRACT

The toxicity of free cyanide (*i.e.* sodium cyanide) to the freshwater fish *Catla catla* was studied using, static bioassay method. The LC50 in 96 h was found to be 30 µg/L. Behavioural changes when exposed to lethal concentration of free cyanide showed increased opercular movement, increased surface behaviour, loss of equilibrium, change in body colour, increased secretion of mucus, irregular swimming activity, rapid jerk movement, 'S' jerk, partial jerk and aggressiveness.

Keywords: *Cyanide, Toxicity, Behaviour, LC50, Catla catla.*

Introduction

Industrial development with a rapid pace in the recent times has taken its toll by causing environmental pollution. The contamination of inland and surface water caused by released of toxic

* Corresponding Author: E-mail: drmsp@rediffmail.com; Phone: (M) 09860129554; (O) 0231-2535706/07; Fax: 0231-2535708.

chemicals can be dangerous to the all classes of living organisms if discharged without proper treatment. One such toxic and lethal chemical known is cyanide. It has the ability to bind heme iron in the oxygen-binding site of mitochondrial enzyme, cytochrome oxidase, thereby blocking of oxidising enzyme metabolism (WHO, 1988). Although notorious, free cyanide (CN^-/HCN) finds large applications in several industries like mining, electroplating, pharmaceutical, organic chemicals, etc. (Patil, 1999). These industries consequently, discharge large quantities of cyanide and metal-cyanide containing wastes. Several physical-chemical methods are employed for the treatment and the levels of total cyanides in discharged liquid wastes are brought down to 0.2 mg/L (200 µg/L). But even 0.2 mg/L of cyanide concentration in aqueous systems is toxic for most of the life forms. There are reports that even cyanide concentration as low as 0.01–0.1 mg/L are able to kill some sensitive animals species present in waters (Blaha, 1976); and fish is one such type of sensitive species. *Catla catla*, one of the Indian major carp is an edible freshwater fish of great economic importance. In the present article, the authors describe the behavioural surveillance of the said species when exposed to free cyanide.

Materials and Methods

Procurement and Maintenance of Fish

Freshwater fish, *Catla catla* (length 7±1 cm; weight 5±1 g) were obtained from Maharashtra State Fisheries Department Fish Farms, Rankala, Kohalpur, India and reared in large cement tank. During acclimatization, the fish were fed with rice bran and oil cake in the ratio of 2 : 1 on alternate days. Water of the tank was changed daily to avoid any fungal and bacterial contamination.

Physico-chemical Characterisation of Water

The physico-chemical characterisation of the water used for fish bioassay was carried out according to the methods described in Standard Methods (APHA, AWWA, WEF, 1998). The water quality parameters were as follows: temperature 28±1°C, salinity 198 mg/L, pH 8±0.2, dissolved oxygen 6.7 to 7.2 mg/L, chloride, 46.3 mg/L, sodium 1.22 mg/L, potassium 30.5 mg/L, calcium 17.04 mg/L, magnesium, 1 mg/L, carbon dioxide 9.0 mg/L, Hardness 115 mg/L, oxygen per cent <R> saturation 57 mg/L (as $CaCO_3$) and specific gravity 1.00374.

Toxicant Selected and Preparation of Stock Solution

For the present study, sodium cyanide (NaCN) was used as a toxicant. The stock solution of cyanide (1000 mg/L) was prepared according to the method prescribed in the Standard Methods (APHA, AWWA, WEF, 1998). For experimental purpose, the required cyanide concentration was drawn from the prepared stock solution.

Toxicity Evaluation

The per cent mortality of fish in different concentrations of free cyanide was determined at 96 h exposure. For this, the experimental fish were divided into batches of ten each, and were exposed to different concentrations of free sodium cyanide ranging from 10 µg/L to 50 µg/L (*i.e.* 0.01 to 0.05 mg/L). This range was obtained on trial and error basis. Toxicity evaluation was carried out in static water (Doudoroff *et al.*, 1951) and mortality rate was observed and recorded for all the concentration after 96 hours. A batch of fish was also maintained simultaneously in freshwater medium without cyanide, which served as negative control. All the experiments were performed in duplicates and repeated thrice to confirm the results. The mean values were derived following the method of Finney. Probit Kill Theory (1971) and Dragstedt and Behren's equation (Carpenter, 1982).

Fixation of Exposure Periods

In order to understand the influence of time over toxicity the effect of lethal concentration of free cyanide on *Catla catla* was studied at different periods of exposure. Before experimentation, the healthy fish collected from the tank were acclimatized to laboratory conditions in glass troughs for fifteen days. Each trough contained 15 L of water with approximately uniform sized fish. They were fed with commercial fish food pallets during acclimatization. The fish were divided into two groups. One group without cyanide served as control and the other group was exposed to lethal concentration of free cyanide for 1, 2, 3 and 4 days were chosen to observe the short-term effects. During this experiment the behavioural changes were critically observed.

Results

Toxicity Studies

The percentage mortality of *Catla catla* was observed to be 0 per cent and 100 per cent at cyanide concentration of 10 μg/L and 50 μg/L, respectively (Table 33.1). The LC_{50} value obtained through sigmoid curve was 31 μg/L and linear curve was found to be 28 μg/L. The LC_{50} value obtained were verified using Dragstedt and Behren's equation and was found to be 29 μg/L. Thus the average LC_{50} for 96 h was found to be 29.33 μg/L.

Table 33.1: Mortality of *Catla catla* in Different of Sodium Cyanide at 96 h Exposure

Sl.No.	Conc. of Toxicant	Log (Conc. of Toxicant)	Number of Fish			% Mortality	Probit Mortality
			Exposed	Alive	Dead		
1.	10	1.0000	10	100	0	0	0
2.	15	1.1761	10	9	1	10	3.72
3.	20	1.3010	10	7	3	30	4.48
4.	25	1.3979	10	6	4	40	4.75
5.	30	1.4771	10	5	5	50	5
6.	35	1.5441	10	4	6	60	5.25
7.	40	1.6021	10	2	8	80	5.84
8.	45	1.6532	10	1	9	90	6.28
9.	50	1.6990	10	0	10	100	8.09

Behavioural Observations

The behavioural changes were observed in control and exposed fish. The fish maintained in normal freshwater behaved in usual manner (*i.e.* they were very active with well-coordinated movements). Also the fish found to be very alert at slightest disturbance and covered one-third area of the bottom. But when, exposed to free cyanide, the fish showed erratic swimming, hyper and hypoactive, imbalance in posture, increase in surfacing activity opecular movement, gradual loss in equilibrium and spreading of excess of mucus all over the surface of the body. The fish occupied twice the area than that of the control group. They were spread out and appeared to be swimming independent of one another. The swimming behaviour was in a cork-screw palter, rotating along horizontal axis and followed by 'S' jerk, partial jerk, sudden rapid, non directed spurt of forward movement. The fish eventually died with their mouth and opecular wide opened. A change in colour of gill lamellae from reddish to light brown with coagulation of mucus on the gill lamellae was seen in dead fish.

Discussion

The behavioural changes are the manifestation of motivational, biochemical physiological and environmentally influenced state of the organism. The migration of fish to the bottom of the tank following the addition of sodium cyanide clearly indicates the avoidance behaviour of the fish, which was reported by Murthy (1987) in trout. The opecular movement of the fish ceases immediately following exposure to cyanide. The increase in opecular movement and corresponding increase in frequency of surfacing of fish clearly indicates that fish adaptively shifts towards aerial respiration (by obtaining atmospheric oxygen surfacing) and the fish tries to avoid contact with the cyanide through gill chamber (Santhakumar and Balaji, 2000). The increased ventilation rate by rapid, repeated opening and closing of mouth and opercular coverings accompanied by partially extended fins (caughing) was observed in the present study. This could be due to clearance of the accumulated mucus debris in the gill region for proper breathing, which was suggested by Prashanth *et al.*, 2005.

The erratic swimming of the treated fish indicates loss of equilibrium. It is likely that the region in the brain, which is associated with the maintenance of equilibrium, should have been affected (Deva Parkasa Raju, 2000; Prashanth *et al.*, 2005). The erratic swimming, jerky movements and convulsions before death were evident and the serenity varied with pesticide concentration. It indicates the signs of esphyxiation as indicated by gasping to death when fish, *Sarotherodan mossambicus* exposed dimethorate (Kalavathy *et al.*, 2001).

The surfacing phenomenon of fish observed under cyanide exposure might either be due to hypoxic condition of the fish as reported by Radhaiah and Jayantha Rao (1988). This fact was clearly evidenced in the present study. The observation on the metabolic shift from aerobic to an anaerobic condition involving glycolytic oxidation with enormous amount of lactic acid accumulation were also seen. Chronic exposure of finfish to aroclor was found to induce surfacing phenomenon of fish as pointed out by Drummond *et al.*, 1986.

Aggressive behaviour such as nudge and nip were increased following exposure to the toxic material. Orientation and locomotor patterns were found to be involved in most aspects of fish behaviour such as migration, mating, courtship and feeding, which were altered under stress conditions of environmental toxicants (David, 1995; Madhab Prasad *et al.*, 2002; Prashanth *et al.*, 2005).

The hyperexcitability of the fish invariably in the lethal exposure to free cyanide may probably be due to the hindrance in the functioning of the enzyme AChE in relation to nervous system as suggested by many authors (Deva Prakasa Raiu, 2000). It leads to accumulation of acetylcholine, which is likely to cause prolonged excitatory post synaptic potential. This may first lead to stimulation and later cause a block in the cholinergic system. The accumulation and increased secretion of mucus in the fish exposed to fee cyanide may be adaptive responses perhaps providing additional protection against corrosive nature of the pesticide and to avoid the absorptions of the toxicant by the general body surface. This agrees to the earlier findings done by Borah and Yadav (1995). In the present study as evidenced the results the abnormal changes in the fish exposed to lethal concentration free cyanide is time dependent.

Behavioural characteristics are obviously sensitive indicators of toxicant effect. It is necessary, however, to select behavioural indices of monitoring that relate to the organisms behaviour in the field in order to derive a more accurate assessment of the hazards that a contaminant may pose in natural system should be considered for species forming social organizational. If social interactions are not considered, only a certain portion of a population may be protected, and the toxicity of contaminant may be underestimated.

Cyanide is notorious for its potentially deadly poisonous characteristics. Its toxicity is critically dependent on speciation and chemical form. Acute toxicity to life forms is generally restricted to the free hydrogen cyanide in the solution. There are at least three mechanisms for the toxicity of cyanide; (*a*) formation of complexes of cyanide with di- or tri-valent metals at the acute site of metallo-enzymes including cyctochrome oxidase; (*b*) formation of Schiff base intermediates during the enzyme catalysis to form stable nitrate derivatives; (*c*) formation of cyanohydrin with metabolic keto groups (Solomonson, 1981). In fish, the early development stages such as embryonic and juvenile stages are the most sensitive to toxic products (Billard and Roubaud, 1985). Studies carried out by Patil and Paknikar (2000) showed that the LC50 of *Poecilia reliculata* when exposed to copper-cyanide and zinc-cyanide was 0.3 mg/L and 0.3 mg/L, respectively at 24 h. Thus, the present study clearly indicates the significant changes in the *Catla catla* when exposed to low quantity of free cyanide.

Acknowledgements

The authors wish to thank the President, Director and Head, Department of Environment Management, SIBER Kolhapur for providing the laboratory facilities. Help offered by the laboratory staff is gratefully acknowledged.

References

APHA, AWWA, WEF, 1998. *Standard Methods for the Examination of Water and Wastewater*, 20th Edn. American Public Health Association, Washington, DC.

Billard, R. and Roubaud, P., 1985. The effect of metals and cyanide on fertilization in rainbow trout (*Salmo gairdneri*). *Water Research*, 19: 209–214.

Blaha, J., 1976. Mathematical analysis of the chemical system cyanide heavy meals in water determination of components and toxicity of the system-I. The theoretical solution. *Water Research*, 10: 815–819.

Carpenter, P.L., 1982. *Immunology and Serology*, 3rd Ed., W.B. Saunders Co., Philadelphia, USA.

Crossland, N.O., 1982. Aquatic toxicology of cypermethrin. II. Fate and biological effects in pond experiment. *Aquatic. Toxicology*, 2: 205–222.

David, M., 1995. Effect of fenvalerate on behavioral, physiological and biochemical aspects of freshwater fish, *Labeo rohita* (Hamilton). *Ph.D. Thesis*, S.K. University, Anantapur, A.P., India.

Deva Parkasa Raju, B., 2002. Fenvalerate induced changes in protein metabolism of fresh water fish, *Tilapia mossambica*. *Ph.D. Thesis*, S.K. University, Anantapur, A.P., India.

Drummond, R.A., Russom, C.L., Diannel Geiger and David L. Defoe., 1986. Behavioural and morphological changes in freshwater minnow *Pimephales promelas*. As diagnostic end point for screening chemicals according to mode of action. *Aqua. Toxicol.*, 9: 415–434.

Finney, D.T., 1971. *Probit Analysis*, 3rd Ed. Cambridge University Press, London.

Kalavathy, K.A., Sirvakumar, A. and Rashmi Chandran, 2001. Toxic Effects of pesticides dimethoate on the fish, *Sarotherodon mossambicus*. *J. Eco. Res. Biocon.*, 2(1 and 2): 27–32.

Madhab Prasad, Bandyopadhaya and Ajit Kumar Aditya, 2002. Xenobiotic impact on sensitivity in *Anabas testudineus* (Bloch). *J. Ecobiol.*, 14(2): 117–124.

Murthy, A.S., 1987. Sub lethal effects of pesticides on fish. In: *Toxicity of Pesticide to Fish*. CRC Press, Boca Roton, Fl., USA, 2(4): 55–100.

Patil, Y.B., 1999. Studies on biological detoxification of metal-cyanides containing industrial effluents. *Ph.D. Thesis*, University of Pune, Pune.

Patil, Y.B. and Paknikar, K.M., 2000. Development of a process for detoxification of metal-cyanides from wastewaters. *Process Biochemistry*, 35: 1139–1151.

Sabita Borah and Yadav, R.N.S., 1995. Static Bioassay and toxicity of two pesticides, rogor and endosulfan to the air-breathing fish, *Heteropneustes fossilis* with special reference to behaviour. *Poll. Res.*, 14(4): 435–438.

Sambasiva Rao and Chandrasekara Rao, 1987. Independent and combined action of carbaryl and phenthoate on snake head fish, *Channa punctatus* (Bloch). *Curr. Sci.*, 56(7): 331–332.

Santhakumar, M. and Balaji, M., 2000. Acute toxicity of an organophours insecticide monocrotophas and its effects on behaviour of an air-breathing fish, *Anabas testudineus* (Bloch). *J. Environ. Biol.*, 21(2): 121–123.

Solomonson, L.P. and Spehar, A.M., 1981. Glyoxylate and cyanide formation. In: *Cyanide in Biology*, (Eds.) Vennesland, B., Conn, E.E., Knowles, C.J. and Wissing, F. Academic Press, London, pp. 363–370.

World Health Organisation, 1988. *Guidelines for Drinking Water Quality*, Vol. 2: Health criteria and other supporting information, Geneva.

Chapter 34

Incidence of Phytophthora Indicant Disease for Citrus Decline in Vidarbha Region (Maharashtra)

S.S. Pande, A.B. Fulke and V.S. Tajane
Rajiv Gandhi Biotechnology Centre, L.I.T. Premises, Amravati Road, Rashtrasant Tukadoji Maharaj Nagpur University, Nagpur – 440 033 (M.S.) India

ABSTRACT

Phytophthora nicotianae var *parasitica P. citrophthora* and *P. palmivora* were found widely distributed in the surveyed area out of total six villages of Nagpur, Amravati and Wardha districts and local nurseries which are main cause of citrus decline in Central India region. Out of the six villages, adjoining area of Morshi was recorded maximum (22.7 CFU/CC) of *Phytophthora* spp. and also intensity of infestation upto 50 per cent in citrus orchard. The number of the propagules in the citrus orchard of Morshi village, found to be much more as compare to other location and was recorded as 56.82 propagules/cc soil, while Rekhi and Selu village of Wardha district recorded comparatively less population of all three different species of *Phytophthora.*

Introduction

India is endowed with a wide range of agro-climatic conditions favorable for the cultivations of temperate, sub-tropical fruit crops. Commercially, the important fruits are mango, banana, citrus, apple, guava, pineapple, litchi, grapes, etc. the increasing area of cultivation and production of fruits to fulfill the domestic demand and to earn foreign exchange by exporting produce, have provided opportunities to many miner pathogens to becomes major ones and to major pathogens for development of frequent epidemics in certain crops.

Gummosis and collar rot epidemics occurred sporadically in Vidarbha region of Maharashtra (Fraser and Singh, 1966; Naqvi, 1988) and it is seem to be one of the major contributing factors for the citrus decline *i.e. Phytophthora* spp. *Phytophthora* causes foot rot, root rot, crown rot, gummosis, leaf fall and brown rot diseases in citrus. Hence, a survey was carried out to elucidate the cause of epidemics of gummosis and collar rot in Vidarbha region. Total six villages and two nurseries in Nagpur, Amravati and Wardha districts were surveyed and soil samples were collected from the rhizosphere of the gummosis and collar rot affected trees of *Citrus reticulata* (Nagpur mandarins) along with the infected roots, leaf and fruits. At least and fruits. At least five orchards were examined in each village.

Materials and Methods

Isolation of *Phytophthora* spp. Symptom less citrus decline and from a heavy diseased orchard where Nagpur, Amravati and Wardha district from 2004–2005. Soil samples were obtained from at least five different locations within the same orchard in September and October month *i.e.* in Mroog Bhar. For isolation, sub-samples of soil were spread onto modified PARPH medium using procedure as described by (Timmer *et al.*, 1988). After two days incubation at 27±1°C, colonies that isolated for further study. The fungal growth on selective medium was again transferred to corn meal agar medium (CMA)/carrot agar medium (CA). The margins of colony growth on CMA/CA medium was cut into 2–3 mm pieces and floated on the surface of sterilized distilled for sporulation. A profuse development of sporangia was absent after 0–36 hrs. *Phytophthora* spp. were identified on the basis of morphological characters of sporangia, sporangiaphores and swelling of the vegetative mycelium.

Similarly the citrus leaf trap technique (Grimm and Alexander, 1973; Naqvi, 1988) was followed to isolate the pathogen from soil.

Results and Discussion

Phytophthora nicotianae var parasitica *P. citrophthora* and *P. palmivora* (Graham *et al.*, 1998) were found widely distributed in the surveyed area out of total six villages of Nagpur, Amravati and Wardha districts and local nurseries, all above three mentioned spp. of *Phytophthora* were isolated from all villages of surveyed area.

Table 34.1: Prevalence of *Phytophthora* spp. Causing Decline to Citrus Orchard in Nagpur Mandarin of Vidarbha Region

Sl.No.	*Name of Orchards*	*Phytophthora spp. Population (CFU/CC)*	*Intensity*
1.	Nagpur District		
	(a) Mohogoan (Zilpi)	10.33	Up to 30 per cent
	(b) Lawha	8.46	Up to 30 per cent
2.	Amravati District		
	(a) Morshi	22.70	Upto 30–50 per cent
	(b) Tiwasa	14.30	Upto 30–40 per cent
3.	Wardha District		
	(a) Rehki	4.00	Upto 10–20 per cent
	(b) Selu	7.3	Upto 10–30 per cent
4.	Local Nurseries	10–63	Upto 10–40 per cent

Out of the six villages, adjoining area of Morshi was recorded maximum (22.7 CFU/CC) of *Phytophthora* spp. and also intensity of infestation up to 50 per cent in Amravati orchard (Table 34.1). The number of the propagules in the citrus orchard of Morshi village, Amravati where found to be much more as compare to other location and was recorded as 56.82 propagules/cc soil. It seems that all three *Phytophthora* spp. were frequently isolated from all locations in Vidarbha region. In Selu, Wardha Dist. *P. palmivora* recorded lowest which was 2 propagules/cc soil similarly *P. citrophthora* recorded 4.24 propagules/cc soil from soil sample of Mohogoan (Zilpi), Nagpur Dist. (Table 34.2). Tiwasa and Morshi region of Amravati Dist. reported predominant prevalence of all three spp. while Rekhi and Selu village of Wardha district recorded comparatively less population of all three different species of *Phytophthora*. While the total number of propagules recorded nearly same *i.e.* 12.67 and 12.01 (CFU/CC) and that of *P. palmivora* recorded only 9.59 CFU/CC (Figure 34.1) in the vicinity of the surveyed area of six villages in Wardha, Amravarti, and Nagpur district of Vidarbha region.

Table 34.2: Prevalence of Different Population of *Phytophthora* spp. in Vidarbha Region

Sl.No.	*Locations*	**Total No. of Propagules (CFU/CC)*	*P. nicotianae*	*P. citrophthora*	*P. palmivora*
1.	Mohogoan (Zilpi) Nagpur Dist.	24.18	12.10	4.24	7.69
2.	Lawha, Nagpur Dist.	32.46	6.08	12.34	14.04
3.	Morshi, Amravati Dist.	56.82	22.96	15.96	20.40
4.	Tiwasa, Amravati Dist.	48.14	21.00	20.00	7.00
5.	Rekhi, Wardha Dist.	26.12	4.00	10.00	12.00
6.	Selu, Wardha Dist.	16.48	10.00	4.47	2.00
7.	Local Nurseries	29.00	12.60	17.10	4.00

*: Average of five replication.

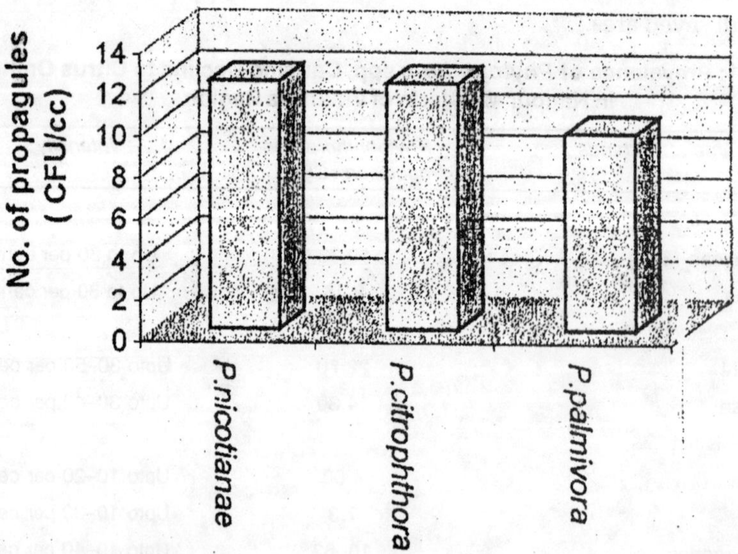

Figure 34.1: Distribution of *Phytophthora* spp. in Vidarbha Region

From this survey it was confirmed that *P. nicotianae* and *P. citrophthora* are dominant in the region of Vidarbha.

References

Fraser, L.R. and Daljit Singh, 1986. *Indian Hort.*, 11: 15–16 and 26.

Graham, J.H., L.W. Timmer, D.L. Drouillard and T.L. Peever, 1998. Characterization of *Phytophthora* spp. causing outbreaks of citrus brown rot in Florida. *Phytopathology*, 88: 724–729.

Grimm, S.R. and F. Alexander, 1973. *Phytopathology*, 63: 540–541.

Naqvi, S.A.M.H., 1988. Prevalence of *Phytophthora* spp. pathogenic to citrus in orange grower of Vidarbha, Maharashtra. *Indian J. Mycol. Pl. Pathol.*, 18(3): 274–276.

Timmer, L.W., Sandler, H.A., Graham, J.H. and Zitko, S.E., 1988. Sampling citrus orchards in Florida to estimate populations of *Phytophthora* parasitica. *Phytopathology*, 78: 940–944.

Chapter 35

Drinking Water Quality and its Impact on the Health of Users in Kattumannarkoil Taluk, Cuddalore District, Tamil Nadu, India

P. Sivagurunathan and B. Dhinakaran
Department of Physics, Annamalai University, Annamalainagar – 608 002, Tamil Nadu, India

ABSTRACT

Observations on the physico-chemical characteristics of undergroundwater used for drinking collected from twenty selected villages of Kattumannarkoil Taluk have been presented in, this article. Water samples collected from ordinary hand pumps and bore-well were analysed for certain physico-chemical properties like pH, Electrical Conductivity (EC), Turbidity, Alkalinity, Total Dissolved Solids (TDS), Total Hardness, Calcium, Magnesium, Sodium, Chloride, Fluoride and Sulphate. The concentration of Total Hardness and Calcium was found to be beyond the WHO and ISI permissible limits. This study indicates that the quality of drinking water in the certain areas of Kattumannarkoil taluk is not suitable for drinking purpose due to TDS and Total Hardness in groundwater.

Keywords: *Physico-chemical Properties, Drinking water, Total hardness.*

Introduction

Water is an excellent solvent and essential to living things. Rapid growth of industrialization, human activities, agriculture, mining and waste disposal practices affect the availability and quality of water (Biswal *et al.*, 2001). Keeping this view, government of India constituted technology mission for drinking water in 1987. In our country 70 per cent of the water is seriously polluted and 75 per cent of illness and 80 per cent of the child mortality is attributed to water pollution (Ravichandran *et al.*, 2002). A number of research reports (Chinoy, 1991 and Susheela, 1985) show that the fluoride content in drinking water results in skeletal disorders, damaging the kidney and nervous system. Excess nitrate content results in methemoglobinemia, gastric cancer, goister, birthmal deformation and hypertension. For consuming excess calcium in drinking water causes hypertension, hypercalciuria and produces pathological changes in the brain tissues (Heilberg, 2000). In this context the present study is an attempt to estimate the physico-chemical characteristics of drinking water in and around Kattumanarkoil Taluk, Cuddalore District, Tamil Nadu.

Materials and Methods

Water samples were collected in plastic containers from different water sources of bore wells and hand pumps in summer and winter seasons of the year 2003 in the 20 selected villages of Kattumannarkoil Taluk, Cuddalore District, Tamil Nadu. The names of sampling stations, depth and type of sources are given in Table 35.1 and their sampling sites are shown in Figure 35.1. Physico-chemical parameters like Colour, Odour, pH, Electrical Conductivity (EC), Turbidity, Alkalinity, Total Dissolved Solids (TDS), Total Hardness, Calcium, Magnesium, Sodium, Chloride, Fluoride and Sulphate content have been estimated by using standard methods (APHA, 1998).

Table 35.1: List of Sampling Sites, Nature of Source and Depth

Sl.No.	*Sampling Station*	*Depth (meter)*	*Source*
1.	Nanjalure	64	Bore well
2.	Thunisiramedu	24	Hand pump
3.	Poolamedu	22	Hand Pump
4.	Thavarthampattu	56	Bore well
5.	Nandimangalam	72	Bore well
6.	Sivayam	20	Hand pump
7.	Neivasal	24	Hand pump
8.	Thirunaraiyur	22	Hand pump
9.	Kumaratchi	22	Hand pump
10.	Lalpet	58	Bore well
11.	Elleri	64	Bore well
12.	Solakur	24	Hand pump
13.	T-Puthur	56	Bore well
14.	Kandakumaran	22	Hand Pump
15.	Nathamalai	24	Hand pump
16.	Chettikatalai	21	Hand pump
17.	Sivakkam	26	Hand pump
18.	Atkondanatham	48	Bore well
19.	Nedumbur	66	Bore well
20.	Vadamur	22	Hand pump

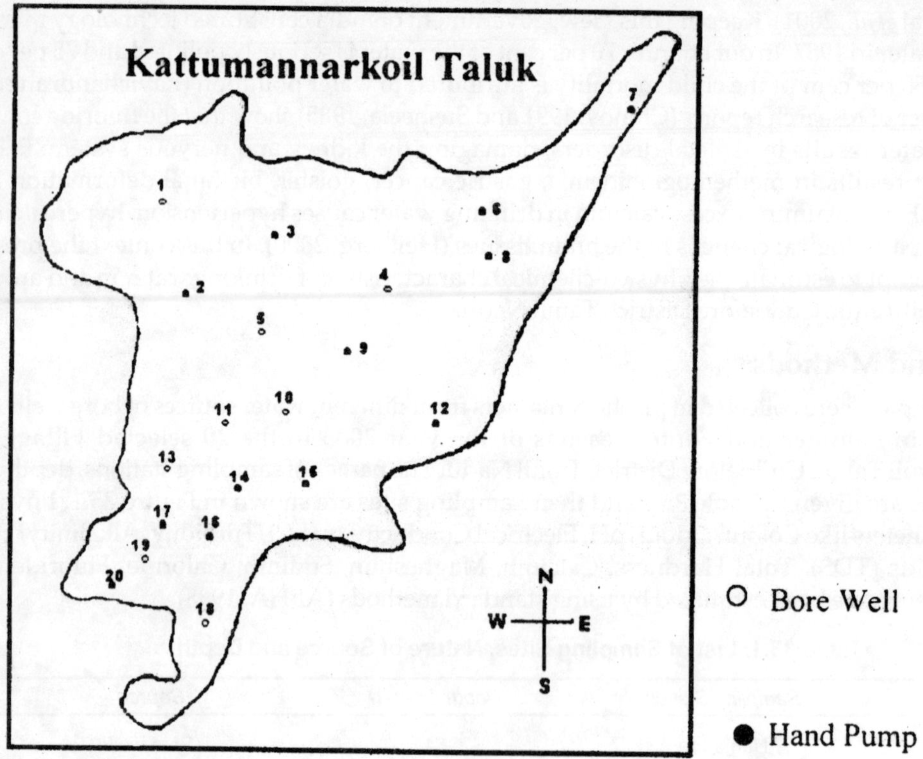

Figure 35.1: Location of Sampling Sites

Results and Discussion

The physico-chemical parameters like pH, EC, Turbidity, Alkalinity, TDS, Total Hardness, Calcium, Magnesium, Sodium, Chloride, Fluoride and Sulphate contents of 20 sites of water samples at summer and winter seasons are provided in Tables 35.2 and 35.3.

pH

The pH values of drinking water have been observed in the ranges from 6.8 to 8.5 for summer season and 6.5 to 7.7 for winter season. The observed pH value in the present study is within the permissible limit of WHO and ISI. The seasonal variation shows that the pH value fluctuates maximum during summer and minimum in winter at all locations. The seasonal fluctuation in drinking water quality in the region shows maximum pH value in summer (8.8) and minimum in winter (6.5). This may be attributed to the leaching of surface soil from surrounding areas in rain water (Garg, 2003).

Electrical Conductivity (EC)

The Electrical conductivity is a measure of the salt content of water in the form of ions. The EC values vary from 650 µmho/cm to 9070 µmho/cm in summer season and 610 µmho/cm to 4250 µmho/cm in winter season. Conductivity depends on the presence of ions, their total concentration, mobility and valance in groundwater. The seasonal fluctuation has been maximum in summer and minimum in winter. Similar results are reported by Garg (2003).

Table 35.2: Physico-chemical Properties of Drinking Water at Different Sampling Stations at Different Seasons

Sl.No.	Summer Season						Winter Season					
	pH	EC µmho/cm	Turbidity mg/l	Alkalinity mg/l	TDS mg/l	Total Hardness mg/l	pH	EC µmho/cm	Turbidity mg/l	Alkalinity mg/l	TDS mg/l	Total Hardness mg/l
1.	7.0	1630	5	500	1141	520	6.8	1150	3	440	980	450
2.	6.8	1420	7	440	994	328	7.2	1060	5	320	780	295
3.	7.4	1750	8	424	1225	750	7.6	1415	8	408	1010	580
4.	6.8	2450	1	624	1715	950	7.6	2100	1	428	1618	570
5.	6.8	1600	2	176	420	260	6.5	1485	1	165	310	180
6.	7.4	1980	5	480	1386	900	7.7	2060	4	444	1120	620
7.	6.8	2210	6	440	1547	710	7.1	2020	4	380	1220	592
8.	6.8	1410	7	352	987	650	7.1	1135	6	224	782	560
9.	7.4	2240	3	600	1568	680	7.4	2300	2	516	1220	568
10.	7.2	1925	8	326	648	652	7.2	1210	5	310	520	532
11.	6.9	1700	4	320	1190	600	6.8	1520	3	308	780	564
12.	7.4	2210	10	440	1547	710	6.7	1810	9	302	1218	660
13.	6.8	2460	5	616	1722	880	7.1	2180	4	540	1320	600
14.	7.6	1640	4	198	448	460	7.4	1310	3	154	322	392
15.	7.4	650	3	186	455	204	7	610	3	132	392	140
16.	8.5	4350	8	680	3045	1000	7.1	2685	8	572	1985	888
17.	6.8	3950	5	648	2765	1400	7	2450	4	558	1315	950
18.	7.4	9070	8	872	5048	3100	7.2	4250	8	660	3880	1510
19.	7.4	1160	7	336	482	600	7.2	988	6	302	430	418
20.	6.9	2188	9	392	1467	728	6.8	1912	7	318	922	592

Turbidity

The turbidity of water samples varies from 1 mg/l to 10 mg/l and 1 mg/l to 9 mg/l for summer and winter seasons respectively. The higher value of 10 mg/l and 9 mg/l were observed for hand pump water source of village Solakur in summer and winter seasons respectively. The observation shows that the turbidity values indicate the clarity of water and is caused by living and non-living suspended matter and colour producing substances. Similar results are reported by Tiwari *et al.* (2003).

Alkalinity

The alkalinity of the water samples vary from 176 mg/l to 872 mg/l and 132 mg/l to 660 mg/l for summer and winter seasons respectively. The observed results in the sampling sites namely Nandimangalam, Kandakumaran and Nathamalai fall with in the permissible limit and the remaining sites have resulted above the permissible limit in both seasons. Alkalinity has public health significance. Highly alkaline water is unpalatable and is not used for domestic water supply. The maximum permissible level of alkalinity in drinking water is 200 mg/l according to WHO and ISI.

Table 35.3: Physico-chemical Properties of Drinking Water at Different Sampling Stations at Different Seasons

Sl.No.	Summer Season (mg/l)						Winter Season (mg/l)					
	Calcium	Magnesium	Sodium	Chloride	Fluoride	Sulphate	Calcium	Magnesium	Sodium	Chloride	Fluoride	Sulphate
1.	140	41	196	552	0.1	24	120	36	122	512	0.1	14
2.	194	22	185	225	0.1	20	124	21	176	210	0.5	13
3.	192	65	162	347	0.1	29	164	48	136	309	0.1	19
4.	180	96	112	81	0.1	24	160	56	78	48	0.1	14
5.	170	20	86	212	0.2	12	120	19	72	195	0.1	8
6.	228	79	178	351	0.3	28	180	49	138	242	0.3	17
7.	208	46	181	495	0.2	30	160	38	147	388	0.1	13
8.	244	46	156	242	0.8	23	174	24	112	184	0.4	8
9.	260	67	184	360	0.2	30	220	48	122	268	0.2	24
10.	267	20	122	175	0.1	22	224	12	88	108	0.1	20
11.	224	46	188	252	0.2	33	160	42	142	182	0.1	21
12.	220	50	194	527	0.3	29	184	44	172	458	0.2	24
13.	256	58	222	477	0.1	46	202	48	176	427	0.1	27
14.	120	38	81	221	0.1	29	96	32	65	180	0.1	21
15.	48	20	80	94	0.3	39	32	19	56	80	0.2	5
16.	264	82	338	1143	0.1	39	218	68	211	768	0.1	25
17.	368	115	212	1053	0.1	37	284	86	164	807	0.1	19
18.	760	288	686	3285	0.3	29	432	250	438	1592	0.2	20
19.	160	48	92	387	0.3	27	128	36	66	266	0.2	21
20.	168	43	76	405	0.1	28	132	31	62	212	0.1	27

Total Dissolved Solids

The experimental values of total dissolved solids of groundwater samples vary from 420 mg/l to 5048 mg/l and 310 mg/l to 3880 mg/l for summer and winter seasons respectively. The higher TDS values for the bore well source of Atkondanatham are 5048 mg/l during summer and 3880 mg/l during winter. As recommended by WHO and ISI, maximum permissible limit of TDS in drinking water is 500 mg/l. From the analysis, four sampling sites namely Nandimangalam, Kandakumaran, Nathamalai and Nedumbur fall within the limit prescribed by WHO and remaining sites exceed the permissible limit. Excess TDS contents in drinking water affect the strength, solidify of concrete, palatability of food cooked, and also cause intestinal irritation. Similar results are reported by Patil *et al.* (2002).

Total Hardness

Hardness of water samples are observed in the present study is from 204 mg/l to 3100 mg/l in summer season and 140 mg/l to 1510 mg/l in winter season. The maximum value of 3100 mg/l was observed in bore well source of Village Atkondanatham and minimum value of 140 mg/l was recorded for hand pump water samples of village Nathamalai. The standard prescribed for total hardness was 600 mg/l as $CaCO_3$, by ISI. The observed result in five sampling sites namely Nanjalure, Thunisiramedu, Nandimanglam, Kandakumaran and Nathamalai fall within the permissible limit in summer season, and remaining sites have above the prescribed limit of ISI. But for winter season, 14 sampling stations fall within the prescribed limit and six sampling sites namely Sivayam, Solakur, T-Puthur, Chettikattali, Sivakkam and Atkondanatham were found to have above the prescribed limit. The present observation suggests that this is mainly due to presence of high calcium content in these places (Ravichandran *et al.*, 2002 and Chun-Yun *et al.*, 1999).

Calcium

The calcium content in the study area ranges from 48 mg/l to 760 mg/l and 32 mg/l to 432 mg/l for summer and winter seasons respectively. The maximum value of 760 mg/l has been observed for bore well source of village Atkondanatham, while the minimum value of 32 mg/l is recorded for hand pump source of Nathamalai. The prescribed limit of calcium values for drinking water quality is 200 mg/l. Kumaratchi, Lalpet, T-Puthur, Chetettikatalai, Sivakkam and Atkondanatham have higher than the permissible limits of WHO and ISI in both seasons. The samples collected from Sivayam, Neivasal, Thirunaraiyur, Elleri and solakur are found to have higher than the permissible limits of calcium in summer season only. The excess calcium in the human body causes hypercalcemia, coma and death (Dasgupta Adk and Purohit, 2001).

Magnesium

The concentration of magnesium in the study area ranges from 20 mg/l to 288 mg/l in summer season and 12 mg/l to 250 mg/l in winter season. The maximum permissible limit of magnesium in drinking water is 150 mg/l, according to WHO and ISI. Magnesium contents in all sampling sites are less than the permissible limit except Atkondanaatham, which recorded 288 mg/l and 250 mg/l in summer and winter season respectively. Excess magnesium present in the drinking water causes nauscea, muscular weakness and paralysis in human body when it reaches up to the level of about 400 mg/l (Nirmalkumar *et al.*, 2003).

Sodium

The sodium content in drinking water was found to be 64 mg/l to 548 mg/l in summer season and 56 mg/l to 431 mg/l in winter season. The prescribed limit of sodium (250 mg/l) was suggested by WHO. In the present study 2 sampling sites out of 20 namely Chetettikatalai and Atkondanatham were found to have high sodium content in summer season. The excess sodium in drinking water causes cardiovascular disease and kidney failure (Pomevanz *et al.*, 2002).

Chloride

The maximum value of chloride 3285 mg/l has been observed in the bore well source of village Atkondanatham, while the minimum value 81 mg/l is observed for bore well source of Thavarthampattu in summer season. In winter season, it is in the range of 48 mg/l to 1592 mg/l. The permissible limit of chloride in drinking water is 250 mg/l as suggested by WHO and ISI. The water samples from Thunisiramedu, Thavarthampattu, Nandimangalam, Thirunaraiyur, Lalpet, Kandakumaran and Nathamalai were found to fall with in the permissible limit in summer season. The higher concentration of chloride may affect heart and kidney diseased persons (Patil *et al.*, 2002).

Fluoride

Fluoride contents are observed in the ranges from 0.1 mg/l to 0.8 mg/l for summer and 0.1 to 0.5 mg/l for winter season. Fluoride content observed in all stations is less than the prescribed limit (2 mg/l) suggested by WHO and ISI.

Sulphate

The sulphate contents of 20 drinking water samples range from 12 mg/l to 46 mg/l in summer season and 5 mg/l to 27 mg/l in winter season. All the samples have been found within the desirable limits of (200 mg/l) prescribed by ISI.

Conclusion

Physico-chemical parameters like pH, EC, Turbidity, Alkalinity, TDS, Total Hardness, Calcium, Magnesium, Sodium, Chloride, Fluoride and Sulphate of 20 water samples at 20 sites were carried out. The water samples from Nandhimangalam, Kandakumaran and Nathamalai are suitable for drinking purpose without purification. But the sampling sources at Nanjalure, Thunisiramedu. Poolamedu, Thavarthampattu Sivayam, Neivasal, Thirunaraiyur, Kumaratchi, Lalapet, Elleri, Solakur, T-Puthur, Chettikatalai, Sivakkam, Atkondanatham, Nedumbur and Vadamur recorded higher TDS and total hardness values. High TDS and total hardness in the drinking water causes health hazards. Consumption of high TDS and total hardness in drinking water necessitates reverse osmosis and demineralization process.

References

APHA, 1998. *Standard Methods for Examination of Water and Wastewater*. American Public Health Association. Washington, D.C.

Biswal, S.K., Majhi, B. and Behera, J.P., 2001.Groundwater quality near ashpond of thermal power plant. *Poll. Res.*, 20(3): 457–490.

Chinoy, N.J., 1991. Effect of fluoride on physiology of animals and human beings. *Indian J. Env. Prot.*, 1(1): 17–32.

Chun Yuh, Y., Shang Shyne, T., Aco Cheng, L. and Chung Feng, H., 1999. Rectal cancer mortality and total hardness levels in Taiwan's drinking water. *Environ. Res. Sec. Ass.*, p. 311–316.

Dasgupta Adak, M. and K.M. Purohit, 2000. Correlation coefficient of some physico-chemical characteristics of surface and groundwater of Rajangpur, Part I. *Indian J. Env. Prot.*, 20(9): 681–687.

Garg, S.S., 2003. Water quality of well and bore well of 10 selected locations of Chitrakoot Region. *Indian J. Environ. Prot.*, 23(9): 966–974.

Heilberg, I.P., 2000. Update dietary recommendations and medical treatment of renal stone disease. *Nephorl Dial Transplant*, 15: 117–123.

ISI, 1984. *India Standard Specification for Drinking Water*. Indian Standard Institution, New Delhi.

Jain, C.K., Bhatia, K.K.S., Kumar, C.P. and Purandare, B.K., 2003. Groundwater quality in Malalprabha Sub-basin Karnataka. *Indian J. Env. Prot.*, 23(3): 321–329.

Nirmalkumar, J.I., Bhatt, I., Kumar, N.E. and Patal, M., 2003. Physico-chemical properties of groundwater in an around Anand District. *Indian J. Env. Prot.*, 23: 1248–1253.

Patil, P.R., Patil, S.K. and Dhande, A.D., 2002. Studies on drinking water quality in Bhusawal corporation water supply. *Indian J. Env. Prot.* 22(2): 161–164.

Pomevanz, T. Doltin, Korzets, Z. and B. Wolach, 2002. Increase sodium concentration in drinking water increases blood pressure in neonates. *J. Hypertension*, 20: 203–207.

Ravichandran, C., Alagappa Moses, A., Girija K. and Chakravarthy, P., 2002. Drinking water quality assessment in few selected pilgrim centres and tourist spots in Tamil Nadu. *Indian J. Env. Prot.*, 22(2): 129–136.

Sharma, S., Jain, P.C. and Mathur, R., 1995. Quality assessment of groundwater municipal and fringe areas near Gwalior. *Indian J. Env. Prot.*, 15(7): 534–538.

Susheela, A.K., 1985. Fluoride content in drinking water and groundwater quality in rural areas of Tamil Nadu. District Chitrakoot. *Proc. Conf. Inter. Soc. Fluoride Res.*, pp. 12–17.

Tiwari, A.K., Dikshit, R.P., Tripathi, E.P. and Chaturvedi, S.K., 2003. Fluoride content in drinking water and groundwater quality in rural areas of Tashil Mau, District Chitrakoot. *Indian J. Env. Prot.*, 23: 1045–1050.

WHO, 1984. *Guideline for Drinking Water Quality*, Vol. 2.

Chapter 36

Impact of Mercuric Chloride on the Biochemical Composition of Blood of *Clarius batracus* (Linn)

A. Parithabhanu and M.A. Subramanian
P.G and Research Department of Zoology C.N. College, Erode – 368 004, Tamil Nadu

ABSTRACT

On exposure to sublethal concentrations of mercuric chloride, the total free amino acids and total free sugars increased in concentration, in the blood of *C. batrachus* whereas the glycogen content, total proteins and total lipids decreased. The stress caused by the metal could inflict an excessive energy demand which is fulfilled by glycogen reserve so that the glycogen content in the blood diminished with an elevated blood sugar, level in the blood. The decrement of free amino acids, proteins and lipids could be due to utilization of these biochemical parameters as energy source under the condition of emergency during prolonged stress. Thus, the heavy metal would bring alterations in the fundamental biochemical mechanism in the present experimental animals leading to mortality in them.

Keywords: *Blood, Biochemical parameters, Mercuric chloride, Clarius batracus.*

Introduction

At present, there is an increasing concern over heavy metal pollution as it endangers aquatic ecosystem (Ramakrishna Rao, 1992). Rapidly expanding industrial potential in India has led to various research works in the field of heavy metal pollution in "Indian rivers (Bhuwaneswaran *et al.*,

1999). Of all the heavy metals, mercury is the most toxic one and many mercuric compounds are included in the blacklist by all International capitals (Tayler *et al.*, 1972). Many fresh water bodies in India are shown to be polluted by mercury (Agarwal and Kumar, 1978). Metelev *et al.* (1971) have opined that even traces of mercuric chloride in water bodies may be highly toxic to the animals. The heavy metals are found to affect the physiology and biochemistry of fishes (Jha, 1991; Janarthana Reddy *et al.*, 1992; Dhana Packiam *et al.*, 1998). Histological changes caused by heavy metals have also been studied in fishes by Richardson *et al.* (1974), Bergen and Patel (1982) and Pavlos and Mahagen (1987).

The impact of pollutants can be well ascertained by analyzing the blood, as the blood is the reflector of the whole body pathophysiology. Therefore the haematological parameters are of great value in understanding the functional status of the animals exposed to toxicants. Investigators like Pandey *et al.* (1984), Sambasiva Rao and Ramana Rao (1985), Sakthivel and Sampath (1987) and Seker and Christy (1996) have made studies on hematological parameters with reference to pesticides in fishes. The adverse effects of industrial effluents on various blood parameters are let known by Haniffa *et al.* (1986), Subramanian *et al.* (1988) and Varadaraj *et al.* (1993). Much importance has also been given to show the adverse effects of heavy metals on the blood of many fresh water fishes. (Balabanova 1999; Mitra *et al.*, 2001; Ramalingam *et al.*, 2001; Joshi *et al.*, 2002). However, studies pertaining to the toxicity of mercuric chloride on the biochemical composition of blood in fishes are scanty. Therefore the present work has been carried out to evaluate the toxicity of mercury on various biochemical parameters in the blood of common fresh water teleost, *Clarius batracus*.

Materials and Methods

The fishes measuring 15–18 cm in length and weighing about 10–5 gms procured from a local fish farm and inorganic mercuric chloride supplied by Sarabai Chemical Ltd, Bombay were used. A standard mercuric chloride solution was prepared by dissolving 10 mg of salt in 10 liter of water. From this standard solution, different concentrations (0.02, 0.04, 0.06, 0.08 and 0.10 ppm) were prepared. After determining the LC_{50} 96 hr value of the heavy metal to the fishes, a group of 10 laboratory-acclimatized fishes were introduced into each concentration for 3 weeks with appropriate controls. The blood from the fishes were collected and subjected to biochemical estimations, such as total free amino acids (Yemm and Cocking, 1957), total proteins (Gornall *et al.*, 1949), total free sugars (Roe, 1955), glycogen (Kemp and Kits, 1975) and total lipids (Barnes, and Black Stock, 1973).

Results and Discussion

As per Table 36.1, the glycogen content, total proteins and total lipids decrease in the blood of *C. batracus* under the toxicity of mercuric chloride whereas the levels of total free sugars and total free amino acids increase. In animals, any stress could inflict excessive energy demand, which is immediately fulfilled by blood glucose. As a result, the blood sugar level could increase as observed in the present experimental fishes. As a consequence, the glycogen reserve in the blood could be subjected to glycogenolysis with the resultant depletion. This is in conformity with the findings of Mukhopadhyay and Dehadrai (1980) and Natarajan (1989) in pesticide-treated fishes. The reduction in the glycogen content of in the blood of present experimental fishes could also be due to the inhibition of glycogenesis, as also observed by Kabeer *et al.* (1978) in bivalves treated with pesticide.

In the present investigation, mercuric chloride causes reduction in total proteins in the blood of *C. batracus*. This is suggestive of degradation of proteins with the resultant increase of total free amino acids. This corraborates the findings of Kabeer *et al.* (1978) in pesticide-treated molluscs and Joyce

Shoba Rani and Janaiah (1991) in *C. batracus* under pesticide toxicity. Nagabhushanam *et al.* (1983) has reported that the free amino acids serve as supplementary energy source under the condition of emergency during chronic stress.

Table 36.1: Biochemical Constituents in the Blood of Control and Treated *C. batracus*

Concentrations of $Hgcl_2$ (ppm)	Biochemical Constituents (mg/l00 ml)				
	Total Free Amino Acids	Total Proteins	Glycogen	Total Free Sugars	Total Lipids
Control	450.00±16.73	69.00±6.63	16.80±1.49	149.00±6.63	392.68±15.43
0.02	472.00±13.29 NS (+4.88; r=0.72)	59.80±1.32* (13.33; r=1.00)	13.6C±1.35* (–19.04; r=0.58)	164.00±10.48NS (+10.06; r=0.35)	355.01±25.11 NS (–9.59; r=0.88)
0.04	512.00±11.66* (+13.77; r=0.61)	51.60±1.35* (–25.21; r=0.51)	11.00±0.89* (–34.52; r=0.69)	196.00±10.74* (+31.54; r=0.08)	328.34±22.24* (–16.38; r=0.75)
0.06	572.00±13.26* (+27.11; r=0.90)	43.00±1.67* (–37.68; r=0.54)	9.20±1.16* (–45.23; r=0.98)	236.00±10.62* (+58.38; r=0.38)	276.47±19.64* (–29.59; r=0.70)
0.08	68.00±7.48* (+35.11; r=0.63)	31.80±1.16* (–53.91; r=0.10)	8.00±1.26* (–52.38; r=0.89)	246.00±11.12* (–65.10; r=0.97)	194.19±20.04* (–50.54; r=0.99)
0.10	628.00±13.26* (+85.55; r=0.90)	29.00±2.36* (–57.97; r=0.95)	6.60±1.01* (–60.71; r=0.93)	266.00±10.19* (–78.52; r=0.08)	179.86±16.87* (–54.19; r=0.80)

*: Significant at P > 0.05; NS: Not Significant.

+: Indicates increase over control; –: Indicates decrease from control.

Each value is mean±SD of 5 observations.

The decrement of total lipids in the blood of the species under study indicates the utilization of lipids to meet the energy demand during the stress caused by mercuric chloride. According to Srinivas *et al.* (1991), the endogenous fat is the only source for energy requirements in animals during prolonged stress. This falls in agreement with the findings of Jebakumar (1993) and Govindan (1994) who have reported the decrement of lipids in the tissues of fishes during stress caused by insecticides.

Thus the metal intoxication has disturbed the normal functioning of cells with the resultant alterations in the fundamental biochemical mechanisms in fishes. This would in turn result in the mortality of fishes on chronic exposure to the heavy metal.

References

Agarwal, M. and Kumar, H.D., 1978. Physio-chemical and physiological assessment of two mercury polluted effluents. *Ind. J. Environ. Health*, 20: 141–155.

Balabanova, L.V., 1999. Toxic effects of heavy metals on the immuno competent cells of fish. *Report of World Fish Convention*, 1999.

Barnes, and Black Stock, J., 1973. Estimation of lipids in marine animals and fish. Detailed investigation of the Sulphophosphovanillin method for total lipids. *J. Exp. Mar. Biol. Ecol.*, 12: 103–118.

Bergeri, K.V. and Patil, H.S., 1982. Behaviour responses of freshwater fishes *Barbus aurilus* and Lepidocephalichithy ountes to heavy metals. *Proc. Ind. Sym. on Physiological Response of Animal to Pollution*.

Bhuvaneswaran, N., Santhalakshmi, G. and Rajeswari, S., 1999. Water quality of River Adyar in Chennai city. The river a boon or a bane. *Indian J. Env. Protection*, 19(6): 412–415.

Dhanapackiam, P., Manohar, L. and Sampoorani, V., Median lethal effects of mercury and cadmium on oxygen consumption and gill histopathology of common crab, *Cyprinus carpio. J. Ecobiol.*, 10(3): 193–198.

Gornall, A.G., Bardawill, C.J and David, M.M., 1949. Determination of total serum proteins by means of biuret reaction. *J. Biol. Chem.*, 177: 551–556.

Govindan, V.S., Jacob, L. and Devika, R., 1994. Toxicity and metaboilic changes in *Gambusia affinis* exposed to Phosphamidon. *J. Ecotoxicol. Env. Monit.*, 4(1): 1–6.

Haniffa, M.A., Murugasan, A.G and Porchelvi, M., 1986. Haematological effects of distillery and paper mill effluent on *Heteropnustes fossilis* (Blood) and *Sarotherodon mossambicus* (Peters). *Proc. Indian Acad. Sci. (Anim. Sci.)*, 95: 155–161.

Janardhana Reddy, S., Kalarani, V., Therakanandha, B., Reddy, D.C. and Ramamurthi, R., 1992. Changes in energy metabolism of the fish *Labeo rohita* in relation to prolonged lead exposure and recóvery. *J. Ecotoxicol. Environ. Monit.*, 8: 43–45.

Jebakumar, S.R.D., Kumaraguru, A.K and Jayaraman, J., 1993. Accumulation and dissipation of phosphomidon in the tissues of freshwater fish *Oreochromis mossambicus. Comp. Physiol. Ecol.*, 18(1): 12–17.

Jha, B.S., 1991. Alterations in the protein and lipid contents of intestine, liver and gonads in the lead exposed fresh water murrel, *Channa puntatus* (Bloch). *J. Eco. Biol.*, 3(1): 29–34.

Joshi, P., Deepa Harish and Manjushri Bose, 2002. Effect of lindane and malathian exposure to certain blood parameters in a freshwater teleost fish *Clarias batrachus. Poll. Res.*, 21(1): 55–57.

Joyce Shoba Ram, V. and Janalah, C., 1991. Ammonia metabolism in freshwater teleost *Clarias batrachus* (Linn) on exposure to trichlorofom. *Bull. Environ. Contam. Toxicol.*, 46: 731–737.

Kabeer, A.I., Begum, M.D., Sivaiah, S. and Ramana Rao, K.V., 1984. Sub-lethal toxicity of malathian on the proteases and free amino acid composition in the liver of teleost, *Tilapa mossambica* (Peters). *Toxicol. Letter*, 20: 59–62.

Kemp, A. and Kits, V.A.J.M., 1975. A colorimetric micromethod for the determination of glygogen in tissues. *J. Biochem.*, 56: 646–648.

Metelev, V.V., Kanrev, A.I and Dzasokhova, N.G., 1971. *Water Toxicology*. Vodnaya Tokshikologiyhs Kolos Publishers, pp. 67–101.

Mitra, J., Samir Banerjee and Nirmal Kumar Sarkar, 2001. A study of food-chain transport and haematotoxicity of Lead. *Poll. Res.*, 20(2): 163–166.

Mukhopadhyay, P.K. and Dehadrai, 1980. Biochemical changes in the air breathing eat fish *Clarias batrachus* (Linn) exposed to malathion. *Environ. Pollut.* 22(A): 149–158.

Nagabhushanam, R., Kedarkar, M.S and Sarojini, R., 1983. *Textbook of Animal Physiology*, Oxford and IBH Publishing Co., New Delhi.

Natarajan, G.M., 1989. Changes in the carbohydrate metabolism during acute and chronic exposure of air breathing fish *Channa straistus* to xydemeton methyl (Metasystax). *Comp. Physiol. Ecol.*, 14(14): 181–184.

Pandey, B.N., Prasad, S.S. Chanchal, A.K. and Singh, S.B., 1984. Effect of some pesticides on the oxidative metabolism and haematology of *Clarias batrachus* (Bloch). *Proc. Sem. Eff. Pest. Agri. Fau.*, p. 13–18.

Paulose, P.V and Mahajan, C.L., 1987. Effect of temperature on toxicity and accumulation of mercury compounds in a fish *Gambusia affinis. Com. Physiol. Ecol.*, 12(1): 52–54.

Ramalingam, V., Vimaladevi, V., Narmadaraji, R. and Prabakaran, P., 2000. Effect of lead on haemotological and biochemical parameters in freshwater fish *Cirrhina mrigala. Poll. Res.*, 19(1): 81–84.

Ramakrishna Rao, T., 1992. Missing the words for the trees: An over views of aquatic ecotoxicology in India. Abstract No. 81. *Loth European Conference on Ecotoxicology*, Amsterdans, The Netherlands.

Richardson, M.E., Fox, M.R.S. Jr. and B.E. Fry. 1974. Pathological changes produced in Japanese guail by ingestion of cadmium. *J.* Nutr., 104: 323–338.

Roe, J.H., 1955. The determination of sugar in blood and spinal fluid with anthroma reagent. *J. Biol. Chem.*, 212: 335–343.

Sakthivel, M. and Sampath, K., 1987. Effects of dye-stuff effluents on different levels of protein and blood parameters of *Cyprinus carpio. Proc. Nat. Symp. of Environ. Biol.*, Department of Zoology, St. Xavier's College, Palayamkottai, T.N., India.

Sambasiva Rao, K.R. and Ramana Rao, K.V., 1986. Independent and combined toxicity of cabaryl and phenthoate on some haemotological parameters of murrel *C. Punctatus* (Bloch). *Ind. Zool.*, 10: 29–31.

Sekar, P. and Christy, I., 1996. Haematological changes in the freshwater cat fish *Mystus vittatus* exposed to sublethal concentration of phosphomidon. *J. Ecobiol.*, 8(1): 25–28.

Srinivas, Md. Rafi G. Prasad T.A.V and Reddy, D.C., 1991. Effect of thermal acclimation on the lipid metabolism in a freshwater field crab, *Ozoitelphusa senex senex* (Fabricius). *J. Ecolbiol.*, 3(2): 134–147.

Subramanian, M.A., Shahul Hameed, M. and Varatharaj, G., 1988. Haematological changes in the fish *Macrones keletius* (Gunther) after exposure to the tannery effluent. *The Indian Zoologist*, 12(162): 71–74.

Tayler, 1972. A review of the lethal and sublethal effects of mercury on aquatic life. *Residue Rev.*, 72: 33–70.

Varatharaj, G., Subramanian, M.A. and Nagarajan, B., 1993. The effect of sublethal concentration of paper and pulp mill effluents on the haematological parameters of *O. mossambicus* (Peters). *J. Environ. Biol.*, 14(4): 321–326.

Yemm, E.W. and Cocking, E.C., 1957. The determination of aminoacid in the ninhydrin. *Analyst*, 80: 209–213.

Chapter 37

Experiments on Protein Requirements of Young Silver Carp *H. molitrix* (Val.)

Surabhi Gupta, A.P. Sharma and M. Das
College of Fisheries, G.B. Pant University of Agriculture and Technology, Pantnagar, Distt. U.S. Nagar, Uttaranchal – 263 145

Introduction

The production of fish under culture conditions greatly depends upon provision of the artificial feed. However, due to lack of feeding with suitable artificial feed, the fish production in inland aquaculture is very low, averaging only 1750 kg/ha/year in India (Sharma and Bhadula, 1995). Among the various nutrients, protein constitutes the costliest and most important dietary component for the growth of fish (Tacon and De-Silva, 1997). If protein in diet is insufficient, it is withdrawn from the tissues to carry out vital life functions, thereby resulting into rapid reduction in growth.

Among cultivable carps, the maximum investigations on requirement of protein and other nutrients have been done on common carp (Ogino and Satia, 1970; Sin, 1973; Takeuchi *et al.*, 1979). In comparison, the availability of information on protein requirement of Indian major carps and cultivable Chinese carps is scarce (Dabrowski, 1977; Das and Tripathi, 1979; Singh and Sinha, 1981; Singh *et al.*, 1986; Swamy and Mohanty, 1988).

Silver carp is an important component in carp culture system in Asia. However, the information on protein requirement of silver carp is inadequate (Bitterlich, 1985; Tacon, 1995; Singh, 1998). Taking into consideration the above facts the present investigation was undertaken to study the protein requirement of young silver carp.

Materials and Methods

Induced bred advanced fingerlings of *H. molitrix* (79-80 g) produced from the brood stock maintained at instructional fish farm of College of Fisheries, Pantnagar were used for the present experiment. The experiment was conducted in sixteen rectangular pools (10 m × 5 m × 1 m), each having individual water supply. Replicates of three were set for each diet and one pool was kept as control. Physico-chemical parameters (temperature, turbidity, dissolved oxygen, free CO_2, alkalinity, pH, NH_3-N etc.) were monitored at weekly intervals. Five isocaloric, diets containing 20, 25, 30, 35 and 40 per cent protein levels were formulated using locally available feed ingredients in varying proportions and combinations (Table 37.2). Proximate composition of raw ingredients is given in Table 37.1. Biochemical composition of diets was estimated following AOAC, 1980 (Table 37.3).

Table 37.1: Biochemical Composition of Raw Ingredients

Ingredients	*Nutrients (%)*				
	Crude Protein	*Total Lipid*	*Crude Fibre*	*Total Ash*	*NFE (by Differences)*
Fish meal	56	8.5	2.50	30.0	3.00
Meat meal	48	9.5	3.00	27.5	12.00
Deoiled soybean cake	45	4.0	7.00	6.0	38.00
Deoiled mustard cake	42	5.5	7.25	7.0	38.25
Deoiled ground cake	40	8.5	6.50	6.0	39.00
Deoiled rice bran	12	7.0	17.50	21.0	45.50

Table 37.2: Proportion of Ingredients in Experimental Diets

Ingredient (%)	*Diets*				
	Diet A (25% CP)	*Diet B (30% CP)*	*Diet C (35% CP)*	*Diet D (40% CP)*	*Control (20% CP)*
Fish meal	50	50	100	163.7	–
Meat meal	50	75	100	163.7	–
Deoiled soybean cake	100	150	100	163.7	–
Deoiled mustard cake	100	150	200	163.7	500
Deoiled ground cake	100	150	200	163.7	–
Deoiled rice bran	600	425	300	218.0	500

Table 37.3: Biochemical Composition of Experimental Diets (on per cent dry weight basis)

Nutrients (%)	*Diets*				
	Diet A (25% CP)	*Diet B (30% CP)*	*Diet C (35% CP)*	*Diet D (40% CP)*	*Control (20% CP)*
Crude protein	25.39	30.00	35.35	40.44	20.00
Crude fibre	4.40	5.30	4.51	5.14	5.54
Ether extract	1.25	1.19	1.26	1.22	1.04
Total ash	22.40	15.8	20.83	17.16	12.05
Nitrogen free extract	46.56	47.71	28.05	36.04	59.81
P/E ratio (mg P/k cal E)	98.14	113.04	127.11	141.41	79.74

Well ground feed ingredients were mixed and made into dough and the balls made out of dough were served in flat earthen pots hanged at fixed sites (2 for each pool). The feed was supplied once (9–10 am) @ 5 per cent of the wet body weight of the fishes. Growth of fishes was measured weekly (Figure 37.1).

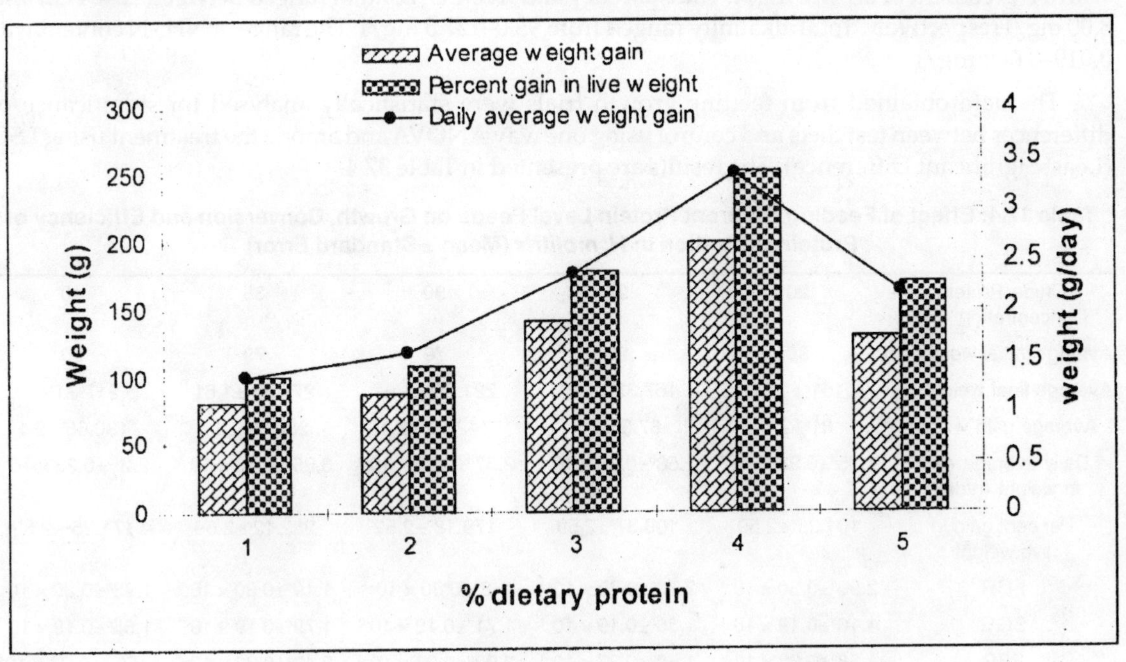

Figure 37.1: Variations in Growth During Experimental Period

Nutritional indices were calculated using the following standard definitions:

$$\text{Per cent gain in live weight} = \frac{\text{Final wet body weight (g)} - \text{initial wet body weight (g)}}{\text{Initial wet body weight (g)}}$$

$$\text{FCR} = \frac{\text{Feed intake (g)}}{\text{Wet weight gain (g)}}$$

$$\text{Per cent SGR} = \frac{\text{Log}_e \text{ final weight} - \text{Log}_e \text{ initial weight}}{\text{Experimental duration (days)}} \times 100$$

$$\text{PER} = \frac{\text{Gain in wet weight of fish (g)}}{\text{Quantity of protein fed (g)}}$$

$$\text{Gross growth efficiency (GGE)} = \frac{\text{Wet weight gain (g)}}{\text{Feed intake (g)}}$$

Results

During the present study the water quality parameters valued between optimum range. Water temperature over the 60 days feeding trials based on weekly measurements ranged between 16–28°C with a DO content of 6.9–9.0 mg/l. Transparency and free CO_2 content ranged between 20–64 cm and 8.00 mg/l respectively. Total alkalinity ranged from 95.0–127.5 mg/l. The range of NH_3-N content was 0.019–0.040 mg/l.

The data obtained from feeding growth trials were statistically analysed for significance of differences between test diets and control using one way ANOVA and among the treatment using LSD (Least Significant Difference). The results are presented in Table 37.4.

Table 37.4: Effect of Feeding Different Protein Level Feeds on Growth, Conversion and Efficiency of Protein Utilization in *H. molitrix* (Mean ± Standard Error)

Crude Protein Concentration (%)	20	25	30	35	40
Average initial weight (g)	80	80	79	79	80
Average final weight (g)	161^{a}±1.61	167.33^{b}±1.61	221.66^{c}±1.61	279.33^{d}±1.61	217^{c}±1.61
Average gain weight	81^{a}±2.1	87.5^{b}±2.1	142.33^{c}±2.1	200.00^{d}±2.1	130.66^{b}±2.1
Daily average gain in weight (g/day)	$1.35^{a}±0.23 \times 10^{-1}$	$1.58^{b}±0.23 \times 10^{-1}$	$2.37^{d}±0.23 \times 10^{-1}$	$3.35^{c}±0.23 \times 10^{-1}$	$2.2^{cd}±0.23 \times 10^{-1}$
Per cent gain in live weight	101.25^{a}±2.59	109.37^{a}±2.59	179.18^{b}±2.59	252.12^{c}±2.59	171.25^{b}±2.59
FCR	$2.96^{d}±0.30 \times 10^{-1}$	$2.73^{c}±0.30 \times 10^{-1}$	$1.8^{b}±0.30 \times 10^{-1}$	$1.19^{a}±0.30 \times 10^{-1}$	$1.83^{b}±0.30 \times 10^{-1}$
SGR	$1.16^{a}±0.19 \times 10^{-1}$	$1.30^{b}±0.19 \times 10^{-1}$	$1.71^{d}±0.19 \times 10^{-1}$	$1.79^{c}±0.19 \times 10^{-1}$	$1.65^{c}±0.19 \times 10^{-1}$
PER	$1.68^{c}±0.23 \times 10^{-1}$	$1.58^{b}±0.23 \times 10^{-1}$	$2.00^{d}±0.23 \times 10^{-1}$	$2.42^{c}±0.23 \times 10^{-1}$	$1.375^{a}±0.23 \times 10^{-1}$
Gross growth efficiency	$0.337^{a}±0.97 \times 10^{-2}$	$0.353^{a}±0.97 \times 10^{-2}$	$0.592^{c}±0.97 \times 10^{-2}$	$0.832^{d}±0.97 \times 10^{-2}$	$0.44^{b}±0.97 \times 10^{-2}$

Figures having same superscripts in the same row are statistically indifferent ($p = 0.05$).

Nearly per cent survival was registered during the experiment. Per cent gain in live weight significantly increased up to 35 per cent dietary protein incorporation ($p < 0.05$) beyond which a fall in growth was observed. Statistical analysis of growth data revealed that diet C (35 per cent dietary protein level) was most effective in comparison to other test diets and control. However other diets A, B and D were more effective than control diet but among themselves these were statistically indifferent ($P < 0.05$). A positive correlation existed between dietary protein and per cent gain in live weight ($r = 0.69$).

Dietary P/E ratio (mg protein/k cal energy) was lowest (79.74) in control feed with 20 per cent protein and 250.8 k cal energy/100 g feed and was highest (141.4.1) at 40 per cent protein and 282.66 k cal energy/100 g feed.

Table 37.4 indicates mean values of growth and nutrient utilization. Out of five experimental diets, diet with 35 per cent protein showed best growth performance (200.00 g) compared to others ($r = 0.69$) without showing any significant correlation.

The SGR of fish fed with 35 per cent protein diet was higher (1.79) than other four diets however no significant correlation ($r = 0.76$, $p > 0.05$) could be established in the present study. A significant correlation ($r = 0.91$, $p < 0.05$) between average weight gain and SGR could be established.

Table 37.5: Pearson's Correlation Matrix (One Tailed)

Parameters	*% Protein*	*Average Weight Gain*	*Daily Average Weight Gain*	*% Gain in Live Weight*	*FCR*	*SGR*	*PER*	*GGE*	*P/E Ratio*
% Protein	1	0.697	0.701	0.731	–0.824*	0.843*	0.085	0.526	0.998**
Average weight gain		1	0.998**	0.999**	–0.973**	0.912*	0.773	0.976**	0.703
Daily average weight gain			1	0.996**	–0.971**	0.910*	0.769	0.974**	0.709
% gain in live weight				1	–0.981**	0.923*	0.740	0.963**	0.736
FCR					1	–0.978**	–0.627	–0.909*	–0.831*
SGR						1	0.533	0.837*	0.856*
PER							1	0.892*	0.098
GGE								1	0.536
P/E ratio									1

*: Significant at 0.05 level; **: Significant at 0.01 level.

The best FCR (1.19) was observed with 35 per cent protein diet. Significant negative correlation (r = –0.82, p < 0.01) between dietary protein and FCR could be set. Whereas control diet (20 per cent protein) showed least food conversion ratio (2.96).

A positive correlation (r = 0.08, p > 0.05) was observed between dietary protein and PER. Fish fed with 35 per cent protein diet demonstrated highest PER (2.42) (Figure 37.2). Whereas, lowest PER (1.37) was observed at 40 per cent protein diet.

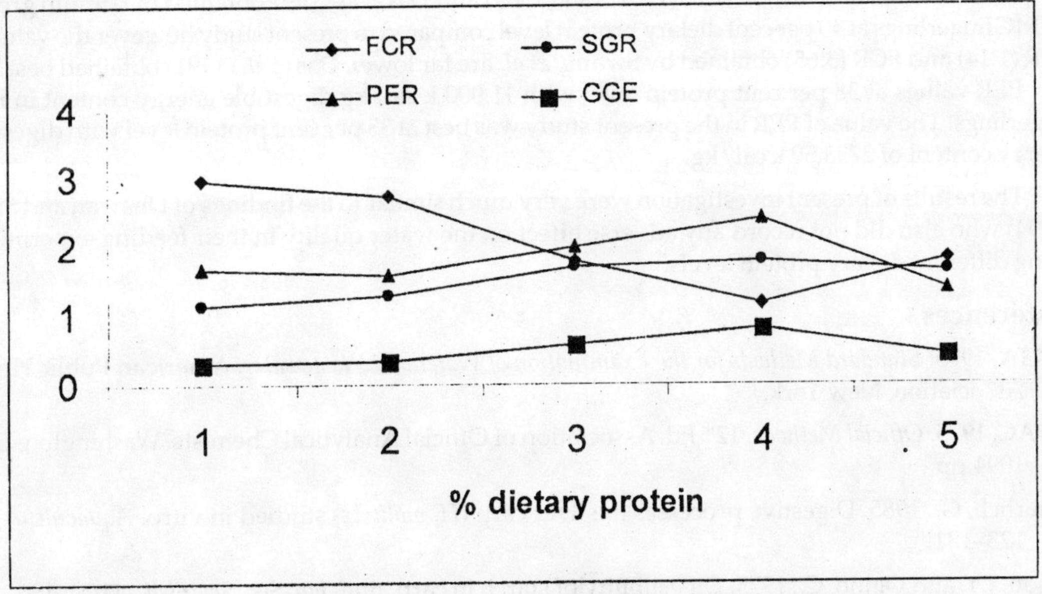

Figure 37.2: Growth and Nutritional Indices

A positive relationship (r = 0.52, p > 0.05) was observed between dietary protein level and GGE. Highest GGE value (0.832) with 35 per cent protein diet compared to 0.337 of control diet.

Discussion

Dietary protein intake of fishes is related to their nutritional status, temperature, age and growth. The data of the investigation revealed optimum growth of *Hypophthalmichthys molitrix* in terms of live weight gain at 35 per cent protein level. Pattern of growth at different protein level was similar to those reported for grass carp (Dabrowsky, 1977) and Indian major carps (Singh and Bhanot, 1988). In comparison to IMC (30 per cent) dietary protein requirement of sub adult of silver carp (35 per cent) appears to be a bit on higher side.

Satpathy *et al.* (2003) in their experimental trial on rohu fingerlings concluded that the fish could utilize lipid to spare protein but there is no significant advantage from this beyond the dietary protein level of 35 per cent. Various investigations have proved it possible to reduce the protein content of diet with higher lipid level inclusions without decrease in weight gain (Jauncy, 1979; Takeuchi *et al.*, 1979). Compared to other studies fat content in the experimental diets of present study was very low (1.04–1.26 per cent) owing to the use of deoiled rice bran and oil cakes, still the optimum protein requirement was reasonable.

Best food conversion ratio (1.19) was revealed by 35 per cent protein diet showing efficient utilization of carbohydrate as energy source by silver carp thereby sparing protein for growth. Chiou and Ogino (1985) have also found that carps were able to digest on an average of 85 per cent of ingested starch at dietary levels of 19–48 per cent.

Significant correlations existed between protein levels and growth parameters. Khan and Jafri (1991) obtained 61.18 per cent live weight increase, 1.130 SGR and 2.58 FCR for catla over eight week experimental period. Comparison of these results with present study shows that silver carp is more efficient in utilization of 35 per cent protein diet than catla.

On the other hand Swami *et al.* (1981, 1988,1989) and Das *et al.*, 1991 obtained maximum growth of IMC fingerlings at 40 per cent dietary protein level compared to present study however the values of SGR (1.14) and FCR (3.65) obtained by Swamy *et al.* are far lower. Das *et al.* (1991) obtained best FCR and PER values at 38 per cent protein level with 11,000 k cal/kg digestible energy content in rohu fingerlings. The value of PER in the present study was best at 35 per cent protein level with digestible energy content of 2753.50 k cal/kg.

The results of present investigation were very much similar to the findings of Dhawan and Singh (1994) who also did not record any adverse effect on the water quality in their feeding experiments using different dietary protein levels.

References

APHA, 1975. *Standard Methods for the Examination of Water and Wastewater*. American Public Health Association, New York.

AOAC, 1975. *Official Methods*, 12th Ed. Association of Official Analytical Chemists. Washington, D.C., 1094 pp.

Bitterlich, G., 1985. Digestive processes in silver carp (*H. molitrix*) studied in vitro. *Aquaculture*, 50: 123–131.

Chiou, J.Y. and Ogino, C., 1975. Digestibility of starch in carp. *Bull. Jap. Soc. Sci. Fish.*, 41(4): 456–466.

Dabrowsky, K., 1977. Protein requirement of grass carp *Ctenopharyngodon idella* (*Val.*) *Aquaculture*, 7: 125–132.

Das, K.M. and Mohanty, S.N. and Sarkar, S., 1991. Optimum dietary protein to energy ratio for *Labeo rohita* fingerlings. In: *Proc. of Fourth Asian Fish. Soc. Spec. Publ.*, 5: 205.

Das, K.M. and Tripathi, S.D., 1979. Feed protein utilization and growth of *Ctenopharyngodon idella* (Val.). In: *Proc. Indian Sci. Congr.*, 66: 85.

Dhawan, A. and Singh, R., 1994. Efficacy of different protein ingredients used in feeds on the growth and reproductive potential of common carp. In: *Proc. of 3rd Indian Fish. Forum*, October 11–14, p. 23–26.

Jauncey, K., 1979. Growth and nutrition of carps in heated effluents. *Ph.D. Thesis*, University of Aston in Birmingham, 200 pp.

Mohanty, S.N., Swamy, D.N. and Tripathi, S.D., 1990. Growth, nutritional indices and carcass composition of Indian major carp fry *Catla catla*, *Labeo rohita* and *Cirrhinus mrigala* fed four dietary protein levels. *Aquaculture Hungarica* (Szaruas), 6: 211–217.

Pandian, T.J., 1989. Protein requirement of fish and prawns cultured in Asia. In: *Proc. of 3rd Asian Fish. Nutr. Netr. Network Meeting*, pp. 11–22.

Satpathy, B.B., Mukherjee, D.D. and Ray, A.K., 2003. Effects of dietary protein and lipid levels on growth, feed conversion and body composition in rohu, *Labeo rohita* (Ham.), fingerlings. *Aquaculture*, 9(1): 17.

Singh, B.N. and Sinha, V.R.P., 198l. Observations on the nutrition of Indian Major Carp. *C. mrigala* (Ham.). In: *All India Seminar on Fish Biology*, November 26–28, p. 21–23.

Singh, B.N., Sinha, V.R.P., Kumar, K. and Swamy, D.N., 1986. Observations on the feed formulation and fortification of conventional fish feed for rohu and mrigal fry and fingerlings. *Int. J. Acad. Icthol.*, 7(1): 31–34.

Swamy, D.N. and Mohanty, S.N., 1990. Response of rohu, *L. rohita* (Ham.) fingerlings to diet of different protein content. *J. Zool. Res.*, 3: 2–4.

Swamy, D.N., Mohanty, S.N. and Tripathi, S.D., 1988. Growth of mrigala (*C. mrigala*) fingerlings fed on fish meal based formulated diets. In: *Proc. of First Indian Fish Forum*, (Ed.) M. Mohan Jeseph. Asian Fisheries Society, Indian Branch, Mangalore, p. 81–83.

Swamy, D.N., Mohanty, S.N. and Tripathi, S.D., 1989. Dietary protein requirement of *Catla catla* (Ham.) fingerlings. *Proc. Nat. Sem. Freshwater Aqua.*, pp. 109–111.

Tacon, A.G.J., 1995. Application of nutrient requirement data under practical conditions: Special problem of semi-intensive and intensive farming system. *J. Appl. Icthyol.*, 11: 205–214.

Tacon, A.G.J. and De Silva, S.S., 1997. Feed preparation and management strategies within semi-intensive fish farming systems in the tropics. *Aquaculture*, 151: 379–404.

Takeuchi, T., Watanabe, T. and Ogeino, C., 1979. Optimum ratio of dietary energy to protein for carps. *Bull. Jap. Soc. Sci. Fish.*, 14: 983–987.

Yorkowski, M. and Tabachek, J.L., 1979. Proximate and amino acid composition of some natural fish foods. In: *Proc. of World Symp. on Fin Fish Nutr. and Fish Feed Technology*, 1: 435–448.

Chapter 38

Effect of Nitrogen Management on Growth and Micro-climatic Parameters of Bed Planted *durum* Wheat (*Triticum durum* Desf.)

Navneet Kaur Brar and D.S. Kler

Department of Agronomy and Agrometeorology, Punjab Agricultural University, Ludhiana – 141 004, Punjab, India

ABSTRACT

Field trial on *durum* wheat (*Triticum durum* Desf.) was conducted during Rabi 2002–03 at Punjab Agricultural University, Ludhiana. The experiment replicated thrice consisted of 16 treatment combinations arranged in Randomized Block Design with three nitrogen management levels (120 kg N; 120 kg N + 15 t FYM and 150 kg N ha^{-1} and five nitrogen application times [1/2 at sowing (S) + 1/2 at crown root initiation (CRI); 1/3 at S + 1/3 at CRI + 1/3 at tillering (T); 1/4 at S + 1/2 at CRI + 1/4 at T; 1/3 at S + 1/3 at CRI + 1/3 at T + 3 per cent urea spray (US) at flowering (F) and 1/4 at S + 1/2 at CRI + 1/4 at T + 3 per cent US at F] on beds and one flat sown treatment (unidirection 22.5 cm). Mean grain yield of 40.02 and 39.53 q ha^{-1} of bed planted and flat sown crop were statistically at par with each other. One extra treatment where whole N was applied at sowing to bed planted wheat, yielded 38.31 q ha^{-1} which was comparable to 2 split application of N (Rec.) either to flat sown (39.53 q ha^{-1} or to bed planted (38.33 q ha^{-1}) crop indicating that whole N could be applied at sowing to bed planted *durum* wheat. 120 kg N + 15 t FYM ha^{-1} gave 9.9 per cent and 7.9 per cent higher grain yield than 120 kg N ha^{-1} and 150 kg N ha^{-1} respectively. 120 kg N + 15 t FYM ha^{-1} on an average tended to record higher values of growth parameters and favorable values of microclimatic parameters.

***Keywords:** Durum wheat, Bed planting, Split application, Growth parameters, Micro-climatic parameters.*

Introduction

Durum wheat occupies 8 to 10 per cent area in Punjab covering about 250 to 300 thousand hectares with a production of about 800 to 1000 thousand mt annually (Anon, 1995). Introduction of *durum* wheat has helped create varietal diversity for a balanced and stable agro ecosystem to sustain high production, caring the industry needs and reduce the risk of crop failures resulting from pest and disease outbreaks. In view of export for foreign exchange of *durum* wheat it becomes imperative to improve its productivity from the present almost a static level being obtained at recommended levels of inputs. To make *durum* wheat competitive in international market it becomes imperative to improve its productivity and quality by manipulating agronomy of the crop through variation in planting methods and nitrogen management with the emphasis on use of organic manures. Hence, present investigations on *durum* wheat were conducted to analyze crop growth and yield under bed planting and different nitrogen management combinations.

Materials and Methods

The experiment was conducted at the Punjab Agricultural University, Ludhiana during 2002–03 on a loamy sand irrigated soil. The soil of the experimental field tested low in available nitrogen and medium in available phosphorous and potash. The 16 treatment combinations (15 bed planted + 1 flat sown) consisting of 3 nitrogen management levels [120 kg N (Rec.); 120 kg N + 15 t PYM and 150 kg N ha^{-1}] with 5 times of N application [1/2 at sowing (S) + 1/2 at crown root initiation (CRI) (Rec.); 1/3 at S + 1/3 at CRI + 1/3 at tillering (T); 1/4 at S + 1/2 at CRI + 1/4 at T; 1/3 at S + 1/3 at CRI + 1/3 at T + 3 per cent urea spray (US) at flowering (F) and 1/4 at S + 1/2 at CRI + 1/4 at T + 3 per cent US at F to bed sown wheat including 1 recommended treatment of flat sown *durum* wheat (Rec.) and an extra treatment where whole N (120 kg ha^{-1}) was applied at sowing to bed planted wheat for comparison were replicated thrice in a randomized block design having gross plot size of 6.00 m × 4.05 m while net plot size measured 5.00 m × 2.70 m. A seed rate of 100 kg ha^{-1} of *durum* wheat (cv. PDW 274) was used. Two rows per bed (67.5 cm) were sown and row to row spacing of 22.5 cm was kept for flat sowing. All other agronomic management practices of the crop were uniformly adopted. The crop was sown on November 9, 2002 and harvested on April 24, 2003. The data on yield attributes and yield were recorded at harvest. Data on various growth parameters and microclimatic parameters such as tiller count, plant height, dry matter accumulation, canopy temperature, soil temperature, relative humidity and Photosynthetically active radiation interception were recorded at periodic intervals of 45, 75, 105 and 135 days after sowing (DAS).

Assman Psychrometer was used for recording data on canopy temperature and relative humidity between 10.00 am to 11.00 am and 2.00 pm to 3.00 pm. Soil temperature (0°C) was recorded at 5 cm soil depth with a Bimetal soil thermometer between 2.00 pm to 3.00 pm. Penetration of photosynthetically active radiation (PAR) in the range of 0.4 to 0.7 microns was measured between 10.00 am to 11.00 am, 12.00 noon to 1.00 pm and 2.00 pm to 3.00 ppm and mean values for the day were calculated for further use. Line quantum sensor (LI-COR Photometer model LI-191-SA) which has quantum (photon) response through wavelength range of 400–700 nm for photosynthetic photon flux density (PPFD) was used for PAR measurements in µmoles m^{-2} sec^{-1}. The percentage photosynthetically active radiation intercepted (PARI) by the crop was computed for reporting.

$$\text{PARI (\%)} = \frac{\text{PAR above the crop canopy} - \text{PAR at the soil surface}}{\text{PAR above the crop canopy}} \times 100$$

All interaction effects were found to be non significant therefore the mean effects are discussed in results.

Results and Discussion

Tiller Count

Tillering is one of the most important determinant of grain yield in wheat. Good tillering leads to production of more number of ears per unit area leading to higher grain yield. The data on tiller count are presented in Table 38.1. The differences between the bed planted and flat sown crop were non-significant.

Table 38.1: Effect of Nitrogen Management and Time of Nitrogen Application on Number of Tillers in Bed Planted *durum* Wheat

Treatments	*Total Tillers* m^{-2}			*Effective Tillers* m^{-2}
	Days After Sowing			
	45	*75*	*105*	*135*
Nitrogen management (ha^{-1})				
120 kg N	325	364	285	239
120 kg N +15 t FYM	380	397	294	245
150 kg N	340	369	287	243
CD (P = 0.05)	NS	NS	NS	NS
Time of N application				
1/2S + 1/2CRI	344	354	273	231
1/3S + 1/3CRI + 1/3T	352	379	282	240
1/4S + 1/2CRI + 1/4T	348	389	284	242
1/3S + 1/3CRI +1/3T + 3 per cent US	354	380	303	247
1/4S + 1/2CRI +1/4T + 3 per cent US	345	380	303	248
CD (P = 0.05)	NS	NS	NS	NS
Interaction	NS	NS	NS	NS
Flat	535.6	494.6	258.5	250.9
CD (P = 0.05) Bed vs Flat	NS	NS	NS	NS
Bed (S)	340	357	270	234

S: Sowing; CRI: Crown root initiation; T: Tillering; US: Urea spray at flowering stage.

Nitrogen management did not influence the tiller count significantly throughout the crop growth as the results are not significant statistically. Similar results were reported by Dhiman *et al.* (1982) and Kaur (2000). Application of FYM along with 120 kg N ha^{-1} tended to produce higher tiller count throughout the crop growth over 120 kg N ha^{-1} or 150 kg N ha^{-1} when applied alone. Similar results have been reported by Negi and Mahajan (2000) and Sushila and Giri (2000).

Time of application of nitrogen also resulted in non-significant differences in tiller count throughout the crop growth. The lowest tiller count was recorded in two splits.followed by three splits and the highest with three splits combined with urea spray. Tiller count values with whole N applied at sowing were comparable with the various split applications on the beds. Similar results have been reported by Uppal (1999) and Dhuka *et al.* (1991).

Plant Height

Plant height is a useful growth index of the developmental phase of plant to give an idea about the dry matter production leading to ultimate yield. The data on plant height are presented in Table 38.2. The plant height was almost similar between the bed planted and flat sown crop throughout the crop growth as they were statistically at par.

Table 38.2: Effect of Nitrogen Management and Time of Nitrogen Application on Periodic Plant Height in Bed Planted *durum* Wheat

Treatments	*Plant Height (cm)*			
	Days After Sowing			
	45	*75*	*105*	*135*
Nitrogen management (ha^{-1})				
120 kg N	28.8	41.9	64.2	77.8
120 kg N + 15 t FYM	34.2	44.5	70.2	85.0
150 kg N	30.6	42.7	65.2	79.3
CD (P = 0.05)	4.1	2.1	3.2	3.3
Time of N application				
1/2S + 1/2CRI	30.9	41.8	65.6	79.2
1/3S + 1/3CRI + 1/3T	31.3	43.0	66.2	79.5
1/4S + 1/2CRI + 1/4T	31.6	43.0	67.7	79.6
1/3S + 1/3CRI + 1/3T + 3 per cent US	31.1	43.7	67.0	82.4
1/4S + 1/2CRI + 1/4T + 3 per cent US	31.0	43.7	67.0	82.1
CD (P = 0.05)	NS	NS	NS	NS
Interaction	NS	NS	NS	NS
Flat	32.1	40.7	62.0	77.1
CD (P = 0.05) Bed vs Flat	NS	NS	NS	7.6
Bed (S)	30.5	40.5	63.7	78.0

The perusal of data indicated an increase in height of the plants with advancement in age and maximum height was obtained at 135 DAS. The increase in height was much rapid during the period between 75 to 105 DAS.

Nitrogen management significantly influenced plant height throughout crop growth. At 45 DAS minimum height was recorded at 120 kg N ha^{-1} which was significantly lower than 120 kg N + FYM and 150 kg N ha^{-1} both being at par between themselves. Similar trend was followed at 75 DAS. However at 105 and 135 DAS 120 kg N + 15 t FYM ha^{-1} recorded maximum plant height which was significantly higher than other two treatments which were at par statistically. The increase in plant height is attributed to the role of nitrogen in plant metabolism and growth (Tisdale, 1985). Similar results have been reported earlier also (Singh and Agarwal, 2001). Significant increase in plant height with application of FYM at 10 t + 120 kg N ha^{-1} over 120/t kg N ha^{-1} was reported by Negi and Mahajan, 2000).

The split application of nitrogen did not affect plant height significantly. The earlier work done by Dhuka *et al.* (1991), Singh (1996) and Uppal (1999) also revealed that plant height did not differ significantly with different times of nitrogen application.

Dry Matter Accumulation

Accumulation of dry matter is good growth index to express the photosynthetic efficiency of plants. The dry matter accumulation (DMA) was not influenced significantly by sowing methods *i.e.* bed planting and flat sowing of wheat (Table 38.3) throughout crop growth. Similar results were reported by Kaur (2000). The maximum DMA was produced during 75–105 DAS (Table 38.3).

Table 38.3: Effect of Nitrogen Management and Time of Nitrogen Application on Periodic Dry Matter Accumulation in Bed Planted *durum* Wheat

Treatments	*Dry Matter (q ha^{-1})*			
	Days After Sowing			
	45	*75*	*105*	*135*
Nitrogen management (ha^{-1})				
120 kg N	6.01	31.44	75.29	90.95
120 kg N + 15 t FYM	6.85	35.12	79.68	97.66
150 kg N	6.21	34.41	77.83	94.99
CD (P = 0.05)	NS	NS	NS	NS
Time of N application				
1/2S + 1/2CRI	6.25	30.85	72.44	84.71
1/3S + 1/3CRI + 1/3T	6.34	34.16	76.29	93.81
1/4S + 1/2CRI + 1/4T	6.47	35.25	75.77	93.97
1/3S + 1/3CRI + 1/3 T + 3 per cent US	6.36	34.01	82.75	99.23
1/4S + 1/2CRI + 1/4 T + 3 per cent US	6.47	34.01	80.76	100.94
CD (P = 0.05)	NS	NS	NS	NS
Interaction	NS	NS	NS	NS
Flat	5.48	31.90	81.98	99.98
CD (P = 0.05) Bed vs Flat	2.28	NS	NS	NS
Bed (S)	6.00	30.20	72.00	82.50

Nitrogen management did not affect the dry matter accumulation by the crop significantly throughout the growth. Application of FYM along with chemical fertilizers (120 kg N + 15 t FYM ha^{-1}) tended to result in higher dry matter accumulation than 120 kg N and 150 kg N ha^{-1} when applied only through chemical fertilizers. This effect of FYM may be due to increased availability of nitrogen to crop for longer period. However, significantly higher dry matter production by application of FYM at 15 t ha^{-1} along with recommended dose of fertilizer (120 kg N ha^{-1}) have been reported by Nehra *et al.* (2001).

The effect of split application of nitrogen was also found non-significant on dry matter accumulation throughout the crop growth. Application of nitrogen through three splits and three splits + urea spray tended to accumulate more dry matter than two splits and whole N applied at the sowing time due to the reason that the application of nitrogen in three splits in light textured soils made the nitrogen available to plants for longer period and synthesis of photosynthates continued and thus increased the dry matter accumulation CGR followed the dry matter accumulation pattern of the crop. Similar results have been reported by Uppal *et al.* (2000) and Singh (1996).

Canopy Temperature

The data on canopy temperature are given in Table 38.4. In general the canopy temperature increased with increase in age of crop. At 45 and 75 DAS the canopy temperature on an average was lower in flat sown as compared to bed sown crop except at 135 DAS. At 135 DAS LAI was less due to more senescence of leaves. Similar results have also been reported by Kaur *et al.* (2001). More LAI captured higher radiation and resulted in lower temperature.

Table 38.4: Effect of Nitrogen Management and Time of Nitrogen Application on Periodic Canopy Air Temperature in Bed Planted *durum* Wheat

Treatments	*Canopy Air Temperature (°C)*			
	Days After Sowing			
	45	*75*	*105*	*135*
Nitrogen management (ha^{-1})				
120 kg N	16.5	20.1	24.1	28.6
120 kg N + 15 t FYM	15.8	19.4	22.1	27.4
150 kg N	15.6	19.8	22.8	28.0
Time of N application				
1/2S + 1/2CRI	17.2	20.6	23.3	28.6
1/3S + 1/3CRI + 1/3T	15.5	20.3	23.1	28.0
1/4S + 1/2CRI + 1/4T	15.9	19.6	23.0	28.4
1/3S + 1/3CRI +1/3T + 3 per cent US	15.6	19.0	22.8	27.6
1/4S + 1/2CRI + 1/4T + 3 per cent US	15.6	19.3	22.9	27.3
Flat	14.5	18.6	22.0	28.9
Bed (S)	15.0	18.9	22.9	28.0

Among nitrogen management, with increase in nitrogen level from 120 kg N to 150 kg N ha^{-1} the canopy air temperature decreased and the lowest was recorded in 120 kg N + 15 t FYM ha^{-1} throughout the crop growth. It may be due to more nitrogen availability which helped in sustaining higher LAI and interception of high solar radiation resulting in lower temperature. Grewal (1987) has reported decreased canopy temperature with increased nitrogen level from 40 to 200 kg N ha^{-1}.

With the increase in the number of splits the canopy air temperature decreased. The maximum value throughout the crop growth was recorded with two split application with a general order of two splits > three splits. Almost same canopy temperature values were obtained with whole N dose applied at sowing on beds. While after 3 per cent urea spray at flowering stage, the air temperature recorded at 105 and 135 DAS showed the trend as two splits > three splits > three splits + urea spray.

Soil Temperature

The data pertaining to soil temperature are given in Table 38.5. In general, the soil temperature remained lower in flat sown as compared to bed planted wheat. Light was falling due to border effect on all rows through vacant space in furrows and keeping the soil warm in bed sown crop. Similar results have been reported by Kaur (2000).

Table 38.5: Effect of Nitrogen Management and Time of Nitrogen Application on Periodic Soil Temperature in Bed Planted *durum* Wheat

Treatments	*Soil Temperature (°C)*			
	Days After Sowing			
	45	*75*	*105*	*135*
Nitrogen management (ha⁻¹)				
120 kg N	15.0	17.7	24.1	28.4
120 kg N + 15 t FYM	14.8	17.2	23.2	27.7
150 kg N	14.9	17.4	23.5	28.2
Time of N application				
1/2S + 1/2CRI	15.0	18.6	25.0	29.0
1/3S + 1/3CRI + 1/3T	14.9	17.2	24.0	28.7
1/4S + 1/2CRI + 1/4T	14.9	17.2	23.8	28.6
1/3S + 1/3CRI +1/3T + 3 per cent US	14.9	17.1	22.8	27.5
1/4S + 1/2CRI + 1/4T + 3 per cent US	14.9	17.1	22.7	27.4
Flat	14.2	16.8	23.0	28.0
Bed (S)	14.0	16.1	23.5	29.0

With increase in nitrogen level from 120 to 150 kg ha^{-1} the soil temperature tended to decrease and it further declined under 120 kg N + 15 t FYM ha^{-1}. The highest temperature was recorded under 120 kg N ha^{-1} and the lowest with 120 kg N + 15t FYM ha^{-1}. The general order was 120 kg N ha^{-1} > 150 kg N ha^{-1} > 120 kg N + 15 t FYM ha^{-1}. This is due to better crop growth and interception of more solar radiation by the crop canopy. Grewal (1987) has reported decrease in soil temperature with increase in nitrogen level from 40 to 200 kg N ha^{-1}. The soil temperature recorded was lower where light interception was more because incoming radiation was intercepted by crop and lesser of it reached the ground level. Such results have been reported by Kler and Bains (1985), Kler *et al.* (1987), Dhaliwal and Kler (1995), where light interception showed inverse relation with soil temperature and direct correlation with yield of various crops.

Among the time of N application the soil temperature decreased with increase in number of splits. The value of soil temperature tended to be higher in two split than three split application. After 3 per cent urea spray the trend was two splits > three splits > three splits + 3 per cent urea spray. The application of urea as spray further tended to increase the canopy growth and PAR interception resulting in lower temperature. However, soil temperature on the beds was slightly lower than two and three splits at 45 and 75 DAS and almost same at 75 and 135 DAS where whole N was applied at sowing, there were no differences statistically.

Relative Humidity

The periodic data on relative humidity was taken at regular intervals and are presented in Table 38.6. The relative humidity tended to be higher in case of flat sown compared to bed planted crop. Among the nitrogen management treatments, with increase in nitrogen level from 120 to 150 kg ha^{-1} there was increase in relative humidity and it further increased under 120 kg N + 15 t FYM ha^{-1} through out the crop growth. Increase in relative humidity with increase in nitrogen level has been reported by Kaur *et al.* (2001). As PARI (per cent) increased by the crop, relative humidity also increased.

Table 38.6. Effect of Nitrogen Management and Time of Nitrogen Application on Periodic Relative Humidity in Bed Planted *durum* Wheat

Treatments	*Relative Humidity (%)*			
	Days After Sowing			
	45	*75*	*105*	*135*
Nitrogen management (ha^{-1})				
120 kg N	50.4	79.8	77.0	54.5
120 kg N + 15 t FYM	54.2	86.4	79.2	61.0
150 kg N	51.2	83.0	77.8	57.0
Time of N application				
1/2S + 1/2CRI	48.0	81.0	76.3	53.0
1/3S + 1/3CRI + 1/3T	52.7	83.7	78.1	56.8
1/4S + 1/2CRI + 1/4T	53.0	83.3	78.2	57.0
1/3S + 1/3CRI + 1/3T + 3 per cent US	52.8	83.6	78.7	57.1
1/4S + 1/2CRI + 1/4T + 3 per cent US	53	83.3	78.7	57.2
Flat	53	83.5	79.5	54.5
Bed (S)	48.5	81.0	77.5	53.2

With increase in the number of N splits there was increase in the relative humidity. Minimum values of relative humidity were recorded under two split application of nitrogen followed by three split doses at 45 and 75 DAS where as at 105 DAS and 135 DAS after 3 per cent urea spray it was further reduced in the plots receiving urea spray. Whole N on beds applied at sowing recorded almost similar values of relative humidity as two splits.

Photosynthetically Active Radiation Interception

The visible light interception (PARI per cent) increased with increase in crop age (Table 38.7) upto 105 days and found slightly less at 135 days of crop age. The effect of sowing methods on PARI (per cent) was found significant. The PARI (per cent) was recorded significantly higher in flat sown crop than bed planted crop at 45 and 75 DAS and tended to be higher at 105 DAS whereas at 135 DAS PARI (per cent) was slightly less in flat sown crop. Significantly higher PARI (per cent) by flat sown over bed planted crop throughout the crop growth has been reported by Kaur (2000). Lower interception of visible radiation by the bed planted crop supports the fact that the bed planting system of sowing has border effect on all the rows of the crop.

Table 38.7: Effect of Nitrogen Management and Time of Nitrogen Application on Photosynthetically Active Radiation Interception in Bed Planted *durum* Wheat

Treatments	*Photosynthetically Active Radiation Interception (PARI per cent)*			
	Days After Sowing			
	45	*75*	*105*	*135*
Nitrogen management (ha^{-1})				
120 kg N	40.2	58.7	74.2	56.2
120 kg N + 15 t FYM	48.4	66.3	75.6	62.2
150 kg N	41.1	62.1	74.6	58.5
CD (P = 0.05)	5.4	6.75	NS	NS
Time of N application				
1/2S + 1/2CRI	40.4	56.5	72.0	55.2
1/3S + 1/3CRI + 1/3T	43.6	63.2	73.3	57.1
1/4S + 1/2CRI + 1/4T	44.7	64.2	75.2	60.0
1/3S + 1/3CRI + 1/3T + 3 per cent US	44.0	63.5	75.5	60.9
1/4S +1/2CRI + 1/4T + 3 per cent US	44.1	64.3	77.0	61.6
CD (P = 0.05)	NS	NS	NS	NS
Interaction	NS	NS	NS	NS
Flat	74.4	98.0	98.4	88.4
CD (P = 0.05) Bed vs Flat	11.9	1.46	NS	13.3
Bed (S)	49.9	60.0	71.0	54.9

The nitrogen management had a significant effect on PARI (per cent) by the crop at initial stages of crop growth at 45 and 75 DAS. The highest values for PARI (per cent) were recorded at 120 kg N + 15 t FYM ha^{-1} throughout the crop growth. At 45 DAS the PARI (per cent) by the crop with 120 kg N + 15 t FYM ha^{-1} (48.4 per cent) was significantly higher than both 120 (40.2 per cent) and 150 kg N ha^{-1} (41.1 per cent). However at 75 DAS the 120 kg N + 15 t FYM intercepted significantly higher PAR (66.3 per cent) than 120 kg N ha^{-1} (58.7 per cent) but was statistically at par with 150 kg N ha^{-1} (62.1 per cent). While at 105 and 135 DAS the PARI (per cent) was not affected significantly by nitrogen management and the observed order of values was 120 kg N + 15 t FYM > 150 kg N ha^{-1} > 120 kg N ha^{-1}. Addition of FYM continued the available supply of nutrients to plant photosynthetic apparatus showing more LAI and evidently more capture of radiant energy.

The time of nitrogen application did not influence the PARI (per cent). However with increase in number of splits from two to three the PARI (per cent) tended to increase. After the 3 per cent urea spray the observed order was two splits < three splits < three splits + 3 per cent urea spray. More LAI captured higher light and hence higher PARI (per cent) values. Thus showing that urea spray helped in maintaining the leaf area for longer period preventing senescence. Application of whole N at sowing on beds also resulted in comparable value of PARI (per cent). With increase in number of splits, nitrogen was available to plants for longer period helping delayed senescence of leaves for trapping more PARI (per cent) at later stages of crop growth.

Grain Yield

The ultimate effect of experimental variables is reflected in the final yield of the crop and thus it is a major criterion to identify the efficiency of various treatments in a given situation. The efficiency of different factors can be judged mainly by the addition they make to yield which is the ultimate aim of the field research investigations. The data on the grain yield as affected by various treatments are given in Table 38.8. The grain yield of bed planted *durum* wheat was statistically on par with the flat sown crop.

Table 38.8: Effect of Nitrogen Management and Time of Nitrogen Application on Grain Yield and Harvest Index in Bed Planted *durum* Wheat

Treatments	*Grain Yield (q ha^{-1})*	*Harvest Index*
Nitrogen management (ha^{-1})		
120 kg N	38.33	0.379
170 kg N +15 t FYM	42.54	0.380
150 kg N	39.19	0.372
CD (P = 0.05)	NS	–
Time of N application		
1/2S + 1/2CRI	38.33	0.373
1/3S + 1/3CRI + 1/3T	39.00	0.372
1/4S + 1/2CRI + 1/4T	39.78	0.379
1/3S + 1/3CRI + 1/3T + 3 per cent US	41.12	0.378
1/4S + 1/2CRI + 1/4T + 3 per cent US	41.88	0.383
CD (P = 0.05)	NS	–
Interaction	NS	–
Flat (Rec.)	39.53	0.377
CD (P = 0.05) Bed vs Flat	NS	–
Bed (S)	38.31	0.373

Nitrogen management levels did not influence the grain yield significantly. However, the application of combined use of 15 t FYM with 120 kg N ha^{-1} as fertilizer in bed planted wheat tended to increase the grain yield over both 120 kg N ha^{-1} (4.21 q ha^{-1} or 9.9 per cent) and 150 kg N ha^{-1} (3.35 q ha^{-1} or 7.9 per cent) as fertilizers. This increase in the yield was due to cumulative increased effect of various yield attributing characters in 120 kg N + 15 t FYM ha^{-1} treatment (Negi and Gulshan, 2000) also reported 10.5 per cent increase in wheat yield with FYM application over no FYM application. Similar results were reported by (Nehra *et al.*, 2001; Thakur *et al.*, 1999) and (Singh and Agarwal, 2001). Time of nitrogen application did not influence the grain yield significantly and all treatments were at par statistically. Similar results were reported by (Robinson *et al.*, 1979; Dhuka *et al.*, 1991; Mairona *et al.*, 1992; Singh, 1996; Uppal, 1999).

Harvest Index

The data regarding harvest index were presented in Table 38.8. The harvest index values for bed planted and flat sown crop were almost same. Among nitrogen management treatments the highest harvest index value was obtained with 120 kg N + 15t FYM ha^{-1} (0.380) followed by 120 kg Nha^{-1} (0.379) and lowest was recorded in 150 kg N ha^{-1} (0.372). Thus showing that increasing the level of nitrogen through FYM manure increased or at least maintained harvest index when compared with 120 kg N ha^{-1} whereas when nitrogen level was increased only through fertilizer at 150 kg N ha^{-1} the harvest index decreased because there was more increase in straw yield than grain yield. Slow availability of nitrogen from FYM improved the yield attributes and thus grain yield.

Time of N application did not influence the harvest index value and 3 per cent urea spray tended to increase the harvest index marginally. When whole dose of N was applied at sowing on beds almost similar harvest index value was obtained as compared with two split application on beds and flat.

Thus it may be concluded that the performance of *durum* wheat with respect of level of nitrogen management and time of nitrogen application under bed and flat sowing was statistically on par between each other. Under bed sown, management of nitrogen through integration of chemical fertilizers with FYM (120 kg N + 15t FYM ha^{-1}) tended to improve yield attributes and yield over alone chemical fertilizers in *durum* wheat. With increase in times of nitrogen application from whole dose of N applied at sowing to two splits, three splits and three splits + 3 per cent urea spray, there was tendency of beneficial effect on growth and microclimatic parameters tending to improve grain yield in bed sown *durum* wheat.

References

Dhaliwal, G. S. and Kler, D. S., 1995. *Principles of Agricultural Ecology.* Himalaya Publishing House, Bombay.

Dhiman, S.D., Mohan, R.D.S. and Sharma, H.C., 1982. Evaluation of nitrogen requirement for triple gene dwarf wheat. *Indian J. Agron.*, 27: 383–393.

Dhuka, A.K., Sadaria, G.S., Patel, J.C. and Vyas, M.N., 1991. Response of irrigated late sown wheat to nitrogen application. *Indian J. Agron.*, 36: 276–277.

Grewal, D.S., 1987. Microenvironmental parameters in wheat field as influenced by geometric pattern and nitrogen fertility. *Indian J.Ecol.*, 14: 72–77.

Kaur, G., 2000. Studies on planting techniques at higher nitrogen nutrition for microclimate, lodging, growth and yield of wheat (*Triticum aestivum* L.). *M.Sc. Thesis*, Punjab Agricultural University, Ludhiana, India.

Kaur, G., Kler, D.S. and Singh, S., 2001. Influence of planting techniques and nitrogen levels on leaf area index, dry matter, canopy air temperature and relative humidity of wheat (*Triticum aestivum* L.). *Indian J. Environ. Ecoplan.*, 5(2): 265–269.

Kler, D.S. and Bains, D.S., 1985. Relationship between soil temperature and light interception in *durum* wheat (*Triticum durum* Desf.) under different sowing pattern. *Haryana J. Agron.*, 1: 107–112.

Kler, D.S., Dhaliwal, J.S. and Atwal, A.S., 1987. Effect of ploughing frequency, stand geometry and nitrogen levels on the performance of *durum* wheat in Bajra-wheat rotation. *Transactions of Indian Society of Desert Technology,* 12: 21–26.

Kler, D.S., Walia, A.S. and Singh, S., 1987. Utility of light interception and soil temperature relations for some crops as influenced by directional orientations of sowing. *Environ. Ecol.*, 5: 422–429.

Negi, S.C. and Mahajan, G., 2000. Effect of FYM, planting methods and fertilizer levels on rainfed wheat. *Crop Res.*, Hisar, 20: 534–36.

Nehra, A.S., Hooda, I.S. and Singh, K.P., 2001. Effect of integrated nutrient management on growth and yield of wheat (*Triticum aestivum*). *Indian J. Agron.*, 46(1): 112–117.

Singh, K., 1996. Effect of irrigation regime and nitrogen levels on growth, yield and quality of *durum* wheat (*Triticum durum* desf.). *M.Sc. Thesis*, Punjab Agricultural University, Ludhiana, India.

Singh, R. and Agarwal, S.K., 2001. Growth and yield of wheat (*Triticum aestivum*) as influenced by levels of farmyard manure and nitrogen. *Indian J. Agron.*, 46(3): 462–467.

Sushila, R. and Giri, G., 2000. Influence of farmyard manure, nitrogen and biofertilizers on growth, yield attributes and yield of wheat (*Triticum aestivum*) under limited water supply. *Indian J. Agron.*, 45(3): 590–595.

Tisdale, S.L., Nelson, W.L. and Beaton, J.D., 1985. *Soil Fertility and Fertilizer.* MacMillan Publishing Company, New York.

Uppal, R.S., 1999. Effect of levels of NPK and time of nitrogen application on yield and quality of *durum* wheat (*Triticum durum* Desf.). *Ph.D. Dissertation*, Punjab Agricultural University, Ludhiana, India.

Uppal, R.S., Singh, J. and Kler, D.S., 2000. Effect of split application of nitrogen on growth and yield of *Durum* wheat. *Indian J. Environ. Ecoplan.*, 3: 527–534.

Chapter 39

Occupational Health Problems Observed Among the Saltpan Workers of Thoothukudi District, South East Coast of India

R. Sornaraj, A. Amutha Jaisheeba and A. Pushparaj
Research Department of Zoology, Kamaraj College, Thoothukudi – 628 003

ABSTRACT

A survey was made extensively to study the saltpan environmental impact on the workers. The information were collected from 490 respondents working in different areas of saltpan located in and around Thoothukudi. The various physical and physiological disorders of the workers were screened in cooperation with physician. Among the various disorders observed, the eye disorder ranks first and its percentage about 45.91. Next to eye disorder, the other common disorders observed are respiratory disorder (23.46 per cent), Circulatory disorder (20.40 per cent), skin disorder (18.36 per cent) and renal disorder (9.18 per cent). The number of workers get affected in relation to the increase of age and also the year of experience of the workers. Female workers are more affected than the male workers.

Keywords: *Saltpan environment, Physiological disorder, Eye disorder, Respiratory disorder, Skin disorder, Circulatory disorder.*

Introduction

To obtain the daily bread and to lead a healthy comfortable life, every individual in the globe engages himself in any one of the available jobs. Most of the jobs, the people are doing are not providing sufficient money and satisfaction. Most of the jobs are not even providing a healthy condition to the workers. Knowingly or unknowingly the workers are suffering a lot due to the bad environmental conditions in theirs working place. Occupational health is an important one in the working spot. The prevention of disease and maintenance of the highest degree of physical, mental and social well-being of workers in all occupations is essential (Park 1997).

India with about 61,000 km long coastal line stands fourth in production of salt in the world with an average annual production of 13.5 million tones. In India Gujarat, Tamil Nadu and Rajasthan contribute 95 per cent of the total salt production. Thoothukudi district constitutes 70 per cent of the total salt production of Tamil Nadu and meets 30 per cent of salt requirement of our country.

Thoothukudi center produces an average of 25 lakh tones per year. It stands next to Gujarat in salt production in the country. This production comes from about 3100 manufactures working in an area of 4100 acres including one public sector Unit (Tamil Nadu Salt Corporation) and one joint sector unit. Both sea and the subsoil brine are the chief sources in Thoothukudi. Being an intensive labour oriented industry; the salt industry has vast employment potential. In India, it provides employment to the tune of 20 million people.

Salt industry in Thoothukudi district provides large-scale employment to laborers. Every year during January and February the saltpans are prepared for the salt production *i.e.* mending of bunds, desilting of channels, preparation of crystallizer bed. All the salt manufacture related works are attended only manually. As the manual works are tough, no children is employed as child labourer in the salt field.

To excavate the various types of occupational hazards found among the saltpan workers of Thoothukudi, an observation was made and valuable finding were obtained.

Materials and Methods

General physiological conditions of the saltpan workers were collected by randomly selecting 490 workers and interviewing them with a prepared standardized questionnaire. The collected information was counter checked and confirmed with the help of a medical practitioner. The major disorders observed among the workers are skin disease, eye disorder, respiratory disorder, renal disease and circulatory disorders. The information were collected from 5 major salt producing spots namely Thoothukudi, Mullakkadu, Muthaiahpuram, Pullavezhi, Mathagiri and Mudukkukadu situated within the 25 km radius of Thoothukudi. Since the workers are illiterates, the information were collected by interviewing them and the questionnaire was filled up by the researcher in the interview spot itself. Every day around 20 respondents were met and the information were collected. Eye disorders are identified using the ophthalmic kit and the blood pressure of the workers are identified using sphygmomanometer. The statistical tool chi-square was adopted to test the validity of the results.

Results

In the saltpan Industry, workers of different age groups are involved. Their work is very hard and unsafe. They are working for about 10–12 hours in a scorching sun. The environmental condition in the working place is very poor. The job causes lot of stress and unhappiness to the workers.

Physical disorder is a notorious feature among saltpan workers. Cent per cent of the workers are affected by anyone type of pain. Apart from general body pain the workers are also affected by anyone of the other types of physical disorders. The various disorders observed among the saltpan workers are given in Table 39.1 and Figures 39.1 to 39.4. The general disorders observed was general body pain, eye disorder, skin disorder, respiratory disorder, renal disorder and circulatory disorder. Regarding eye disorder the long sight, short sight and reddening of the eye was observed. Regarding skin disorder most of them are affected by dry and scaly skin (keratosin). As far as circulatory disorder is concerned the BP of the patients alone identified and it was found that most of them are having low blood pressure. Regarding the respiratory disorder, most of the workers are frequently affected by cold and some of them are having breathing problems.

Table 39.1: Showing the Physically Affected Saltpan Workers in Relation to their Age. The Chi-Square values of various diseases in relation to age of the workers are also represented in the table.

Age	Eye Disease	Percentage	Skin Disease	Percentage	Circulatory Disorder	Percentage	Renal Disorder	Percentage	Respiratory Disorder	Percentage
15–25	20	4.08	–	–	–	–	–	–	10	2.04
26–35	35	7.14	5	1.02	5	1.02	5	1.02	25	5.10
36–45	65	13.27	55	11.22	10	2.04	10	2.04	45	9.18
46–55	50	10.20	10	2.04	35	7.14	5	1.02	20	4.08
56–65	55	11.22	20	4.08	50	10.20	25	5.10	15	3.06
Total	225	45.91	90	18.36	100	20.40	45	9.18	115	23.46
Chi-square value	9.27 P > 0.05 S		37.37 P > 0.05 S		5.84 P > 0.05 S		17.37 P > 0.05 S		20.56 P > 0.05 S	

Figure 39.1 clearly tells of the physical disorders that commonly occur among the saltpan workers. The common physical disorders observed among the saltpan workers are eye disease, skin disease, circulatory disorder, renal disorders and respiratory disorders in the percentage of 45.95, 18.36, 20.40, 9.18 and 23.00 respectively. Among the various disorders observed, the eye disease is in the apex position. The second position is occupied by the respiratory disorders and the third position is occupied by the circulatory disorder. The lowest position is occupied by the renal disorders.

Figure 39.2 clearly states the relationship between the disorders and years of experience of the workers. The most experienced workers are more affected by eye and circulatory disorder in the percentage of 24.49 and 13.26 respectively. The workers who are having the experience of 10 to 15 years occupy the second position. Among them nearly 10.2 per cent is affected by eye disease.

Figure 39.3 shows the sex-wise distribution of physically affected persons. The workers in salt industry are affected by more than one disease. The Figure 39.2 clearly shows the relationship between the year of experience and the disease appeared among the workers. Age old and experienced workers are affected by more than one disease. By visualizing the Figure 39.3 we can clearly recognize the different types of diseases and the number of individuals affected and so on. Majority of the workers (45.91 per cent) are having eye disorders. The eye disease equally affects both the sex. The percentage of eye defect among male is 20.4 per cent and among female is 25.51 per cent. The respiratory diseases take the second position among saltpan workers. The percentage is about 23.46 per cent. The other diseases such as skin, heart and kidney represents 18.36, 20.40 and 9.18 percentage respectively.

Among the total observed population 54.08 percentage of workers are having good vision. The remaining 45.9 per cent is affected by anyone of the eye disorder. Most of the affected workers are

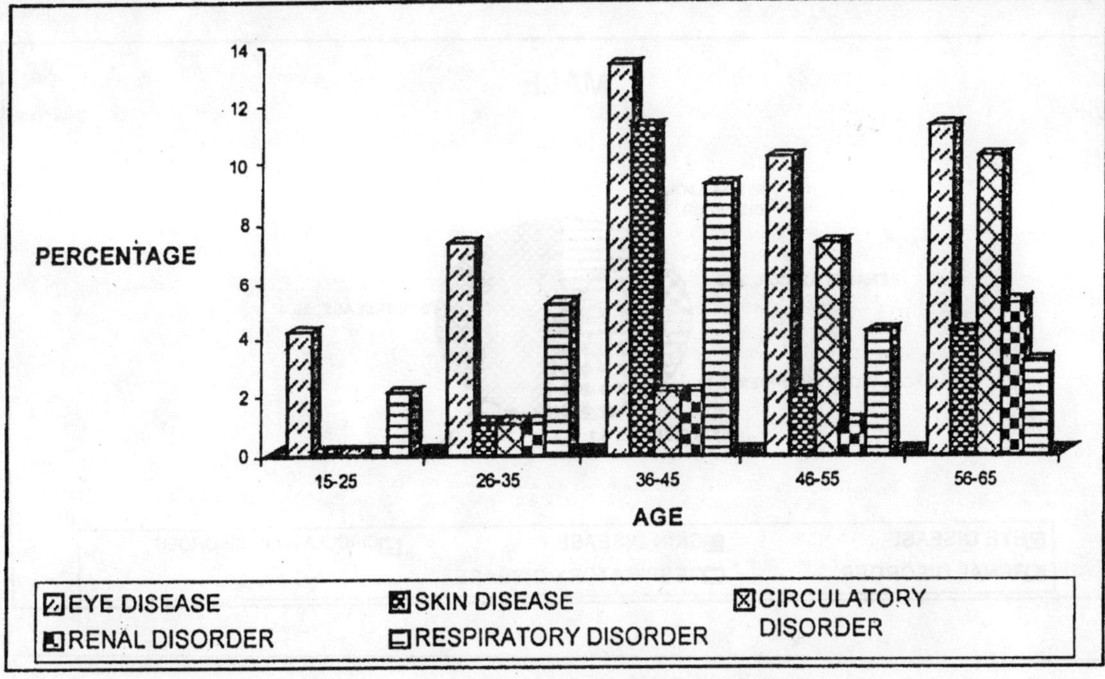

Figure 39.1: Showing the Physically Affected Saltpan Workers in Relation to their Age in the Thoothukudi Area

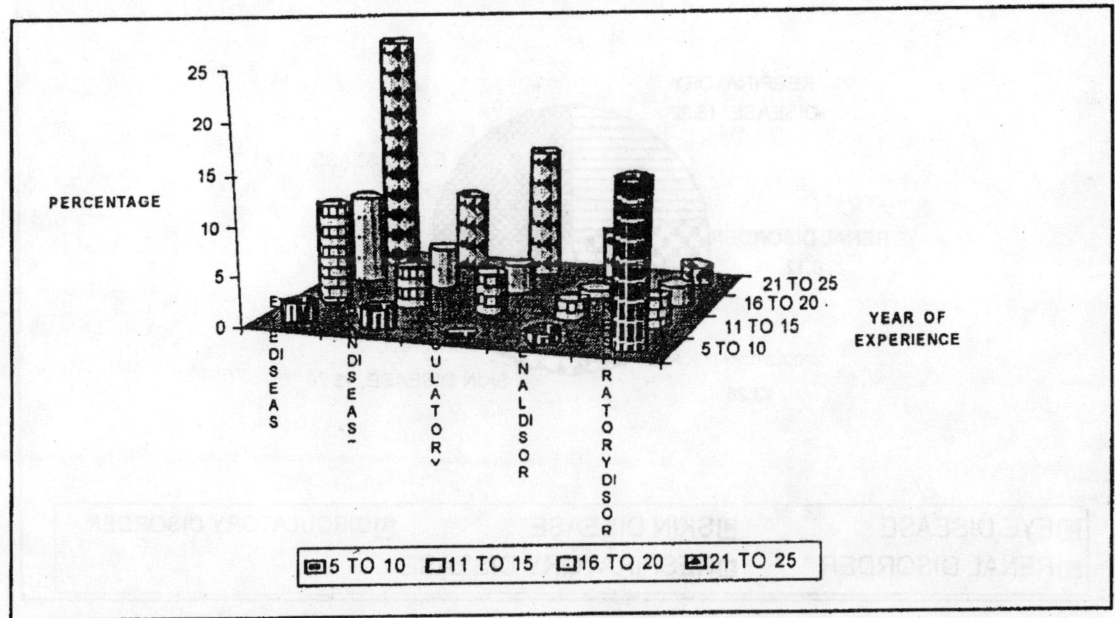

Figure 39.2: Showing the Relationship Between the Experience of the Saltpan Workers and their Nature of Physical Illness in the Thoothukudi Area

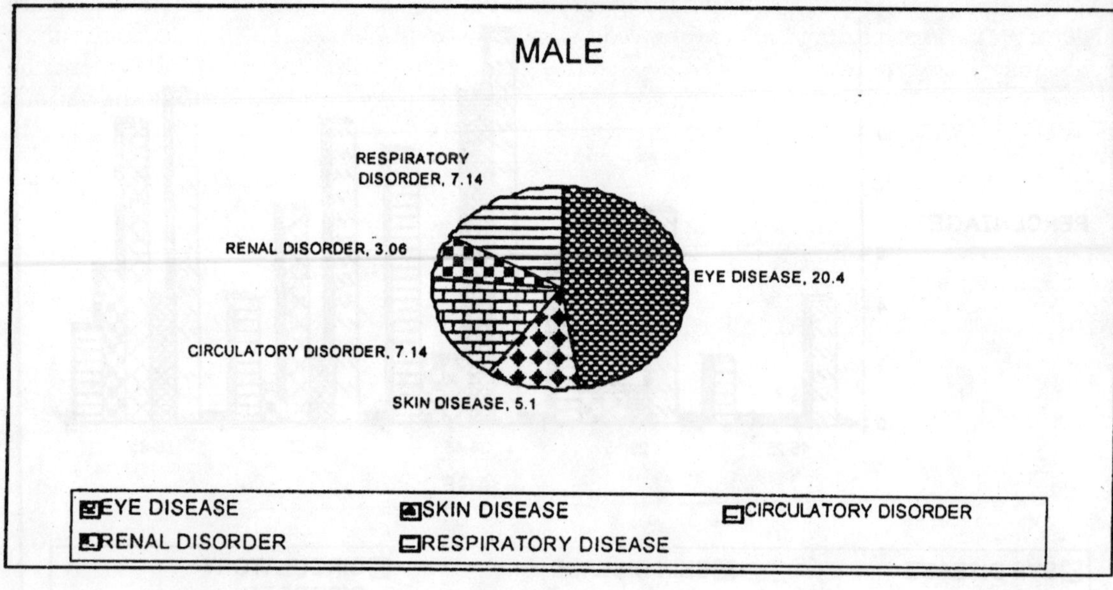

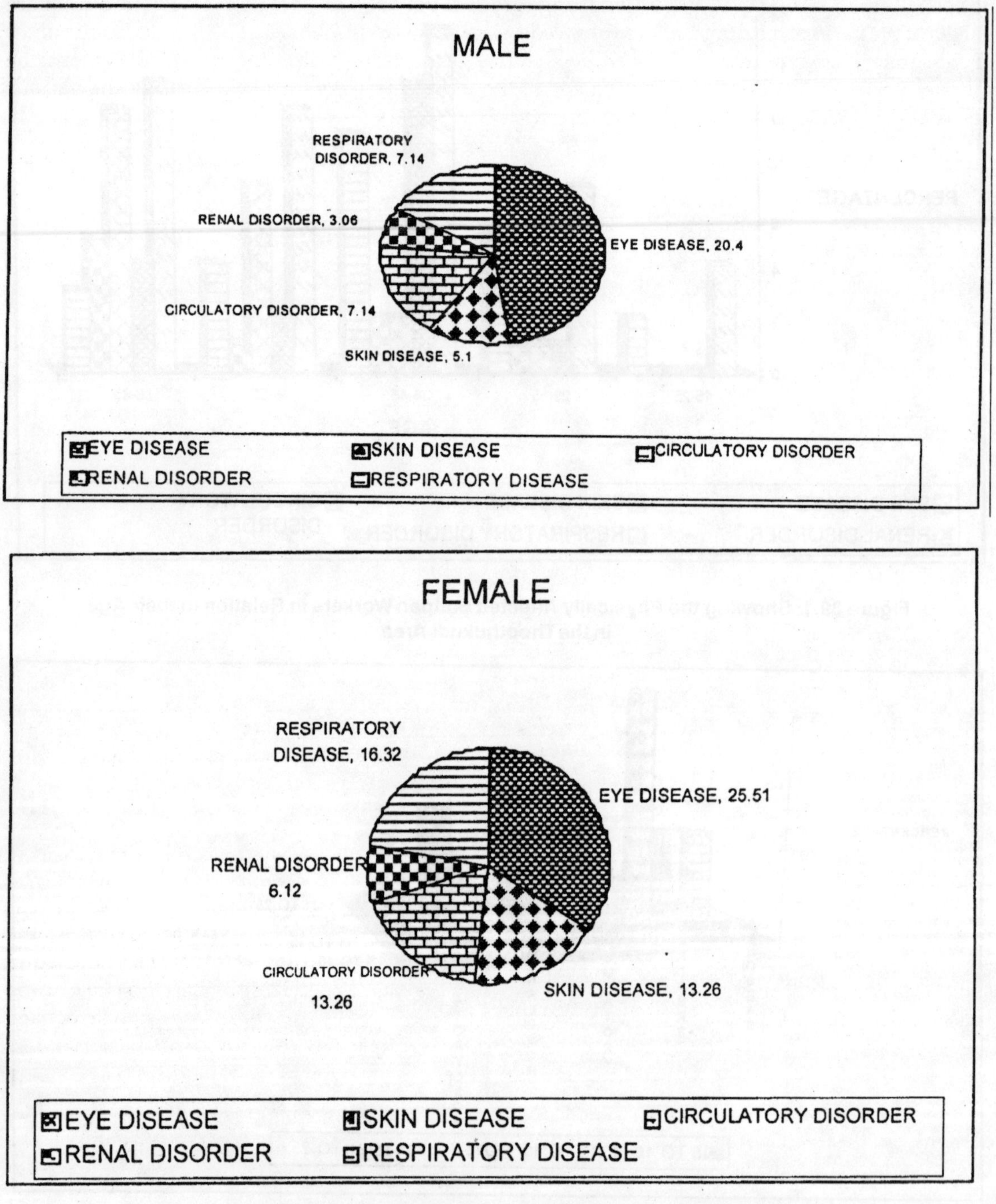

MALE
RESPIRATORY DISORDER, 7.14
RENAL DISORDER, 3.06
EYE DISEASE, 20.4
CIRCULATORY DISORDER, 7.14
SKIN DISEASE, 5.1
EYE DISEASE
SKIN DISEASE
CIRCULATORY DISORDER
RENAL DISORDER
RESPIRATORY DISEASE
FEMALE
RESPIRATORY DISEASE, 16.32
EYE DISEASE, 25.51
RENAL DISORDER 6.12
CIRCULATORY DISORDER 13.26
SKIN DISEASE, 13.26
EYE DISEASE
SKIN DISEASE
CIRCULATORY DISORDER
RENAL DISORDER
RESPIRATORY DISEASE

Figure 39.3: Showing the Sex-wise Distribution of Physically Affected Saltpan Workers in Thoothukudi Area

having short sight (41.83 per cent) only 4 per cent alone are affected by long sight. Eye defect occupies its apex position among the workers whose age group is between 35–45 (Figure 39.4). The chi-square values of the various results calculated showed a significant trend (Table 39.1). The calculated values of chi-square is greater than its table value, that means the various disorders developed among the saltpan workers have relationship with age of the individuals. As the age increased, the various disorders developed among the workers are also increased.

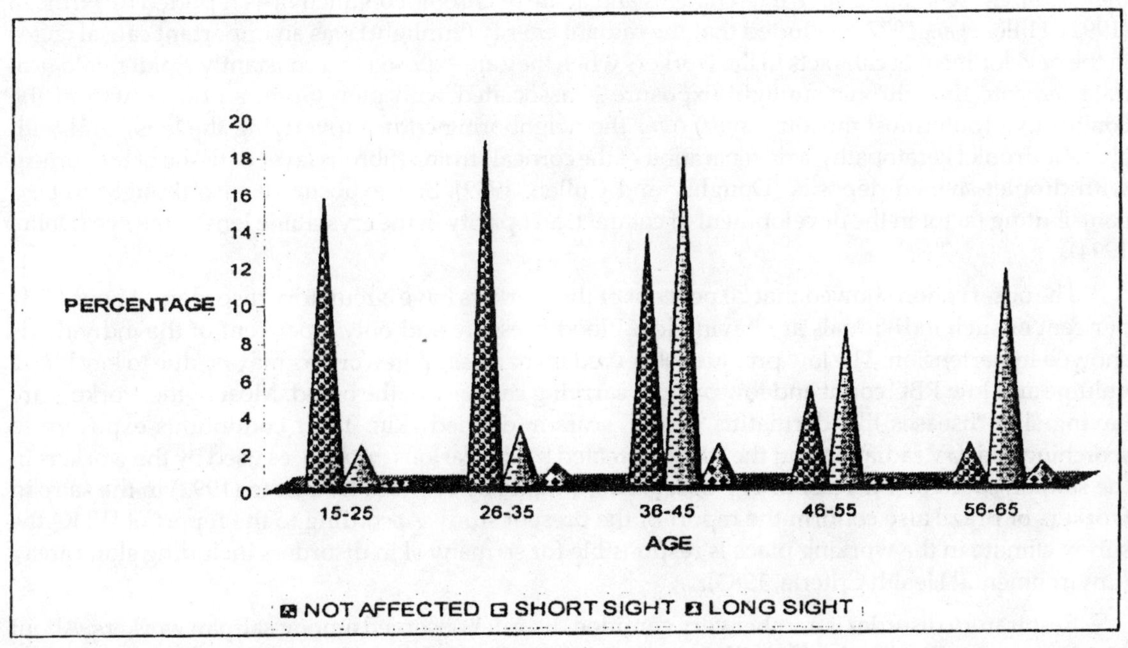

Figure 39.4: Showing the Saltpan Workers who are Affected by Various Eye Disorders

Discussion

The World Health Organization (WHO) has defined health as "physical, mental and social well-being". Well-built physical condition is essential for every individual in a society. There is a common saying, "Health is Wealth". The health of an individual is affected due to so many reasons such as malnutrition, environment in which he is living or working, nature of work and the like. As the saltpan workers are doing a hard and much painful work and are constantly exposed to the salt and saltpan environment, defects in their physical and mental health is inevitable. Generally speaking the physical disorder is a common occurrence in all kind of workers including the women who engage themselves in kitchen and other household works. In case of saltpan workers also the physical disorders are common. But the degree of disorder is entirely different from other kinds of workers.

Irrespective of the age, sex and year of experience the saltpan workers are affected by certain ailment. But the degree of damage corresponds to the year of experience in saltpan environment. The common physical disorders are the eye disease, skin disease, circulatory disorder, renal disorder and respiratory disorders. Most of the workers are very severely affected by the eye, heart and respiratory disorders.

Nearly 41.83 per cent of workers are having short sight and 4 per cent of the workers are having long sight the high percentage of short sight observed among the saltpan workers may be due to the excessive radiation from the saltpan and the heaps of salt crystals. The opinion of the author is supported by other researchers. Finch and Moriyama, (1980) has the opinion that due to severe radiation in the environment the nature of the eye lens and other accessory structures are affected resulted in eye defect. The workers exposed to excessive light rays comprised of ultra violet and infrared rays affect the vision, cause retinal injury, lens changes and acute or chronic conjunctiva as reported by Firmeza (1992). Hiller *et al.*, 1977 concluded that the radiant energy (Sunlight) was an important causal factor in the development of cataracts in the workers when they are exposed to it constantly. Epidemiological data indicate that chronic sunlight exposure is associated with pterygium, an outgrowth of the conjunctiva (outermost mucous layer) over the neighboring cornea (overlying the lens), and with climatic droplet keratopathy, a degeneration of the corneal stroma (fibrous layer of tissue of the cornea) with droplet-shaped deposits (Doughty and Cullen, 1989). Sun exposure is also thought to be a contributing factor in the development of cataract, an opacity in the crystalline lens of the eye (Dolin, 1994).

The observation showed that 20 per cent of the workers have circulatory disorders. About 17.34 per cent of such individuals are having low blood pressure and only 3 per cent of the individuals showed hypertension. The low pressure observed in many saltpan workers may be due to low blood volume and low RBC count and low oxygen carrying capacity of the blood. Most of the workers are having skin diseases like dermatitis, lesion, scars and dried skin. Their continuous exposure to scorching sun, UV radiation and the friction created by the various appliances used by the workers in the saltpan could be the reason. The observations made by Firmeza De Brit to (1992) in the saltpan workers of Brazil also confirm the report of the present study. According to the report of WHO the sunny climate in the working place is responsible for so many skin disorders including skin cancer (Environmental Health Criteria, 1983).

Respiratory disorder is another very common disorder observed among saltpan workers. About 23.46 per cent are frequently affected by common cold and sneezing and breathing problem irrespective of age and sex. The respiratory disorders among the saltpan workers may be due to the deposition of suspended micro particles and salt dust in the nasal cavity and upper respiratory tract during their normal work. During windy days, the inhalation of the salt dust may be very high and that may result in serious respiratory disorders and the deposits may affect the epithelial layers of the respiratory passage and it also affects the easy inhalation resulting in wheezing and other respiratory disorders. Chatterjee (1998) reported that an excessive deposition of dust particles in the nasal passage provokes the fibrous tissue causing permanent disability.

The various physical disorders observed among the saltpan workers in the study area could be due to the continuous exposure of the workers to the saltpan environment, their hard and restless work, under nutrition and absence of awareness about their environment and work. Sidorenko, (1978) has the opinion that the population's health could be an integral index of environmental quality and the effects of various environmental agents. According to Bustueva and Sluchanko, (1979) the effect of some environmental agents is specific, but most are non-specific in their action.

Acknowledgement

The financial support provided by UGC, Government of India, New Delhi is gratefully acknowledged by the authors.

References

Bustueva, K.A. and I.S. Slucanko, 1979. *Methods and Criteria of the Evaluation of the Health of the Population in Relation to Environmental Pollution*. Meditsyna, Moscow.

Chatterjee, C.C., 1998. *Human Physiology*, 11th edn. Medical Allied Agency, Kolkata, Special Reprint, pp. 366–367.

Dolin, P.J., 1994. Ultraviolet radiation and cataract: A review of the epidemiological evidence. *Brit. J. Ophthalmol.*, 78: 4788–482.

Doughty, M.J. and A.P. Cullen, 1989. Long-term effects of a single dose of ultraviolet-B on albino rabbit cornea–I *in vivo* analysis, *Photochem. Photobiol.*, 49: 185– 96.

Environmental Health Criteria, 1983. *Guidelines on Studies in Environmental Epidemiology*. World Health Organization, Geneva.

Finch, S.C. and I.M. Moriyama, 1980. The delayed effects of radiation exposure among atomic bomb survivors, Hiroshima and Nagasaki, 1945–79: A brief summary. Radiation effects research foundation technical report, pp. 16–78

Firmeza De Brito, H., 1992. Pathogenic activity of Sodium chloride dust in eye. In: *Encyclopedia of Occupational Health and Safety*, 3rd edition. International Labour Office, Geneva.

Hiller, R., Giacometti, L., and K. Yuen, 1977. Sun light and cataract, an epidemiological investigation *Am. J. Epidemiol.*, 105: 450–459.

Sidorenko, G.I., 1978. Modern problems of environmental health science. *Gigi. Sanit.*, 10: 9–15.

Chapter 40

Lentic Fishery: Fishery of a Tropical Wetland (Beel) in Assam

Devashish Kar*

Department of Life Science, Assam (Central) University, Silchar – 11, India

ABSTRACT

Fish is an item of staple food of the people of this region. It is also taken of prosperity and peace among many social customs of this region. The North-eastern region of India in general and Assam in particular is very rich in both running and stagnant water bodies. They serve as potential habitats of fish for breeding and feeding. The present communication is an in-depth analysis of the fisheries aspects of the biggest tropical wetland in Assam, *viz.*, the Sone Beel situated in the Barak valley region of Assam bordering Bangladesh.

Keywords: *Wetland, Beel, Fishery, Assam.*

Introduction

Fishes mean different things to different people. To many, fish is best known as an item of food. To a fisherman, fish is a source of earning livelihood. To an amateur, it is an article of hobby or sport. To an ichthyologist, a true fish is a cold-blooded vertebrate having gills for respiration and paired and unpaired fins for locomotion and balancing.

The occurrence of fish had been known to the Indians since three millennium BC (Hora, 1956). Remains of fish, with indications of their use as food, have been obtained from excavations at

* President, Conservation Forum, Silchar; E-mail: devashishkar@sify.com.

Mohenjodaro and Harappa, which dates back to 2500–1500 BC. King Somesvara (1127 AD) was the first to record the common sport fishes of India. Bloch's (1785) work was the first modern record on Indian fishes. This was followed by Lacepede's book (1798–1903). Hamilton (1822) described 269 ichthyospecies from the Ganges and its tributaries. Day's work (1878, 1889) eclipses the taxonomy and distribution of almost all the fish groups of India. Recent contributions on the fish and fish groups of India, in general, have been made by Hora (1920, 1924, 1930, 1935, 1940, 1945, 1956), followed and extended by Misra (1959), Menon (1951, 1974), Jayaram (1981), Jhingran (1993). On the other hand, significant works on the lentic and lotic water fishes of North-East (NE) India, in particular, have been done by Dey (1973, 1981), Dey and Kar (1989 a,b,c, 1990), Sen (1982), Lahon (1983), Kar (1990, 1999, 2000a,b, 2002, 2003a,b, 2004, 2005a,b,c,d,e), Kar and Dey (1982a,b, 1986, 1987, 1990a,b,c, 1991, 1993, 1996, 2000, 2002), Kar *et al.* (1995, 1996, 1998, 2002a,b,c, 2003, 2004, 2005), Kar and Barbhuiya (2000a,b, 2004), Dhar *et al.* (2004), Sen (1982), Lahon (1983), Goswami (1985), Nath (1987), Yadava (1987), Nath (1989), Yadava *et al.* (1989).

Sampling and data collection were done for two years (January 2002–December 2003). Fish identification and the Biodiversity of fishes have been studied after Menon (1974, 1999), and Jayaram (1981, 1999). Data on annual fish yield have been extrapolated from daily catch statistics recorded at all the landing stations of the Beel (FAO, 1974) while, the trend, population dynamics and cyclic variations have been constructed by 12 month's moving average method (Coxton and Cowden, 1950). The fish catching devices have been studied after Brandt (1980) and Dey (1981). The physico-chemical characteristics of water and soil have been analysed after APHA (1995) and Jackson (1973) respectively. Other limnological parameters have been studied after Welch (1948).

Lentic Potential of Assam

The simple aspect of ichthyology is the logical unity of theoretical and practical knowledge of fish to produce efficient and rational methods of raising more fish. Contrary to such background, the naturally-formed lakes and wetlands constitute great potential of fishery resources in India. Of the 0.72×10^6 ha lake coverage in India, the North-eastern (NE) region of the country, particularly, Assam, portrays significant potentialities of such water bodies for productive fisheries. Besides lotic territories, the region is gifted with myriads of different kinds of wetlands, which could be broadly of three categories, *viz.* (a) perennial ones (locally called 'Beel') and (b) seasonal ones (locally called 'Haor') and (c) river-formed oxbow types (locally called 'Anua') (Kar, 2002). All these constitute *c* 81 per cent of the total lentic area ($> 0.12 \times 10^6$ ha) in Assam (Kar, 2005c).

In Assam, there are *c* 1392 numbers of such wetlands having a total of *c* 22,896 number of fisheries of different categories. Out of this, the number of registered wetlands is only 394 (30.38 per cent) covering an area of *c* 70,000 ha; of which, *c* 19,000 ha is in good condition; *c* 15,000 ha is in semi-derelict condition and *c* 35,000 ha is in derelict condition (Anon, 1980). The remaining wetlands are mostly unregistered and under the control of both public and private sectors (Kar, 2003a).

In this context, it is interesting to note that the district of Cachar, Karimganj and Hailakandi covers the highest number of such unregistered wetlands in Assam. But, inspite of the vast resources, the existing fish production rate in Assam is < 50 per cent of the total requirement based on a minimal consumption of 5 kg/person/year. Absence of adequate ichthyological data is one of the reasons why fish culture is not encouraged in these vast water bodies although the fish prone water area constituted by such wetlands, otherwise, justifies the tremendous potentialities of the Beel fisheries of the region (Kar, 2000a). The importance of these Beels is also judged from the fact that the Beels directly act as prospective spawning grounds of large growing riverine fishes, particularly, the coveted Indian major carps (IMC). It is with this basic aim that a detailed pilot study had been conducted on the Limnology and fishery of Sone Beel, the biggest wetland in Assam in order to portray various aspect of its limnology and fishery (Kar, 1990, 2005a). The following is an account of the fishery of Sone Beel.

Fishery, Fish Yield, Fisherfolk and Fishing Efforts of Sone Beel Wetland

The Sone Beel

The 'Sone Beel', situated in Karimganj district (92°24'50" E–24°44'30" N), is the biggest (Δ 3458.12 ha at Full Storage Level) tectonic (Mio-pliocene) Beel in Assam. It is unique in having continuous inflowing and out flowing channels. The Beel has a long shoreline of 35.4 km, a shore development of 1.69, a gross of 101.54 × $10^6 m^3$/sec and a mean depth of 0.29 m. The area of the Beel diminishes to 409.37 ha at the Dead Storage Level (DSL). The continuous inlet and outlet channels of the Beel recorded maximum discharges of 33.91 and 87.07 m^3/sec respectively. The Beel water is circumneutral with pH varying from 6.0 to 7.9, Dissolved Oxygen (DO) ranging from 2.6 to 5.9 mg/lit, Free carbon dioxide (FCO_2) fluctuating from 0.9 to 14.5 mg/lit and Total Alkalinity (TA) lying between 25.0 to 76.0 mg/lit (Dey and Kar 1987; Dey and Kar, 1989c; Kar, 2005e).

Ichthyospecies of the Beel and their Zoogeography

In Sone Beel, 70 valid ichthyospecies belonging to 49 genera under 24 families and 11 orders have been recorded (Dey and Kar 1990; Kar, 2005b). Of these, 59 species (*Chitala chitala, Notopterus notopterus, Amblypharyngodon mola, Aspidoparia morar, Barilius bendelisis, Catla catla, Cirrhinus mrigala, C. reba, Cyprinus carpio, Danio devario, Esomus danricus, Labeo bata, L. calbasu, L. gonius, L. nondino, L. rohita, Puntius chola, P. conchonius, P. sarana sarana, P. ticto, Rasbora daniconius, R. elanga, Salmostoma bacaila, Botia dario, Lepidocephalus guntea, Acanthocobitis botia, Somileptes gongota, Mystus cavasius, M corsula, M tengara, M vititatus, Aorichthys. seenghala, Rita rita, Ompok bimaculatus, Wallago attu, Ailia coila, Pseudeutropius atherinoides, Clupisoma garua, Eutropiichthys vacha, Silonia silondia, Pangasius pangasius, Gagata nangra, Glyptothorax telchitta, Clarius batrachus, Heteropneustes fossilis, Chaca chaca, Channa gachua, C. marulius, C. punctatus, C. striatus, Badis badis, Nandus nondus, Anobas testudineus, Colisa fasciatus, C. lalius, Macrognathus aculeatus, M pancalus, Mastocembelus armatus* and *Tetraodon cutcutia*) belong to the Primary Freshwater Group (Myers, 1949); while, the remaining 11 species (*Pisodonophis boro, Gudusia chapra, Hilsa ilisha, Xenentodon cancila, Aplocheilus panchax, Amphipnous cuchia, Chanda baculis, C. nama, Rhinomugil corsula, Sicamugil cascasia* and *Glossogobius giuris*) belong to the Peripheral Class (Nichols, 1928; Darlington, 1957). In other words, the Primary class formed *c* 84.28 per cent while the Peripheral class made 15.72 per cent of the total fish composition. Significantly, the cyprinids contained 35.59 per cent of the Primary forms. On the basis of distributional pattern, it is more appropriate to categorise families Notopteridae, Nandidae, Anabantidae and Mastocembelidae as Primary forms (Mirza, 1975) rather than as Secondary types (Jayaram, 1974) although these fishes have better salinity tolerance (Sen, 1982). On the basis of Indian and Extra-Indian territorial distribution (Motwani *et al.*, 1962), the fish denizens of Sone Beel include 20 'Widely distributed species' of little significance and eight species of 'Northern India'. Ecomorphologically (Dey, 1973), *G. telchitta* is the only true 'Rheophilic species' occurring in Sone Beel. The 'Semi-torrential group' contained five species with weak and minimal body adaptations. A critical analysis of the migratory propensity of the fishes of Sone Beel revealed that *A. mala, A. testudineus, C. baculis, C. nama, C. punctata, C. striata, G. chapra, H fassilis, L. guntea, M. aculeatus, M. pancalus, M. armatus, A. seenghala, M. vittatus, N. nandus, O. bimaculatus, P. chala, P. canchanius, S. bacaila, W. attu* and *X. cancila* are the 'resident forms' with perennial species *P. chala*, constituting, on average, 24.60 per cent of the total fish yield of the Beel. Among the 'casual visitors', the IMCs enter the Beel for breeding, while *C. garua, E. danrica, E. vacha, M. corsula, M. tengara, R. corsula, R. rita* and *S. gangota* are of 'seasonal occurrence'. Lastly, the 'Plain water group' includes 57 species of the Beel with least corporal modification and insignificant migratory habit (Kar and Dey, 1986; 2005e).

Commercial Production and Composition

The annual average fish yield from Sone Beel has been recorded as 335.18 metric tones (mt), of which, IMC constituted 0.73 per cent, other carps and minnows 29.93 per cent, clupeids 8.31 per cent, featherbacks 0.27 per cent, loaches 1.68 per cent, bagrids 7.94 per cent, siluroids 3.72 per cent, schilbeids 0.88 per cent, pangasiids 0.00112 per cent, sisorids 0.015 per cent, live cat fish 1.92 per cent, chacids 0.025 per cent, needle fish 0.18 per cent, killi fish 0.03 per cent, murrels 7.53 per cent, glass fish 13.91 per cent, perches 7.45 per cent, mullets 0.045 per cent, gobies 6.31 per cent, eels 7.43 per cent, and puffer fish 0.94 per cent (Kar and Dey, 1987; Kar, 2005d).

Average per hectare fish yield from Sone has been estimated as 96.8 kg/ha obtained by Vaas and Schurman (1949) in some tropical freshwaters of Java. The poor fish yield could be assigned due to siltation of the Beel (Welcomme, 1979) exposing the fish in the shallower areas to shortage of living space through an eventual diminution of the water spread area. However, the Beel soil with pH 5.0 to 5.9, organic carbon (°C) 0.25 to 1.74 per cent, available phosphorus 0.15 to 1.93 mg/100 g; and, water with pH 6.0 to 7.9, DO 2.4 to 5.9 mg/lit are not considered very suitable for rich fish production (Banerjea, 1967; Michael, 1969).

IMC Fishery

In Sone Beel, the IMC consists of all the five species (*Catla catla, Cirrhinus mrigala, Labeo calbasu, L. gonius* and *L. rohita*) and its fishery is found not be a detached and isolated one as the crop is dependent on the stocks in the adjoining rivers (Kar, 1990; Dey and Kar, 1990; Kar and Dey, 2000). The landing of IMC from Sone has been recorded to be 2571.52 kg with a per hectare yield of 0.74 kg on the average (Kar, 2005e). The species-wise average monthly relative yield (MRY) have been 0.017, 0.130, 0.26, 0.076, 0.36 per cent respectively. Concomitantly, the species-wise annual relative yield (ARY) successively have been 0.015, 0.02, 0.21, 0.39, and 0.23 per cent during the first year; and, 0.01, 0.13, 0.16, 0.07 and 0.24 per cent during the second year. The IMCs entering Sone Beel varied in size from 19.0 to 66.8 cm for *C. catla;* 12.9 to 65.0 cm for *C. mrigala;* 15.0 to 54.3 cm for *L. calbasu;* 6.8 to 50.0 cm for *L. gonius;* and, 12.0 to 78.5 cm for *L. rohita.* As these fishes prefer a lotic territory to lentic one, in Sone Beel too, they tend to occur in regions which are more fluviatile in nature. Indeed, direction, coupled with ingress of water through the perennial channels could be the influential factors for recruitment of these coveted species during monsoon. Likewise, temperature is not found to have any direct impact on the movement of the IMCs. Their occurrence in Sone Beel almost throughout the year, without much impact being caused by temperature, lends support to this view. Field investigations further revealed that, with the onset of rains, shoals of IMC enter the Beel through the outlet when they are caught in appreciable number without much discrimination in size. Such non-judicious and irrational harvest of the IMC including their juveniles, cause imbalance in the IMC fishery of the Beel culminating in overall poor contribution of this coveted group in the total fish yield of the Beel (Kar, 2005 d,e).

Hilsa

Concomitant to IMC, the Hilsa (locally called 'ilish') also forms an important fishery of Sone Beel and its crop is also related to its stock in the adjoining rivers. In Sone Beel (Kar and Dey, 1982a,b; Dey and Kar, 1989a; Yadava *et al.*, 1989), there is a single run of hilsa with the fish entering the Beel in shoals on the advent of monsoon unlike in other lentic (Jones and Sujansingani, 1951) and lotic (Pillay, 1958) bodies of India. The total landing of hilsa from Sone Beel has been recorded as 163.5 kg during the first year and 220.7 kg during the following year; the trend thereby showed average per hectare yield of 0.055 kg during the investigating period. Concomitantly, the MRY recorded during the first session have been 0.04 (April), 0.23 (July), 0.11 (August), 0.01 (September) and 0.00403 per cent (October); while, during the following session, the MRY values have been 0.07 (May), 0.09 (June), 0.06 (July), 0.04 (August), and 0.37 per cent (September). The ARY have been 0.04 and 0.07 per cent during the first and second sessions respectively. The hilsa entering the Beel varied in size from 30–60 cm and from 23.0–53.4 cm during first and second sessions respectively. Unusual turbidity of water may not always be considered as a criterion for increased rate of natural recruitment of hilsa, because Sone Beel exhibited a good population of the fish within a turbidity of range of 20.5 to 83.6 TU (Dey and Kar, 1987). Migratory behaviour of shad does not seem to be directly influenced by temperature. However, their occurrence in Sone Beel only during April to October indicated that the fish prefer to live in Sone Beel during the hot season when its water is warm. After hilsa enter the Beel through the outlet, they are caught in large number without any discrimination in size. Subsequently, intensive fishing in the Beel does not spare much of the stragglers to perform the return journey.

Concomitantly to Sone Beel, not only hilsa, but also its advance fry stages have been recorded for the first time by Kar and Dey (2002) in Chatla Haor seasonal floodplain wetland in Cachar district of Assam revealing a new vista about the biology and breeding behaviour of the fish.

Conservation of IMC and Hilsa

Protection of both gravid and juvenile fish is the best course for the conservation of IMC and hilsa fisheries. However, the conditions prevalent in Sone Beel exhibit many difficulties for the successful achievement of this objective. The enormity of the Beel coupled with intense fishing throughout the year make it very difficult to restrict fishing and fish capture for any period or to observe a 'closed season' for some months. This would cause immense hardship to the fisher folk in the absence of alternative source of income. Notwithstanding the above limitation, fishing restrictions during the monsoon season along the inlet and outlet belts would probably afford an opportunity for the conservation of these fishes in Sone Beel (Kar, 2002, 2003b, 2005d,e). The Epizootic Ulcerative Fish Disease Syndrome (EUS) has been posing a serious threat to fish sustenance and conservation (Kar, 1999, 2005e; Kar *et al.*, 1995, 2003).

Fishing Centres and Fish Landing Stations

Fishing Centres (FC)

Sone Beel, not being a very deep wetland, the entire waterspread area is used for intensive fishing throughout the year both on domestic (littoral) and commercial (limnetic) basis. While such a trend is more discernible during the Full Storage Level (FSL) of the Beel, eight distinct 'Fishing Centres', locally called 'Bundhs', are leased out principally to the 'Kaibarta' fishing community (Dey and Kar, 1989 b) at Dead Storage Level (DSL) of the Beel every year, mostly during the period November to April. Such fishing zones, which lie at deeper portions of the Beel at DSL, are kept encircled with bamboo mat (height 2–8 m) and fishes are captured and marketed regularly as warranted throughout the period of operation. The names of such 'bundhs' with their fishing areas (FA), months of operation (MO), water level (WL) and fish yield (FY) at the centres, as recorded (Kar, 2005e), are given in Table 40.1.

Table 40.1

Bundh	*FA (ha)*	*MO*	*WL (m)*
Belala	1.12	August–April	0.10–4.00
Chirakhaora	48.50	November–April	0.10–2.00
Nunail	6.50	September–November	2.00–4.00
Putighat	2.00	November–April	0.10–2.00
Kurerkata	15.05	December–March	0.25–1.50
Kanagajar	50.00	October–December	1.50–3.00
Baroitila	105.90	November–April	0.10–2.00
Vashankuri	8.46	December–March	0.25–1.50

Fish Landing Stations

There are three principal fish landing stations (LS) situated along the shoreline of Sone Beel. These are Debodwar LS, Kalibaribazar LS and Phakhoagram LS. The position of LS are shifted with the change in the water level of the Beel. An account of these LS along with the landing area (LA), leasing amount (la), month operation (MO) and quantity of fish landed (QL) in each (Kar, 2005e), is summarized in Table 40.2.

Table 40.2

LS	*LA (m^2)*	*la (Rs)*	*MO*	*QL (mt)*
Debodwar	700	1451.00	May–September	
	2800	6672.00	October–December	159.90
	1900	3500.00	January–April	
Kalibaribazar	400	15937.00	May–September	
	468	16500.00	October–November	
	1491	20000.00	November–April	258.60
Phakhoagram	182	Nil (Railway plot)	Whole year	217.10

Fisher Folk and Fishing Efforts

Fisher Folk

The activities of the fishermen and their socio-economic condition have been influencing the development and utilization of the fishery resources of the wetlands in a significant way. The aspect gained importance in view of the inferior social status and over fishing of the Beels by the unprogressive but simple, poverty-stricken people. In Sone Beel, fisher folk belonging to four principal communities (Dey and Kar, 1989b) inhabit 39 villages around the Beel (Kar, 2000b, 2003b, 2005e). An account of these communities is given in Table 40.3.

Moreover, on the basis of fishing intensity, the fisher folk of Sone Beel have been further classified, after FAO (1962), into three categories, *viz.*, Professional, Part-time and Occasional (Kar, 1990), an account of which, with relevant information (Kar, 2005e), is given in Table 40.4.

Table 40.3

Fishermen Community	Relative Abundance (%) Out of		Fishing Area (ha/fisherman)
	Total Fishermen	Total Inhabitants	
Kaibarta	68.14	47.67	0.32
Patni	14.66	10.24	1.51
Maimal	12.29	8.59	1.81
Namasudra	4.91	3.43	4.53
		30.07 (Non–fishermen)	

Table 40.4

Fisherman Category	Relative Abundance (%) Amongst				Total
	Kaibarta	Patni	Maimal	Namasudra	
Professional	55.04	6.91	5.54	2.51	70.00
Part-time	8.00	3.00	2.60	1.40	15.00
Occasional	4.50	4.75	4.75	1.00	15.00
Total	67.54	14.66	12.89	4.91	100.00

Fishing Efforts and Crafts

Fishing Efforts

24 different kinds of fishing gears and methods under 7 categories (FAO, 1963; Brandt, 1980) have been found in operation in Sone Beel (Kar, 1990, 2005e; Kar and Dey, 1991, 1993, 1996; Kar *et al.*, 1998). An account of these gears along with their CPGH (Catch/Person/Gear/Hour: Dey, 1981) of some of the principal gears is given in Table 40.5.

Table 40.5

Category	Gear (Local Name)	Size (m)	Mesh Size (mm)
I. Hook and Line gears	1. Lar barshi	L. 400–600	—
	2. Kupa barshi	L. 450–500	—
	3. Tanga barshi	L.350–450	—
II. Traps	4. Dori	H. 0.45–1.35 W. 0.30–0.60	Bamboo slits 1 mm apart
	5. Gui	H.0.90–2.00 W. 0.75–1.00	Bamboo slits 10 mm apart
	6. Paran	H. 0.20–0.40 W. 0.10–0.14	Bamboo slits 7 mm apart
	7. Khati bundh	H. 2.00–3.00 L/unit 4.50	Dense bamboo mat

Contd...

Table 40.5–Contd...

Category	*Gear (Local Name)*	*Size (m)*	*Mesh Size (mm)*
III. Trawls	8. Chhatjal	L. 3.50–4.50	Upper part: 4.0
		W. in front	Lower part: 1.0
		2.00–3.00	
	9. Pelain	L. 2.50–3.00	1.0–5.0
		W. in front	
		1.00–2.00	
IV. Scooping gears	10. Dheki jal	L. 8.00–12.00	Wider part: 50.0
		W. in front	Narrower part: 6.0
		7.0–9.00	
	11. Darar jal	L. 4.50–5.00	5.0
		W. 4.00–5.00	
	12. Kuchrung jal	L. 1.25–1.50	5.0
		Dia: 1.00–1.15	
		Cir: 3.00–3.50	
V. Entangling gears	13. Patan jal:		
	Big variety	L. 300–400	46.0
		H. 4.00–4.50	
	Medium variety	L. 300–390	19.0
		H. 3.00–3.50	
	Small variety	L. 300–400	15.0
		H. 1.50–2.00	
	14. Haran jal	L. 4.50–6.50	Upper part: 8.0
		W. of widest portion	Lower part: 6.0
		3.50–4.50	
VI. Encircling gears	15. Maha jal	L. 700–750	Episurface: 5.0
		H. 3.75–6.00	Underwater: 1.0
	16. Dal jal	L. 80–90	1.0
		H. 2.00–5.00	
	17. Chat jal	L. 110–120	1.0–8.0
		H. 2.50–3.50	
	18. Ghuran jal	L. 30.00–40.50	1.0–6.0
		H. 1.50–2.50	
	19. Jhaki jal	L.4.50–5.40	6.0
		Cir.11.25–14.00	
		Dia. 4.00– 4.50	
	20. Tana jal	L. 45.00–67.50	46.0
		H. 3.60– 4.50	

Contd...

Table 40.5–Contd...

Category	Gear (Local Name)	Size (m)	Mesh Size (mm)
	21. Rekh jal	L. 5.00–5.50	35.0–38.0
		Dia. 4.50–5.50	
		Cir. 11.50–15.00	
VII. Miscellaneous	22. Arar jal	L. 3.50–4.50	4.0–5.0
		H. 2.50–2.75	
	23. Fal jal	L. 100–2000	1.0
		H. 3.00–4.00	
	24.Impoundment: with bailing	L. 0.50–5.00	-
	vessel ('Kuin')	W. 0.20–0.45	
	Without bailing Vessel ('Debli')	-	-
	'Dhagamara'	-	-

Crafts and their Salient Features

Craft	Size (m)	Construction	Propulsion	Crew
Plank–built boat		Wooden planks nailed together to form a rigid frame		
Flat bottom:				
Large	L. 25.0–35.0		Manual	20–25
	W. 8.0– 10.0			
	H. 1.5– 2.0			
Small	L. 10.0–20.0		Manual	3–6
	W. 2.0– 4.0			
	H. 0.75– 1.0			
Round bottom:	L.10.0–20.0		Manual	3–6
	W. 2.0– 4.0			
	H. 0.75– 1.0			
Dug-out boat	L. 5.0–10.0	Hollowing-out single	Manual	1–3
(Canoe)	W. 1.0–2.0	log of wood.		
	H. 0.5–0.75			

Concluding Remarks

Beels in Assam are to be considered as gifts of nature, a treasure worthy to adore. Such potential fishery beds are to be properly utilized to the fullest extent through tangible schemes to result in prolific harvest of fish for the region. And, an overall improvement and judicious harnessing of these water bodies will be imperative and have no alternative, if the fish production of the NE region in general and Assam in particular needs to be augmented in real perspective.

References

Anonymous, 1980. *Statistical Handbook of Assam*. Directorate of Economics and Statistics, Government of Assam.

APHA, 1995. *Standard Methods for the Examination of Water and Wastewater*. American Public Health Association (USA), pp. 1193.

Banerjea, S.M., 1967. Water quality and soil conditions of fish ponds in some States of India in relation to fish production. *Indian J. Fish.*, 14: 115–144.

Bloch, M.E., 1785. *Naturgeschichte der auslandischen Fische in 3 Vols*, Berlin.

Brandt, A.V., 1980. *Fish Catching Methods of the World*. Fishing News Books, Ltd., Surrey, UK, pp. 418.

Coxton, F.E. and Cowden, D.J., 1950. *Applied General Statistics*. Sir Isaac Pitman and Sons (London), pp. 944.

Darlington, P.J., 1957. *Zoogeography: The Geographical Distribution of Animals*. John Wiley and Sons, New York, USA, pp. 675.

Day, F., 1878. *The Fishes of India: Being a Natural History of the Fishes Known to Inhabit the Seas and Freshwaters of India, Burma and Ceylon*, 195 pls., Text and Atlas in 4 parts. WM Dawson and Sons Ltd., London, pp. 778.

Day, F., 1889. *The Fauna of British India, including Ceylon and Burma: Fishes*, Vols. I and II, pp. 548 and 509.

Dey, S.C., 1973. Studies on the distribution and taxonomy of the ichthyofauna of the hill streams of Kamrup-Khasi–Garo regions of Assam with special reference to the functional morphology of some rheophillic fishes. *D.Sc. Thesis*, University of Calcutta, India, pp. 299.

Dey, S.C., 1981. Studies on the hydrobiological conditions of some commercially important lakes (Beels) of Kamrup district of Assam and their bearing on fish production, 18 Figs., 8 pls. Final Technical Report, North-Eastern Council, Govt. of India, Shillong, pp. 177.

Dey, S.C. and Kar, D., 1987. Physico-chemical complexes of water and soil in Sone: An ichthyologically potential tectonic lake of Assam. *J. Assam Sci. Soc.*, 30: 1–11.

Dey, S.C. and Kar, D., 1989a. An account of *Hilsa ilisha* (Hamilton) of lake Sone in the Karimganj district of Assam. *Bangladesh J. Zoology*, 17: 69–73.

Dey, S.C. and Kar, D., 1989b. Fishermen of lake Sone in Assam: Their socio-economic status. *Science and Culture*, 55: 395–398.

Dey, S.C. and Kar, D., 1989c. Aquatic macrophytes of lake Sone in Assam. *Environment and Ecology*, 7: 253–254.

Dey, S.C. and Kar, D. 1990. Fish yield trend in Sone, a tectonic lake of Assam. *Matsya*, 15–16: 39–43.

Dhar, N., Goswami, M.M., and, Kar, D., 2004. Cooperative Movement: Need for the hour in capture fishery sector. In: *Cooperative Movement in Assam: Problems and Prospects*, (Eds.) P.J. Goswami, and P.R. Acharjee. G.C. College Employees' Credit and Thrift Cooperative Society Ltd., Silchar, pp. 113–116.

FAO, 1962. Rapport au Government de la Re'publique du Niger sur la situation et' evolution de la pêche au Niger: Base' sur le travail de M. Jaques MA. Daget'. Rapp. FAO/PEAT: 1525: pp. 27

FAO, 1963. Modern fishing gears of the world. In: *Proc. 2nd World Fishing Gear Congress* (London).

FAO, 1974. *Design of Fisheries Statistical Surveys: Inland Waters*. Food and Agricultural Organisation. Fisheries Tech. Pap., 133: 1–122.

Goswami, M.M., 1985. Limnological investigations of a tectonic lake of Assam, India with their bearing on fish production. *Ph.D. Thesis*, University of Gauhati, Assam, pp. 395.

Hamilton, B., 1822. *Account of the Fishes Found in the River Ganges and its Tributaries*. Archibald Constable and Co., Edinburgh, UK, pp. 405.

Hora, S.L., 1920. Revision of the Indian Homalopteridae and of the genus *Psilorhynchus (Cyprinidae)*. *Rec. Indian Mus.*, 19: 195–215.

Hora, S.L., 1924. Fishes of Siju cave, Garo Hills, Assam. *Rec. Indian Mus.*, 26: 27–32.

Hora, S.L., 1930. Ecology, bionomics and evolution of torrential fauna with special reference to the organs of attachment. *Phil. Trans. Roy. Soc. London*, 218(B): 171–282.

Hora, S.L., 1945. Symposium on the factors influencing breeding of carps. *Proc. Nat. Inst. Sci., India*, 11(3).

Hora, S.L., 1956. Fish painting of 3rd Millennium BC from Nal (Baluchistan) and their geographic significance. *Mem. Indian Mus.*, 14: 78–84.

Jackson, M.L., 1973. *Soil Chemical Analysis*. Prentice Hall of India Pvt. Ltd., New Delhi, pp. 498.

Jayaram, K.C., 1974. Ecology and distribution of freshwater fishes, amphibians and reptiles. In: *Ecology and Biogeography in India*, (Ed.) M.S. Mani. W. Junk Publishers, The Hague (The Netherlands), pp. 517–584.

Jayaram, K.C., 1981. *The Freshwater Fishes of India, Pakistan, Bangladesh, Burma, Sri Lanka*. Zoological Survey of India, Calcutta, pp. 475.

Jayaram, K.C., 1999. *The Freshwater Fishes of the Indian Region*. Narendra Publishing House, Delhi, pp. 551.

Jhinghran, V.G., 1997. *Fish and Fisheries of India*. Hindustan Publishing Corporation, Delhi, pp. 727.

Jones, S. and Sujansingani, K.H., 1951. The hilsa fishery of the Chilka Lake. *J. Bombay Nat. Hist. Soc.*, 50: 264–280.

Kar, D., 1990. Limnology and fisheries of Lake Sone in the Cachar district of Assam (India). *Matsya*, 15–16: 209–213.

Kar, D., 1999b. Microbiological and environmental studies in relation to fish Disease in India. *Gordon Research Conference*, Connecticut, USA.

Kar, D., 2000a. Present status of fish Biodiversity in South Assam and Tripura. In: *Fish Biodiversity of North-East India*, (Eds.) A.G. Ponniah and U.K. Sarkar. NBPGR-NATP Publication No. 2. National Bureau of Fish Genetic Resources, Lucknow, pp. 80–82.

Kar, D., 2000b. Socio-economic development of the fisherwomen through aquaculture with emphasis on integrated farming in the villages around Chatla Haor Wetland in Silchar, Assam. Seminar Presented on DBT-sponsored Awarness Workshop on Biotechnology-based Programmes for Women and Rural Development, NEHU, Shillong, 1: 13–14.

Kar, D., 2002. Present status of biodiversity of fishes of Barak valley region of Assam with a note on their management and conservation. In: *Proc. UGC-Sponsored State-level Seminar on Biodiversity of Assam, Session Chairman's Lecture*, 30, January 2000, (Eds.) M.K. Bhattacharjee, M. Dattachoudhury and P.B. Mazumder. Karimganj College, Assam University, Assam, pp. 3–10.

Kar, D., 2003a. Fishes of barak drainage, Mizoram and Tripura. In: *Environment, Pollution and Management* (Eds.) A. Kumar, C. Bohra and L.K. Singh. APH Publishing Corporation, New Delhi, pp. 203–211.

Kar, D., 2003b. Peoples perspective on fish conservation in the water bodies of South Assam, Mizoram and Tripura. In: *Participatory Approach for Fish Biodiversity Conservation in North-East India*, (Eds.) P.C. Mahanta, and L.K. Tyagi, National Bureau of Fish Genetic Resources (ICAR), Lucknow, pp. 325–328.

Kar, D., 2004. A Glimpse into the fish bioresources of North-East India with a note on their management, conservation and biotechnological potential. In: *Proc. (Invited Lecture) DBT-Sponsored National Symposium on Biodiversity Conservation and Sustainable Utilisation of Environmental Resources*, Tripura University, 10–11 January, pp. 9.

Kar, D., 2005a. Fish genetic resources and habitat diversity of the Barak drainage. In: *Aquatic Ecosystems, Conservation, Restoration and Management*, (Eds.) T.V. Ramachandra, N. Ahalya and C. Rajsekara Murthy. Capital Publishing Company, Bangalore, pp. 68–76.

Kar, D., 2005b. Fish fauna of river Barak, of Mizoram and of Tripura with a note on conservation. *J. Freshwater Biol.*, p. 16.

Kar, D., 2005c. Fish diversity in the major Rivers in Southern Assam, Mizoram and Tripura. In: *GIS/ Spatial Analysis in Fishery and Aquatic Sciences*, (Eds.) T. Nishida, P.J. Kailola and C.E. Hollingworth. Proc. 2nd International Symposium on GIS and Spatial Analyses in Fisheries and Aquatic Sciences, 2–6th September 2002, University of Sussex, Brighton (UK) and Fisheries and Aquatic GIS Research Group, Japan, pp. 679–691.

Kar, D., 2005d. Sustainability issues of inland fish biodiversity and fisheries in Barak drainage (Assam), in Mizoram and in Tripura. In: *Proc. International Symposium on Improved Sustainability of Fish Production Systems and Appropriate Technologies for Utilisation*, Cochin University of Science and Technology, School of Industrial Fish and Fisheries, 1.

Kar, D., 2007. *Fundamentals of Limnology and Aquaculture Biotechnology*. Daya Publishing House, New Delhi, pp. 454.

Kar, D. and Dey, S.C., 1982a. *Hilsa ilisha* (Hamilton) from Lake Sone in Cachar, Assam. *Proc. Indian Sci. Congr.*, 69: 77.

Kar, D. and Dey, S.C., 1982b. An account of *Hilsa ilisha* (Hamilton) of Sone Beel (Cachar district, Assam, India). *Proc. All India Sem. Ichthyol.*, 2: 3.

Kar, D. and Dey, S.C., 1986. An account of the ichthyospecies of Lake Sone in the Barak valley of Assam (India). *Proc. All India Sem. Ichthyol.*, 6: 29.

Kar, D. and Dey, S.C., 1987. An account of the fish and fisheries of Lake Sone in the Barak valley of Assam (India). In: *Proc. Workshop Development of Beel Fisheries in Assam*, 1: 13.

Kar, D. and Dey, S.C., 1991. Gill nets in lake Sone of Assam with their economics and impact on fishery. *J. Applied Zool. Res.*, 2: 76–79.

Kar, D. and Dey, S.C., 1993. Variegated encircling gears in lake Sone of Assam. *J. Appl. Zool. Res.*, 4: 171–175.

Kar, D. and Dey, S.C., 1996. Scooping gears of lake Sone in Assam. *J. Applied Zool. Res.*, 7: 65–68.

Kar, D. and Dey, S.C., 2000. Yield and conservation of Indian major carps of lake Sone in Assam. *Environment and Ecology*, 18: 1036–1038.

Kar, D. and Dey, S.C., 2002. On the occurrence of advanced fry of *Hilsa* (*Tenualosa*) *ilisha* (Hamilton-Buchanan) in Chatla Haor Seasonal Wetland of Assam. *Proc. Zool. Soc., Calcutta*, 55: 15–19.

Kar, D., Roy, A., Dey, S.C., Menon, A.G.K. and, Kar, S., 1995. Epizootic ulcerative syndrome in fishes of India. *World Congress of in vitro Biology, In Vitro*, 31: 7.

Kar, D., Dey, S.C., Kar, S., Michael, R.G., and, Gadgil, M., 1996. Ichthyoecology: Management and conservation fish resources of lake Sone in Assam (India). *Tiger Paper* (FAO, UN), 23: 27–32.

Kar, D., Dey, S.C., Kar, S., and Ramachandra, T.V., 1998. Trawls of lake Sone in Assam. *J. Applied Zool. Res.*, 10: 170–172.

Kar, D., Laskar, B.A., Mandal, M., Lalsiamliana, and Nath, D., 2002a. Fish genetic diversity and habitat parameters in Barak drainage, Mizoram and Tripura. *Indian J. Environment and Ecoplanning*, 6: 473–480.

Kar, D., Laskar, B.A, Nath, D., Mandal, M. and Lalsiamliana, 2002b. *Tor progenius* (McClelland) under threat in River Jatinga, Assam. *Science and Culture*, 68: 211.

Kar, D., Laskar, B.A. and Nath, D., 2002c. *Tor* sp. (Mahseer fish) in river Mat in Mizoram. *Aquacult.*, 3: 229–234.

Kar, D., Dey, S.C. and Datta, N.C., 2003. *Welfare Biology in the New Millennium*. Allied Publishers Pvt. Ltd., Bangalore, pp. 97.

Kar, D., Roy, A. and Dey, S.C., 2004. An overview of fish genetic diversity of North-East India. In: *Proc. National Workshop on Rational Use of Water Resources for Aquaculture*, (Eds.) S.K. Garg and K.L. Jain, 18–19th March, CCS Haryana Agricultural University, Hisar, 1: 164–171.

Kar, D., Laskar, B.A., Nath, D. and Barbhuiya, M.H., 2005. Length-weight relationship of *Neolissochilus hexagonolepis* (McClelland) and *Garra lissorhynchus* (McClelland) collected from Jatinga river in Assam. *Indian J. Fish* (in press).

Kar, D. and Barbhuiya, M.H., 2000a. Length-weight relationship and condition factor in *Gudusia chapra* (Ham-Buch) and *Botia dario* (Ham-Buch) from Chatla Haor (flood plain wetland) in Cachar district of Assam. *Environment and Ecology*, 18: 227–229.

Kar, D. and Barbhuiya, M.H., 2000b. Ichthyodiversity of Chatla Haor: A floodplain wetland in Barak valley region of Assam. In: *Advances in Zoology and Environmental Degradation and Biodiversity*, (Eds.) B.N. Pandey and B.K. Singh. Daya Publishing House, New Delhi, pp. 3–6.

Kar, D. and Barbhuiya, M.H., 2004. Abundance and diversity of zooplankton in Chatla Haor, a floodplain wetland in Cachar district of Assam. *Environment and Ecology*, 22: 247–248.

Kar, D. and Barbhuiya, M.H., 2005. Length-weight relationship and relative condition factor in *Hilsa ilisha* (Hamilton) of Barak drainage in Assam. *Indian J. Environment and Ecoplanning*, pp. 10.

Lacepe'de, S., 1798–1803. *Historie Naturelle des Poissons*, Vols. I–V.

Lahon, B., 1983. Limnology and fisheries of some commercial Beels of Assam, India. *Ph.D. Thesis*, University of Gauhati, pp. 349.

Menon, A.G.K., 1951. The role of the Eastern Ghats in the distribution of the Malayan fauna and flora to peninsular India. *Proc. Nat. Inst. Sci., India*, 17: 475–497.

Menon, A.G.K., 1974. A checklist of the fishes of the Himalayan and the Indo-Gangetic plains. *Inland Fish. Soc. India, Barrackpore*, pp. 136.

Menon A.G.K., 1999. *Checklist: Freshwater Fishes of India*. Occasional Paper No. 175. Zoological Survey of India, Calcutta, pp. 366.

Michael, R.G., 1969. Seasonal trends in physico-chemical factors and plankton of freshwater fish pond and their role in fish culture. *Hydrobiologia*, pp. 33.

Mirza, M.R., 1975. Freshwater fishes and zoogeography of Pakistan. *Pakistan J. Zool.*, 2: 249–250.

Misra, K.S., 1959. An aid to the identification of commercial fishes of India and Pakistan. *Rec. Indian Mus.*, 57: 1–320.

Motwani, M.P., Jayaram, K.C. and Sehgal, K.L., 1962. Fish and fisheries of Brahmaputra river system, Assam. I. Fish fauna with observation on their zoogeographical significance. *Trop. Ecol.*, 3: 17–43.

Myers, G.S., 1949. Salt tolerance of freshwater fish groups in relation to geographical problems. *Bijdr tot de Dierk*, 28: 315–322.

Nath, S., 1987. The fish and fisheries of the Barak river system in Assam, India with special reference to the ecomorphology and distribution of its ichthyofauna. *Ph.D. Thesis*, University of Gauhati, Assam, pp. 259.

Nath, P., 1989. Systematics, distribution and ecology of ichthyofauna of Arunachal Pradesh with particular reference to the Limnology of Itanagar Lake. *Ph.D. Thesis*, University of Gauhati, Assam, pp. 362.

Nichols, J.T., 1928. Fishes of the White Nile (with table of world's freshwater fish faunae). *American Mus. Novitates* No. 319.

Pillay, T.V.R., 1958. Biology of hilsa, *Hilsa ilisha* from river Hooghly. *Indian J. Fish.*, 5: 201–257.

Schneider, J.G., 1801. *Systema Ichthyologics.*

Sen, N., 1982. Studies on the systematics, distribution and ecology of the ichthyofauna of Meghalaya and their bearing on the fish and fisheries of the state. *Ph.D. Thesis*, University of Gauhati, Assam, pp. 576.

Vaas, K.F. and Schurman, J.J., 1949. On the ecology and fisheries of some Javanese freshwaters. *Common Gen. Agric. Res. Sta., Buitenzorg Java*, 97: 1–60.

Welcomme, R.L., 1979. *Fisheries Ecology of the Floodplain Rivers*. Longmans, London, pp. 317.

Yadava, Y.S., 1987. Limnology and productivity of Dhir Beel of Assam. *Ph.D. Thesis*, University of Gauhati, Assam, pp. 467.

Yadava, Y.S., Goswami, M.M., Kar, D. and Choudhury, M., 1989. Present status of Hilsa resources in Assam. In: *Fish Genetics in India*, (Eds.) P. Das and A.G. Jhingran. Today and Tomorrow's Printers and Publishers, New Delhi, pp. 252–256.

Chapter 41

Water Quality Index for Groundwater Affected with Bicycle Manufacturing Industrial Wastes: An Environmental Quality Audit

Vineeta Shukla*, Sharda Abusaria, Monika Dhankhar* and K.V. Sastry***

***Department of Biosciences, Maharshi Dayanand University, Rohtak – 124 001, Haryana**

****Lecturer, Department of Zoology, A.I.J.H.M. College, Rohtak**

ABSTRACT

The present study is intended to calculate water quality index for groundwater at Sonepat in Haryana State of India from various parameters, *viz.*, pH, dissolved oxygen, total hardness, chlorides, total alkalinity, biological oxygen demand, sulphates, iron, chromium, cyanide and MPN coliforms to ascertain the quality of water for human consumption and to assess the impact of bicycle manufacturing industrial wastes on groundwater quality. The high values of water quality indices for groundwater samples from various sources located in the industrial premises and vicinity of the industry indicate the influence of industrial wastes on groundwater quality and make it unsuitable for drinking purposes. The trend of variation in water quality index values collected around and inside the industry was irregular while for control samples, water quality index values were well below the permissible limit throughout the study period and did not vary widely. The importance of water quality index and pollution control has also been discussed and emphasized.

Keywords: *Groundwater, Bicycle manufacturing industrial wastes, Water quality index, Pollution control.*

Introduction

Environment is our biological capital. It makes up a huge, enormously complex living machine on which our total productivity depends. Water is one of the basic natural resources needed for survival of life. Nobel laureate A. Szent-Gyorgy, physician and biochemist, called water the "matrix of life". Many organisms can live without air, but none without water. The use of this under ground fluid for agriculture and drinking purposes has a long history. But groundwater is safe only as long as it is free from disease causing organisms and chemicals that are deleterious to human health. Unfortunately, in recent years industrial growth has created severe environmental pollution, and degradation has not spared even groundwater aquifer system. A number of places in the country face serious groundwater contamination problem due to indiscriminate dumping of toxic chemicals by polluting industries (Singh and Parwana, 1999). Water soluble chemicals are not retained by soil matrix, they percolate through porous medium and enter the groundwater. Environmental protection and mankind's survival are closely linked. Groundwater once polluted cannot be treated, therefore, the effect is long lasting and problem is of vital concern. Hence, there is need for protection and management of groundwater quality. Accurate and timely information on water quality is essential to frame a sound public policy and to implement the water quality improvement programs.

Water quality index is the most efficient tool of environmental audit process to convey the overall water quality of a water resource. It is a rating of water quality parameters by a single numerical expression reflecting the complete influence of water quality parameters. Unfortunately in India water quality index determination has been much neglected. Few attempts have been made to calculate water quality for river waters. Studies on groundwater quality index are scanty.

Sonepat in Haryana State is known for bicycle manufacturing industry whose product is available globally. But the industrial effluents and sludge generated as byproduct may cause great environmental stress locally. Spent waters of different operations such as electroplating and heat treatment are normally stored inside the industrial premises before discharge. For the past more than forty five years, surrounding land has been receiving untreated/partially treated effluents of the industry. Sludge is being dumped within the premises and in low lying areas located near the industry. The dumping and holding of such chemical wastes are likely to affect the groundwater quality in these areas. So far, no detailed investigation on the effect of these industrial wastes on groundwater quality has been carried out. The present study has, therefore, been undertaken to calculate the water quality index for groundwater in and around the industry to measure the extent of damage to water quality.

Materials and Methods

In the industry and surrounding regions groundwater is the only source of drinking water and is being tapped by tubewells. Four sampling sites were selected for the present investigation. Details of these sites are given in Table 41.1. In the formulation of water quality index, the importance of various parameters depends upon the intended use of water. In this study eleven water quality parameters of groundwater are studied from the point of view of suitability of water from different sources at Sonepat for human consumption. For this purpose monthly samples of groundwater from selected sites were collected and analysed as per Standard Methods during March, 1999 to February, 2000.

Based on the groundwater analysis data, water quality index was calculated for each sample by a simple method. The 'standards' (permissible values) of various parameters for drinking water, recommended by the ISI are used for comparison. For dissolved oxygen and BOD, European Economic Community (EEC) Standard and United Public Health Service (UPHS) have been considered respectively.

Table 41.1: Particulars of Groundwater Sampling Sites

Reference Number	*Name and Location*	*Description*	*Age (Yrs)*	*Depth (M)*
GW1	Tubewell No. 3 in bicycle industry premises in contaminated area	20M from effluents bodies	40	115 up to May, 1999. June onwards 120. Closed in August
GW2	Tubewell No. 7 in bicycle industry premises in contaminated area	Earlier 30M, now 1000M from sludge dumping area and effluents bodies	8	105
GW3	Sujan Singh Tubewell in agricultural field contaminated with industrial effluents	Near cyanide sludge dumping area of industry and about 1000M from effluents flow channel	8	75
GW4	Public Health Tubewell in unpolluted area	Beyond the influence of industrial wastes and it served as control groundwater	5	83

Calculation of Subindex (SI)

To calculate the Water Quality Index, first of all subindex (SI) i for the ith parameter was calculated which was the product of the quality rating (qi) and the unit weight (wi) of the ith parameter *i.e.*

$$(S)\, i = qi\ wi \qquad \text{.........(1)}$$

Calculation of Quality Rating (qi)

The quality rating, qi for the ith water quality parameter was obtained from the equation:

$$qi = 100\, (Vi/Si) \qquad \text{.........(2)}$$

where,

Vi: Value of the ith parameter at a given sampling station; and

Si: Recommended standard for the ith parameter.

Equation (2) ensures that qi = 0 when pollutant (the ith parameter) is absent in the water, while qi = 100 if the value of this parameter is just equal to its permissible value for drinking water.

Quality ratings for pH, dissolved oxygen and MPN coliforms required special handling.

The permissible range of pH for drinking water is 6.5 to 8.5. Therefore, the quality rating qpH was calculated from the relation:

$$qpH = 100[(V_{pH} \sim 6.5)/(8.5–6.5)] \qquad \text{.........(3)}$$

where,

V_{pH}: Value of pH; and

Symbol "~" means simply the numerical differences between V_{pH} and 6.5 ignoring the algebraic sign.

The situation is slightly complicated in the case of dissolved oxygen, since, in contrast to other pollutants, the quality of water is enhanced if it contains more dissolved oxygen. Therefore quality rating qDO was calculated from the relation.

$$qDO = 100\left[\frac{14.6}{14.6-5.0} - \frac{V_{DO}}{14.6-5.0}\right] \quad \text{.........(4)}$$

where,

V_{DO}: Value of dissolved oxygen.

14.6 is the solubility of oxygen (mg/l) in distilled water at 0°C; and 5.0 is the standard (mg/l) for drinking water.

In case of MPN coliforms, the permissible value for the drinking water is 1/100 ml, while their actual number in untreated water may be several thousands/100 ml. Therefore to get a convenient number for quality rating qMPN was calculated from the relation

$$qMPN = 100\left[\frac{\text{Log}_{10} V_{MPN}^{+1}}{\text{Log}_{10} S_{MPN}^{+1}}\right] \quad \text{.........(5)}$$

where,

V_{MPN}: MPN count/100 ml; and

S_{MPN}: MPN standard for drinking water.

Calculation of Unit Weight (wi)

$$wi = k/Si \quad \text{.........(6)}$$

where,

wi: Unit weight;

Si: Standard for the ith parameter (i = 1, 2..............11); and

k: Constant of proportionality which was determined from the condition

$$\sum_{i=1}^{i=11} wi = k\sum_{i=1}^{i=11}\left(\frac{1}{si}\right) = 1 \quad \text{.........(7)}$$

Final Value of Water Quality Index

Finally the water quality index was obtained by aggregating these eleven subindices (SI) i linearly

$$W.Q.I. = \sum_{i=1}^{i=11}(qiwi) \quad \text{.........(8)}$$

Results and Discussion

Unit weights assigned to water quality parameters and a sample calculation of overall water quality index from the subindices for groundwater samples from different sources have been shown in Table 41.2. Water quality indices for one year study period are depicted in Table 41.3. On extensive comparison with control sample to document the impact of industrial pollution on groundwater quality present study reveals an alarming feature that in 100 per cent samples from affected area, quality of groundwater is much altered with respected to natural and local background indicating pollution status of groundwater and is not fit for human consumption at any time period of the year due to non-conforming water quality index with drinking water limit (100). A critical study of Table 41.2 indicates that for GW1, GW2 and DW3 samples the main dominating contribution to water quality index comes from the existence of un-natural toxic chemicals–chromium and cyanide–which are used in bicycle manufacturing industry. It is further evidenced by the fact that the more harmful a given pollutant is, the smaller is its permissible value for drinking water. So the, unit weights for various water quality parameters are assumed to be inversely proportional to the recommended standards for the corresponding parameters. Maximum value of the unit weight for water quality parameters in the present study was in the order of chromium and cyanide > iron > MPN coliforms > BOD and dissolved oxygen > pH > total alkalinity and sulphates > chlorides > total hardness, consequently due to high values of chromium and cyanide subindices WQI increased drastically. It may thus be, inferred that higher water quality indices for groundwater samples from different sources–in and around the bicycle manufacturing industry–may be due to leaching of chemicals from industrial wastes through soil into groundwater. Further, WQI for GW1, GW2 and GW3 samples varied depending upon the location of sampling site and variation was sizeable. Very high value of WQI for GW1 indicates the influence of proximity of effluents tanks. There was a wide difference in the values of WQI between GW1 and other two samples coinciding with levels of pollutants. Amongst above three samples, GW1 water exhibited a maximum value of WQI in the month of April which is about 39 times greater than the limit for drinking water. However, GW2 and GW3 samples also exhibited deteriorated status of water quality. On the other hand chromium and cyanide were not the constituents of WQI for GW4 sample (control) coinciding with their nil quality rating values. WQI for GW4 was quite less than other three samples and remained below the limit for drinking water. The variability in WQI among different groundwater samples was measured statistically by calculating the coefficient of variation. It confirmed well marked variation in WQI for groundwater samples from different sources (Table 41.3).

Another striking feature emerging from the present investigation is that, different sources do not synchronize in monthly WQI variation. This may be attributed to the fact that an index significantly increases or decreases when some of the critical parameters exceed or become less than the specified permissible value. Over a period of about half a century enormous quantity of pollutants has been added to the soil which may act as a bank of pollutants. Thus, chemical characteristics of percolating effluents, and nature and status of soil may be the possible reasons for irregular variations in WQI for affected sample. However, WQI for control samples indicates uniformly good water quality, reflecting pollution free status of groundwater of area not yet contaminated, without any irregular variation. But if the bicycle industry is allowed to dump its wastes without proper treatment the possibility of spread of pollution of groundwater in this area also cannot be ruled out.

It is clearly brought out by the findings of this study that the groundwater environment of bicycle manufacturing industrial area has already become grossly polluted. Operational phase of the industry establishes beyond doubt the potential threat to groundwater. It is unfortunate that groundwater, a gift of nature, turns out to be a man-made curse. Good quality groundwater, as of now is a rare and scarce commodity in the area. Therefore, there is an urgent need for strict enforcement of Environment (Protection) Act, 1986. Violation of these rules can be disastrous and can have far reaching consequences.

Present study also clearly confirms the utility of WQI as a proper indicator for conveying the idea of the overall quality of water with a single number. This type of information is helpful for public to understand the quality of drinking water and may have an administrative use in the field of water quality management.

Table 41.2: Water Quality Index of Groundwater Samples (March)

Water Quality Parameters	*Drinking Water Standard (Si)*	*Unit Weight (wt)*	*Parameter Sub-index (qiwi)*			
			GW1	*GW2*	*GW3*	*GW4*
pH	6.5–8.5	0.00262	0.1179	0.1048	0.1179	0.1048
Dissolved oxygen	5.0 mg/l	0.00446	0.3836	0.3791	0.4371	0.4371
Total hardness	300.0 mg/l	0.00007	0.0104	0.0238	0.0149	0.0141
Chlorides	250.0 mg/l	0.00008	0.0028	0.0214	0.0071	0.0053
Total alkalinity	200.0 mg/l	0.00011	0.0157	0.0092	0.0109	0.0144
BOD	5.0 mg/l	0.00446	1.6484	0.8028	0.0347	0.2382
Sulphates	200.0 mg/l	0.00011	0.0037	0.0049	0.0037	0.0028
Iron	0.3 mg/l	0.7430	0.2194	3.4178	3.2194	12.4764
Total chromium	0.05 mg/l	0.44580	3584.2320	53.4960	53.4960	0.0000
Cyanide	0.05 mg/l	0.44580	71.3280	53.4960	53.4960	0.0000
MPN coliforms	1/100 mg/l	0.02229	0.0000	0.0000	0.0000	0.0000
Water quality index			3660.9619	111.7528	111.8377	3.2931

Table 41.3: Monthly Variation in Water Quality Index of Groundwater Samples

Months	*Water Quality Index (WQI)*			
	GW1	*GW2*	*GW3*	*GW4*
Mar	3360.96	111.75	111.84	3.29
Apr	3911.12	111.02	120.73	3.43
May	3633.90	114.32	142.60	3.07
Jun	3635.88	110.46	123.77	4.28
Jul	3654.14*	110.69	148.53	9.48
Aug		119.45	113.44	7.43
Sep		132.22	131.19	3.18
Oct		141.64	139.60	3.19
Nov		129.55	122.18	3.74
Dec		119.80	130.35	3.42
Jan		129.87	121.90	3.73
Feb		138.14	130.91	3.42
Average	3639.20	122.41	128.09	4.30

* GWI (Tubewell No.3) was closed by the industry from August, 1999 CV (%) = 182.64.

References

APHA 1985. *Standard Methods for the Examination of Water and Wastewater*. American Public Health Association, Washington, D.C.

Horton, R.K. 1965. An index number system for rating water quality. *J. Water Poll. Cant. Fed.*, 37(3): 300–306.

ISI 1991. *Indian Standard Specification for Drinking Water* (IS: 10500), Indian Standards Institution, New Delhi.

Jeevan Rao, K. and Shantaram, M.V. 1995. Groundwater pollution from refuse dumps at Hyderabad. *Indian J. Environ. Hlth.*, 37(3): 197–204.

Mariappan, P., Vasudevan, T. and Yegnaraman, V. (1999). Surveillançe of groundwater quality in Thiruppathur Block of Sivagangai district. *Ind. J. Environ. Prot.*, 19(4): 250–254.

Singh, A.P. and Ghosh, S.K. 1999. Water quality index for river Yamuna. *Poll. Res.*, 18(4): 435–439.

Singh, K.P. and Parwana, B.K. 1999. Groundwater pollution due to industrial wastewater in Punjab State and strategies for its control. *Ind. J. Environ. Prot.*, 19(4): 241–244.

Tiwari, T.N., Das, S.C. and Bose, P.K. 1986. Water quality index for the river Jhelum in Kashmir and its seasonal variation. *Poll. Res.*, 5(1): 1–5.

Venkata Mohan, S., Rama Mohan, S. and Jayaraman Reddy, S. 1998. Determination of water quality index (GWQI) of groundwater: An application. *J. Environ. Poll.*, 5(3): 219–223.

Chapter 42

Effect of Different Medicinal Plant Extracts Against Root Knot Nematode Penetration into the Roots of Solanacae Crops

L. Joymati Devi and W. Mema Devi
Department of Zoology, G.P. Women's College, Imphal

ABSTRACT

Effect of aqueous leaves extracts of nine medicinal plants *Jatropha curcas, Melothria purpusilla, Mellia azedarach, Andrographis panniculata, Acrorus calamus, Plectranthes ternifolius, Jusminum multiflorum, Plumeria rubra* and *Abrus fruticulosus* against penetration of root knot nematode *Meloidogyne incognita* into the roots of solanacae crops (brinjal) were tested. Among these nine medicinal plants extracts leaves extracts of *A. panniculata* found to be most inhibitory effect followed by *Acrorus calamus* and *Plectranthes ternifolius.* Larval penetration rate of juvenile J_2 was inversely correlated with the concentration of extracts and directly to the duration of time showing decrease with increase in exposure period.

Keywords: Plant extracts, Root knot nematode, Penetration, Brinjal.

Introduction

The use of plant and plant products having nematicidal properties involved low cost which are easy to apply and produced no pollution hazards. The problem of pesticide resistance, environmental degradation of non target organisms including human associated with the use of synthetic broad spectrum pesticide have triggered the interest of scientist in exploring alternate pest control measure. Thus use of plant products is an alternative method for controlling root knot nematode instead of chemical nematicides. Sharma and Trivedi (1992) reported that leaf products of *Aegla marmelos, Withamia somnifer, Hibiscus rosasinensis, Murraya koenigi* and stem of *Cuscuta reflexa* significantly enhanced plant growth and reduced root galling in Fenugreek. Singh *et al.* (2001) also observed inhibitory effect of aqueous extracts of plant product on egg hatching, mortality and larval penetration of *M. incognita* infecting sunflower. Thus the present investigation has been taken up to evaluate the effect of these nine medicinal plants against root knot nematode infecting brinjal.

Materials and Methods

Fresh plant leaves of these nine medicinal plants *i.e. J. curcas, M. purpusilla, M. azedarach, A. panniculata, A. calamus, P. ternifolius, J. multiflorum, P. rubra* and *A. fruticulosus* were taken, cleaned, washed and chopped. All the materials are separately dissolved in 50 g/50 ml of distilled water in a mixer at 25°C for two minutes. The extracts were separately filtered with a muslin cloth and through Whatmann filter paper no 1. The solution was termed as Standard solution and store at 5°C until needed. Different concentration such as S/100, S/1000 and S/10000 were made by adding required amount of distilled water.

Healthy nursery plants of *Solanum melongena* (brinjal) were collected from *M. incognita* free field soil and transplanted singly in treatment pot containing 300 g of sterilized soil. Ten ml of different concentration of aqueous extracts of these nine medicinal plant were added to each pot containing one plant of brinjal, five days after transplantation of plants. Plants treated with water were served as control. All the treatments were replicated five times. Seven days after extract treatment, the plant was inoculated with 100 freshly hatched second stage juveniles of *M. incognita.* The untreated plants were also inoculated with nematodes. Ten ml of each extract concentration were applied to the treatment plants once a week and pots were watered during intervening periods. Penetration of second stage juveniles were recorded at the intervals of 15, 20 and 25 after inoculation of nematodes by staining the roots with lectophenol acid fuchsin.

Results and Discussion

Table 42.1 shows effect of nine different medicinal plants against penetration of root knot nematode into the roots of brinjal. Among those nine, leaves extracts of *A. panniculata* was found to be most inhibitory effect followed by *A. calamus* and *P. ternifolius.* Larval penetration rate of juvenile was inversely correlated with the concentration of extracts and directly to the duration of time showing decrease in exposure period. Thus all the tested plants shows minimum larval penetration rate in comparison with control but in different rate of penetration rate in due to different in the quantity of nemathelmintic properties present in these tested plants. The findings supported the word of Joymati and Dhanachand (1999) who reported efficacy of aqueous extracts of *Parkia javanica* on larval penetration of *M. incognita* into the roots of broad bean and observed maximum reduction in larval penetration at higher concentration treatment plants and among plant parts seeds was found most inhibitory effect followed by leaf and stem. The results can be adjustable conformity with the findings of Dhanger *et al.* (1996) with different parts of *T. erecta* against penetration rate on brinjal, there was reduction in larval penetration with increase in extract concentration and roots was found most inhibitory followed by leaves and stem extracts. In conformity with them the present findings suggest that these medicinal plants which are commonly available in Manipur can be used for controlling root knot trematode *M. incognita* which can infect variety of host plant.

Table 42.1: Effect of Nine Different Medicinal Plants Against Penetration of Root Knot Nematode into the Roots of Brinjal

Plant	*Duration (Days)*	*No. of Larvae Penetrated Plant⁻¹ Extract Concentration (%)*				
		0	*S/1000*	*S/100*	*S/10*	*S*
J. curcas	15	41	26.7	20.7	17	13.3
	20	57	41.7	33	21.7	19
	25	88	51.3	39.3	31.3	23.7
M. purpusilla	15	47	31.2	27	25	20.1
	20	57	44.2	40	37	30.3
	25	88	69.3	60	48	41.2
M. azedarach	15	47	31.2	27	25	20.1
	20	56	46	38.2	26.2	23
	25	87	56	44.2	38.2	26.4
A. paniculata	15	40.3	14.4	9.3	8.5	4.2
	20	45.4	16.2	10.1	9.6	5.4
	25	53.3	18.5	12.1	10.1	7.2
P. ternifolius	15	40.3	19.1	16.1	13.1	8.1
	20	45.4	23.2	21.2	19.3	12.2
	25	53.3	29.3	25.2	21.2	18.3
A. calamus	15	40.3	17.2	14.1	11.2	6.3
	20	45.4	21.2	18.2	16.1	10.1
	25	53.3	25.7	21.1	19.2	13.3
J. multiflorum	15	38	23.1	19.2	14.1	10.4
	20	54	25.2	21.5	17.3	12.6
	25	37	28.6	23.1	21.2	16.1
P. rubra	15	38	29.1	22.1	23.2	16.1
	20	54	33.4	25.1	19.1	19.2
	25	83	38.2	27.6	22.1	21.3
A. fruticulosus	15	38	29	27.6	22.1	13
	20	54	32.4	29.1	26.4	14.1
	25	81	38.4	31.2	27.2	19.4

Acknowledgement

The author greatly acknowledged to Principal, G.P. Women's College for providing laboratory facility and U.G.C. New Delhi for giving financial assistance of Minor research project during the course of studies.

References

Alam, M.M., Masood, A. and Husain, S.T., 1975. Effect of margosa and marigold root exudates on mortality and larval hatch of certain nematodes. *Indian J. Expt. Biol.*, 13: 412–414.

Dhangar, D.S., Gupta, D.C. and Jain, R.K.,, 1996. Studies on aqueous extract of different parts of *Tagetes erecta* on hatching larval penetration of *M. javanica*. *Indian J. Nematol.*, 26(1): 46–51.

Joymati, L. and Dhanachand, Ch., 1999. Effect of aqueous extract of *Parkia javanica* on egg hatching larval mortality and penetration of larvae of *M. incognita*. *Vasundhara*, 4: 75–79.

Salam, M.A. and Sinha, A.R.P., 1990. Effect of some plant extracts on the hatching of *M. incognita*. *Current Nematology*, 1(1): 59–60.

Sharma, R. and Trivedy, P.C., 1992. Effect of root extract of some plants on larval hatching of *M. incognita*. *Current Nematology*, 3(1): 31–34.

Chapter 43

Mercury Chloride Induced Changes in the DNA, RNA, Protein and Alkaline Phosphatase of Kidney of Freshwater Teleost, *Labeo nandina*

M.M. Prakash[1], R.S. Maheshwari[2] and V.K. Sharma[3]*

[1]Department of Zoology, [2]Department of Biochemistry, Government Model Autonomous Holkar Science College, Indore – 452 017, India

[3]Department of Zoology, PMB Gujarati Science College, Indore – 452 001, India

ABSTRACT

Fish *Labeo nandina* were exposed to 0.1 ppm aqueous mercury chloride solution for 96 hrs. They were sacrificed at various autopsy intervals (0 to 96 hrs.) to examine quantitative changes in the kidney for certain biochemical parameters. In experimental fishes no significant change was noted in the DNA content of kidney. However, there was regular decrease in RNA content up to 61.734 per cent in 96 hrs. The activity of alkaline phosphatase was also observed in decreasing manner up to 54.939 per cent in the experimental groups. But protein content of liver was observed increasing gradually up to 16.228 per cent. All these changes showed that in the experimental fishes short term exposure of mercury chloride altered the activity of RNA, Protein and Alkaline phosphatase in the kidney, which disturbed its normal functioning.

Keywords: *Labeo nandina, Kidney, Nucleic acids, Protein and Alkaline phosphatase.*

* Corresponding Author: E-mail: vipulkeerti@yahoo.com.

Introduction

Mercury is viewed as the most toxic inorganic pollutant available in natural water (Ramadevi and Srinivasan, 2005). It has several forms like Hg^{o}, Hg^{++}, Hg^{+2} and is toxic to various degrees. It combines with other element such as chlorine, sulphur or oxygen to form inorganic mercury compounds. Mercurous and mercuric forms of mercury are encountered in some chemical-metal processing units, electric equipments, automotive and building industries and medical and dental services (Lal, 2005). It is usually washed away with rain water and get deposited in river, lake and sea bed etc. Bio-concentration of it takes place through food-chain (Verma *et al.*, 2002). It enters the food-web through bacteria, algae and fishes of freshwater and marine water etc. as compared to land animals (Gupta, 1998). Inorganic form of mercury enters into the human through fish etc (Ayyaduri, 2004).

Inhalation of high concentration of mercury causes severe respiratory irritation, digestive disturbance and marked kidney damage. Repeated or prolonged exposure to mercury vapour is highly toxic to the central nervous system (Lal, 2005). Recent research has demonstrated that even at much lower level mercury exposure to pregnant women through dietary intake of fish and whale meat, an important regional food staple, is associated with decrements in motor functions, language, memory and neural transmission in their offsprings (Murata *et al.*, 1999). Mercury is also capable of bioaccumulation in body tissue and alters physiological and biochemical metabolism (Rogers and Beamish, 1981; Rogers and Beamish, 1982; Muthukrishnan *et al.*, 1986). Looking to these facts the present study was undertaken to find out the effect of mercury chloride on the nucleic acids (DNA and RNA), protein and alkaline phosphatase of the kidney of *Labeo nandina*.

Materials and Methods

Test Animal

Healthy fish *Labeo nandina* were purchased from Naulakkha fish market of Indore. They were acclimatized to laboratory conditions for one week. Fishes were regularly fed with pawn powder and Soya meal once in a day.

Test Chemical

Mercury chloride of m. wt. 271.52 (E. Merck India, Ltd., Mumbai) was used in the present experiment.

Experimental Protocol

In the present study fishes were divided into two groups *i.e.* control and experimental. Experimental fishes were exposed to 0.1 ppm aqueous solution of mercury chloride for 96 hrs. The control was run with distilled water as vehicle. Fishes of control and experimental groups were sacrificed after 0, 6, 12, 24, 72 and 96 hrs of exposure. The kidney was removed, blotted, weighed and then processed for biochemical Analysis.

Biochemical Analysis

Following standard biochemical methods were used, which are described in the laboratory Manual in Biochemistry (Jayaraman, 2000):

1. DNA-Burrton's Diphenylamine Test
2. RNA-Orcinol reagent Test

3. Protein-Lowry's method
4. Alkaline phosphatase-Kind and king Method.

Results and Discussion

Nucleic Acids

Table 43.1 and Figure 43.1 shows than DNA content of kidney remain unchanged after 96 hrs exposure of mercury chloride. However, at 6 hrs 0.925 per cent change was observed which may be due to experimental errors. On the contrary RNA activity was found inhibited in decreasing order due to exposure of mercury chloride. After 6, 12, 24, 72 and 96 hrs there was a decrease in the activity of RNA by 17.040, 32.721, 54.081, 59.039 and 61.734 per cent respectively (Table 43.2 and Figure 43.2). The inhibition in RNA activity was gradual and chemo-exposure-duration-dependent. Similar results were observed by Prakash and Maheshwary (2005). In our opinion the inhibition in RNA activity was due to inhibition of some enzymes which were involved in RNA synthesis. But according to Oomen and Pilo (2003) it is due to release of glucocorticoids, which in excess suppress the RNA level. However, Kaur and Sidhu (2004) stated that it was due to the blockage of macromolecules synthesis.

Table 43.1: DNA Changes in the Kidney of *L. nandina* Exposed to 0.1 ppm Concentration of Mercury Chloride

Sl.No.	*Exposure Duration in Hrs*	*DNA Content in µg/wt. of Tissue*		*Difference*	*% Alter*
		Control	*Experimental*		
1.	0	1.08	1.08	0	0
2.	6	1.09	1.08	0.01	0.925
3.	24	1.08	1.08	0	0
4.	48	1.08	1.08	0	0
5.	72	1.08	1.08	0	0
6.	96	1.08	1.08	0	0

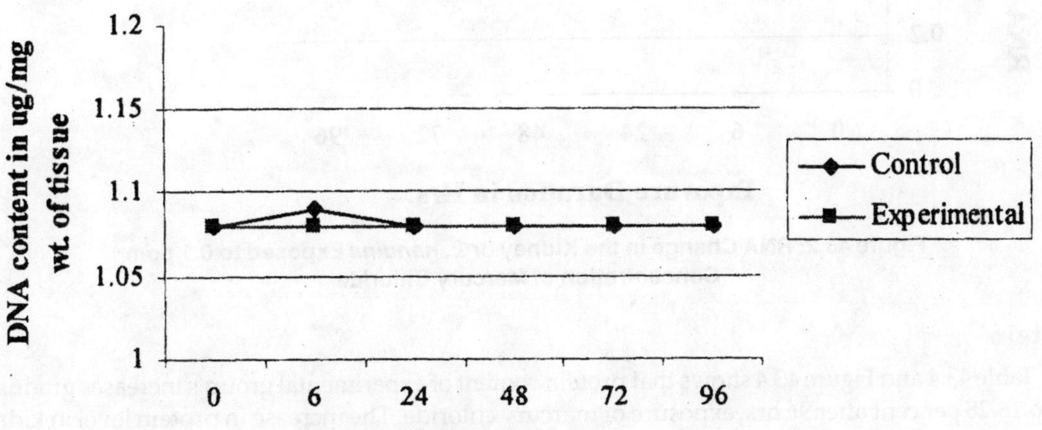

Figure 43.1: DNA Change in the Kidney of *L. nandina* Exposed to 0.1 ppm Concentration of Mercury Chloride

Alkaline Phosphatase

In the present investigation alkaline phosphatase activity was found to be decreasing as the duration of exposure of mercury chloride increases. The total inhibition in alkaline phosphatase activity was 54.939 per cent after 96 hrs of exposure of mercury chloride (Table 43.3 and Figure 43.3). Sastry and Sharma (1981) also observed inhibition in alkaline phosphatase activity in brain of *Heteropneustes fossilis* and Prakash and Maheshwari (2005) in liver of *Labeo nandina* due to exposure of mercury chloride.

Table 43.2: RNA Changes in the Kidney of *L. nandina* Exposed to 0.1 ppm Concentration of Mercury Chloride

Sl.No.	Exposure Duration in Hrs	RNA Content in µg/wt. of Tissue		Difference	% Alter
		Control	Experimental		
1.	0	0.980	0.980	Nil	Nil
2.	6	0.980	0.813	− 0.167	17.040
3.	24	0.980	0.659	− 0.321	32.721
4.	48	0.980	0.450	− 0.530	54.081
5.	72	0.980	0.402	− 0.578	59.039
6.	96	0.980	0.375	− 0.605	61.734

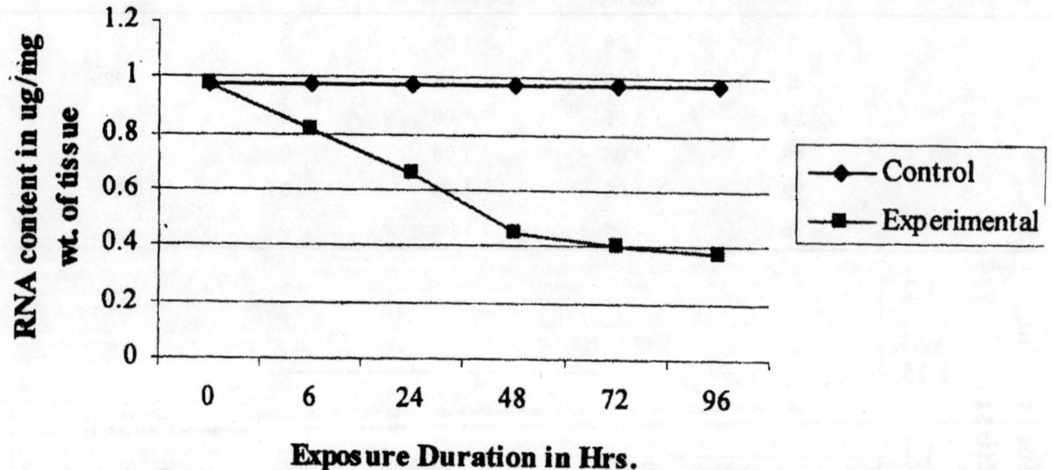

Figure 43.2: RNA Change in the Kidney or *L. nandina* Exposed to 0.1 ppm Concentration of Mercury Chloride

Protein

Table 43.4 and Figure 43.4 shows that protein content of experimental group's increases gradually up to 16.28 per cent after 96 hrs. exposure of mercury chloride. The increase in protein level in kidney may be due to fulfilment of the increased demand of protein for energy supply to fight with the toxicant. Pakash and Mahashwari (2005) and Kaur and Sidhu (2004) have also reported increase in the level of protein due to exposure of heavy metals.

Table 43.3: Protein Changes in the Kidney of *L. nandina* Exposed to 0.1 ppm Concentration of Mercury Chloride

Sl.No.	*Exposure Duration in Hrs*	*Protein in mg/wt of Tissue*		*Difference*	*% Alter*
		Control	*Experimental*		
1.	0	4.55	4.55	Nil	Nil
2.	6	4.55	4.76	0.21	17.040
3.	24	4.55	4.79	0.24	32.721
4.	48	4.55	5.13	0.58	54.081
5.	72	4.55	5.21	0.74	59.039
6.	96	4.55	5.29	0.74	61.228

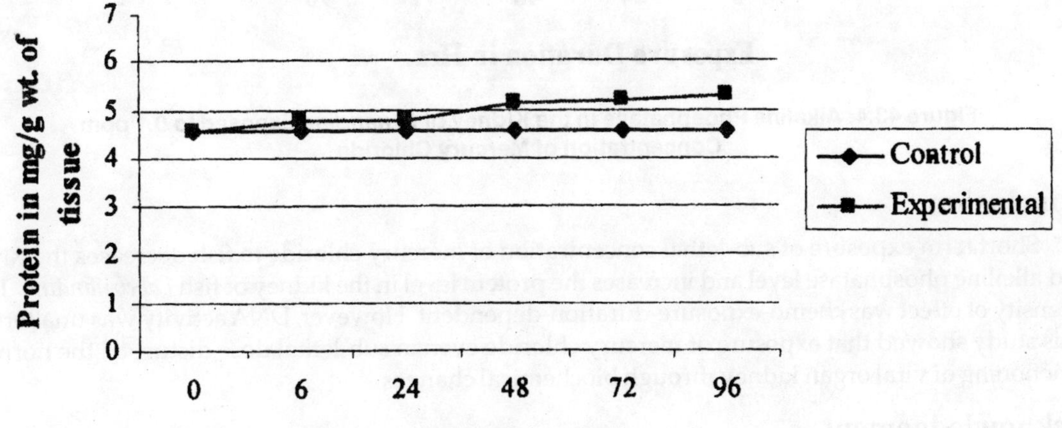

Figure 43.3: Protein Change in the Kidney or *L. nandina* Exposed to 0.1 ppm Concentration of Mercury Chloride

Table 43.4: Alkaline Phosphatase in the Kidney of *L. nandina* Exposed to 0.1 ppm Concentration of Mercury Chloride

Sl.No.	*Exposure Duration in Hrs*	*Alkaline Phosphatase Activity in KA unit/100 ml*		*Difference*	*% Alter*
		Control	*Experimental*		
1.	0	7.39	7.39	Nil	Nil
2.	6	7.39	5.78	– 1.61	21.756
3.	24	7.39	5.08	– 2.31	31.300
4.	48	7.39	4.00	– 3.39	45.872
5.	72	7.39	3.34	– 4.05	54.878
6.	96	7.39	3.33	4.06	54.939

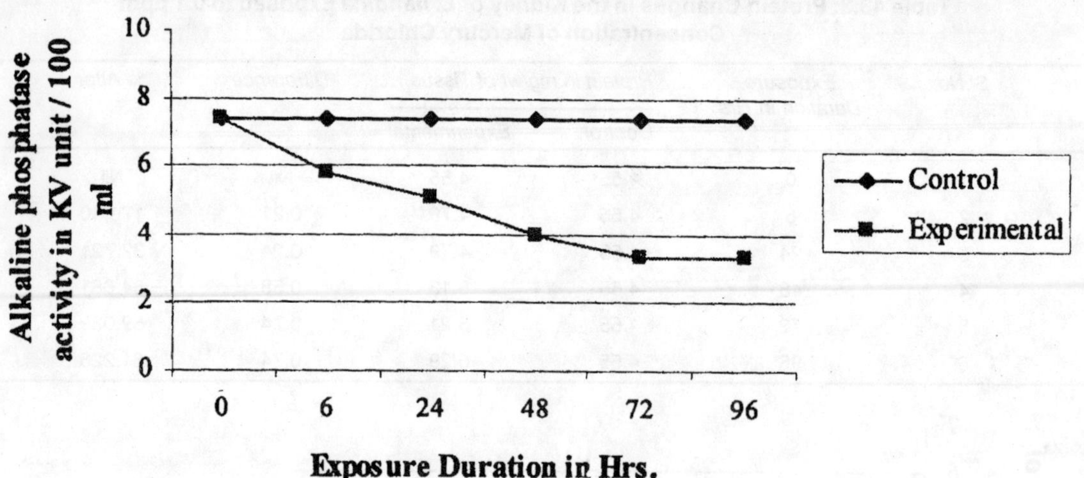

Figure 43.4: Alkaline Phosphatase in the Kidney of *L. nandina* Exposed to 0.1 ppm Concentration of Mercury Chloride

Conclusion

Short term exposure of sub-lethal concentration of mercury chloride to fish decreases the RNA and alkaline phosphatase level and increases the protein level in the kidney of fish *Labeo nandina.* The intensity of effect was chemo-exposure-duration-dependent. However, DNA activity was unaltered. This study showed that exposure of mercury chloride even in sub-lethal dose disturbed the normal functioning of vital organ kidney through biochemical changes.

Acknowledgement

Authors are indebted to Principal, Govt. Model Autonomous Holkar Science College, Indore (M.P.) for encouraging and providing necessary laboratory facilities.

References

Ayyaduri, K., 2004. Mercury pollution in different lotic environment. *Res. J. Chem. Environ.*, 2: 57–61.

Gupta, V.K., 1998. An overview of heavy metals: Impact and remediation. *Res. J. Chem. Environ.*, 2: 63–66.

Jayaraman, J., 2000. *Laboratory manual in Biochemistry*. New Age International Publication, pp. 1–180.

Kaur, S.P. and Sidhu, D.S., 2004. Effect of Cadmium Chloride on the nucleic acid content-DNA/RNA and Protein contents during the adult life of *Henosepilachna vigintioctopunctata* (Fabr.) (Coleoptera). *National Journal of Life Sciences*, 1: 111–115.

Lal, R.B., Sinha, A.P. and Nomani, M.M.R., 2005. Effect of metal toxicity and toxicity exposure on human health. *Indian J. Environ and Ecoplan.*, 10: 533–536.

Muthukrishnan, J., Viswaranjan, S. and Subbulakshmi, R., 1986. Effect of sub-lethal concentration of mercuric chloride on transformation of food by fish *Cyprinus carpio. Environ. Ecol.*, 4: 526–532.

Murata, K., Weighed, P., Araki, S., Budtz, J.E. and Grandjean, P., 1999. Evoked potentials in Faroese children prenatal exposed to methyl mercury Neurotoxicol. *Teratol.*, 21: 471–472.

Oommen, S. and Pilo, B., 2003. Effect of chemical sympathectomy on liver and kidney of rat and pigeon. *Journal of Tissue Research*, 3: 18–24.

Prakash, M.M. and Maheshwari, R.S., 2005. Mercury chloride induced certain biochemical changes in the liver of *Labeo nandina. Adv. Pharmacol. Toxicol.*, 6: 37–43.

Rogers, D.W. and Beamish, F.W.H., 1981. Uptake of water borne methyl mercury by rainbow trout (*Salmo gairdneri*) in relation to oxygen consumption. *J. Fish Aqua. Sci.*, 38: 89–97.

Rogers, D.W. and Beamish, F.W.H., 1982. Dynamics of dietary methyl mercury in rainbow *Salmo gardneri. Aqual. Toxicol.*, 2: 271–290.

Ramadevi, A. and Srinivasan, K., 2005. Agriculture solid waste for the removal of mercury (II) from aqueous solution by Tamarind nut carbon. *Res. J. Chem. Environ.*, 9: 54–65.

Sastry, K.V. and Sharma, Ku. Kamana, 1981. Effect of mercury chloride on the activities of brain enzymes in *heteropneustes fossilis. Matsya*, 7: 66–69.

Verma, B.P., Sinha, A.P. and Nomani, M.M.R., 2002. Effect of methyl mercury chloride (MMC) on ovarian cycle of *Clarias batrachus. Aquacul.*, 3: 17–27.

Chapter 44

Important Medicinal Plants of Agency Tracts of Visakhapatnam District, A.P.

M. Mutyala Naidu and T. Suseela
Agricultural College, Naira – 532 185, Srikakulam District, Andhra Pradesh

ABSTRACT

The agency tract representing the high altitude and tribal zone of Visakhapatnam district holds a promise to its tribal inhabitants in solving their day to day health problems, by the utilization of rich biodiversity of medicinal plants available in the tract.

The important medicinal plants which are the natural inhabitants of this tract and are being widely utilized by the tribals include: *Abrus precatorius, Adhatoda Vasica, Andrographis paniculata, Aristolochia indica, Bixa orellana, Cardiospermum helicacabum, Celastrus paniculatus, Centalla asiatica, Costus speciosus, Curculigo orchiodes, Dioscorea alata, Dioscorea hispida, Eryngium foetidum, Gloriosa superba, Hemidesmus indicus, Oroxylum indicum, Oxalis corniculata, Piper longum, Rhaphidophora decursiva, Rauvolfia serpentina, Tylophora indica* and *Litsea chinensis.* The details of the medicinal uses of the plants are mentioned in the article.

Introduction

Since most of the tribal farmers of using resort to these medicinal plants for solving their health problems, they collect the material as and when needed from their natural inhabitants. However, large scale cultivation of *Piper longum* and *Bixa orellana* is being taken up in the Visakhapatnam agency tract which is facilitated by market facilities from Girijan Corporation. The cultivation practices of these two medicinal plants are detailed in this article.

There has been an increasing trend in the awareness of the utility of medicinal plants in the health care management (Rao *et al.*, 1999) and magnificient diversity of plant resources exists in high altitude agency tract of Visakhapatnam District especially in the medicinal plants. But this wealth is slowly being lost due to continuous loss of forestland; uncontrolled grazing and unscientific collection by local herbal vendors and commercial entrepreneurs. Over 35,000 to 70,000 species of higher plants possess medicinal properties all over the world of which about 7,500 plant species have been used in all the four systems of indigenous medicines *viz.*, Aurveda, Unani, Siddha and Homeopathy. Over 80 per cent of the world's population rely on local health practitioners and traditional medicines for their medicinal needs. Tribal habitants of north coastal districts mostly depend on herbal medicines for solving their day to day health problems.

Looking at the wonderful medicinal diversity and its direct and indirect values it appears that high altitude and tribal zone holds a promise and hope to its habitants. This region due to its variable topographical set up of land forms, variation in climate and soils have resulted into a mosaic of biodiversity in general and medicinal plants in particular. The medicinal uses of different plants have been described in details by Kirtikar and Basu (1987). Depending upon the different altitudes, the diversity, distribution of medicinal wealth of this region varies and the prominent medicinal plants of agency tract of Visakhapatnam District are described below:

Botanical Name: *Abrus precatorius* Linn.

Local Name: Gurivinda

Trade Name: Indian Liquorice

Plant Characters: Deciduous climber growing to 4 mts. Height. The plant has compound leaves, cluster of flowers and seed pods containing scarlet or (rarely) white seeds.

Parts Used: Roots, leaves, seeds

Uses: Seeds are used as contraceptive and also used in treatment of chronic conjunctivitis.

Botanical Name: *Adhatoda vasica* Nees.

Local Name: Addasaramu

Trade Name: Malabarnut tree

Plant Characters: Erect shrub 1–2.5 m height, leaves elliptic, lanceolate, margins entire, flowers white in axillary and terminal spikes, corolla 1–2 cm long, seeds–4.

Parts Used: Leaves, flowers, roots and bark

Uses: The shrub has a considerable reputation as an expectorant and antispasmodic and is prescribed in constipation and other chest infections attended with cough.

Botanical Name: *Andrgraphis paniculata* (Burm. f.) Wall. ex. Nees.

Trade Name: King of Bitters, Kalamegh

Local Name: Nelavemu

Plant Characters: Erect herbs, 50 cm height, leaves 6 × 25 cm, opposite, linear, lanceolate, sessile, flower pink, 0.4 cm long, in terminal, paniculate recemes; capsule sparsely pubescent, seeds about 8.

Parts Used: Whole plant.

Uses: The fresh juice of the leaves is mixed with spices such as cardamom, cloves, cinnamon etc., dried and made into pills for use in household remedy for the minor digestive ailments of children. An infusion and tincture of the dried stems and roots is used for dyspepsia and influenza. An infusion of the plant is used against malaria.

Botanical Name: *Aristolochia indica* Linn.

Trade Name: Serpent root plant, Indian birthwort

Local Name: Nallaeswari

Plant Characters: Unpleasant smelling perennial with heart-shaped leaves and tubular yellow flowers.

Uses: It was formerly used to treat wounds, sores and snake bite. It has been taken after child birth to prevent infection and is also a potent menstruation-inducing herb and a (very dangerous) abortifacient. Decoction is taken to encourage healing of ulcers, asthma and bronchitis.

Botanical Name: *Bixa orellana* Linn.

Trade Name: Bixa or Anota

Local Name: Jaffra

Plant Characters: Evergreen tree growing to a height of 4 to 5 mts. It has large leaves, pink or white flowers and red fruit capsules containing red seeds which are triangular in shape. The seed is of 3–4 mm length. The fruits are available from February to May. The plant starts bearing from third year onwards and continues to yield up to 12th year.

Parts Used: Stem bark, seed, leaves and roots.

Uses: Bark is generally used to treat Genorrhoea, and fever. Seed pulp and leaves are used to treat dysentry and jaundice. Seed is mosquito repellent and also used as colouring agent in food materials.

Botanical Name: *Cardiospermum helicacabum* Linn.

Trade Name: Balloon vine

Local Name: Budda kakara

Plant Characters: Deciduous perennial climbers growing to 3 mts. height with compound leaves, small white flowers and black seeds.

Parts Used: Roots, leaves and seeds.

Uses: Balloon vine root is used to bring on delayed menstruation and to relieve backache and arthritis. The leaves stimulate local circulation and are applied to painful joints to help speed the clearing of toxins. The seeds are also used in the treatment of arthritis. The plant as a whole has sedative properties.

Botanical Name: *Celastrus paniculatus* Willd.

Local Name: Jyotishmati, Malkankini

Trade Name: Black oil tree

Plant Characters: Perennial climbing shrub, branches 4–5 metres length, leaves are thick with toothy margin. Flowers are yellow and hang from tip of the branches. Fruit contains 4–6 seeds. The flowering starts from February and fruits are formed in the month of September.

Parts Used: Leaves, seeds, bark.

Uses: Jyotishmati oil is prepared from leaves, bark and seeds. Oil is used as brain tonic and also improves memory power.

Botanical Name: *Centella asiatica* (Linn.) Urb.

Trade Name: Indian penny wort

Local Name: Saraswathi aku

Plant Characters: A trailing prostrate herb, leaves 4 × 4 cm, orbicular deeply cordate. Flowers 0.2 cm long reddish in axillary umbels.

Parts Used: Whole plant

Uses: Leaves taken as tonic for improving memory and also useful in controlling syphilitic skin diseases both internally and externally.

Botanical Name: *Costus speciosus* Koen. ex Retz.

Local Name: Chengalavakostu

Trade Name: Costus

Plant Characters: Plant grows to a height of 1 metre with underground rhizome. Leaf single at each node, leaf pubscent, flowers white arise from tip of the plant, flower bracts attractive and red in colour. Fruits are red in colour resembling large cardamom.

Parts Used: Rhizomes

Uses: Rhizome has insecticidal properties. It also controls indigestion, fever, skin diseases, Jaundice and antidote for snake bite.

Botanical Name: *Curculiga orchioides* Caertn.

Trade Name: Black Musli

Local Name: Nela taty gadda, Nela tadi

Plant Characters: Common on the gravelly and rocky slopes, usually appear after rains. Erect herb, leaves linear, flowers yellow.

Parts Used: Tubers are used as tonic

Uses: Aphrodisiac, controls piles and jaundice.

Botanical Name: *Dioscorea alata* Linn.

Trade Name: Asiatic yam, Greater Yam

Local Name: Pendalamu

Plant Characters: Climbing perennial herb growing to 5 meters. It has a thick fleshy root, ridged slender stem.

Parts Used: Tubers

Uses: Yam is mostly eaten as a vegetable. It is also useful in the treatment of piles, leprosy, tiredness, weight loss and lack of appetite. Like other yam species its traditional use indicates a hormonal effect.

Botanical Name: *Dioscorea hispida* Linn.

Local Name: Pulidumpa

Plant Characters: Climbing herb growing to 5 mts. It has thick fleshy root. The tuber is like foot print of Tiger and hence the name 'pulidumpa'. The tuber is light yellow in colour.

Parts Used: Tuber

Uses: The tuber is poisonous and its juice is used as antidote for snake poison in agency area.

Botanical Name: *Eryngium foetidum* Linn.

Local Name: Burmis coriander, Kerala Kottimeera

Plant Characters: Spinescent perennial herb, leaves spinous-toothed. Leaves have specific odour similar to that of coriander leaves.

Parts Used: Leaves, roots

Uses: Leaf is stomachic and flavouring agent.

Botanical Name: *Gloriosa superba* Linn.

Trade Name: Malabar glory lilly

Local Name: Adavi Nabhi

Plant Characters: Climbing herb with perennial cylindrical and branched corm, leaves ovate-acuminate, flowers sub-axillary, Capsules oblong.

Parts Used: Tubers, seeds and leaves

Uses: Tubers are poisonous and are used as abortifacient.

Botanical Name: *Hemidesmus indicus* Linn. R.Br. / ex.Schult.

Trade Name: Indian sarsaparilla

Local Name: Sugnadhipala

Plant Characters: Twining herb. Leaves variable in size and shape linear, oblong, lanceolate or broadly elliptic, acuminate glabrous, flowers yellow with purple centre.

Parts Used: Roots

Uses: The roots are used as substitute for sarasaparilla, demulcent, alternative, diaphoretic, diurectic, tonic in loss of appetite. The roots are also soaked in coconut oil and used as hair tonic.

Botanical Name: *Oroxylum indicum* (Linn.) Vent.

Trade Name: Indian Trumpet tree

Local Name: Pampani

Plant Characters: A small tree 7-8 metres height, branched at top, leaves large, flowers numerous light purple in colour.

Parts Used: Roots, leaves, fruits and seeds.

Uses: The bark is acrid, bitter, pungent, astringent to the bowels, cooling, aphrodisiac, tonic. Fruit is expectorant and also improves the appetite. The powder made from the bark is useful to cure the sore-backs of cattle.

Botanical Name: *Oxalis corniculata* Linn.

Trade Name: Indian sorrel

Local Name: Pilli Adugu, Pulichinta

Plant Characters: A common weed of moist and shady places. Small procumbent creeper, rooting at the nodes, leaves trifoliate, leaflets cordate, Flowers in axillary umbels, capsules 1 cm long, linear-oblong, beaked.

Parts Used: Leaves and roots.

Uses: The dried leaves are used in fevers, dysentery, scurvy, piles, biliousness and warts.

Botanical Name: *Piper longum* Linn.

Trade Name: Long Pepper

Local Name: Pippali

Plant Characters: Slender aromatic climber with perennial woody roots. Stems creeping and jointed. Leaves ovate, cordate and oblong.

Parts Used: Roots and fruits

Uses: The fruits and roots are used in respiratory diseases.

Botanical Name: *Rhaphidophora decursiva* (Linn.) Vent.

Local Name: Atukula baddu

Plant Characters: Stem climbing the highest tree-trunks, cylindric, leaves broadly ovate, dark green. Flowers hexagonal.

Parts Used: Stem

Uses: Juice of the stem is applied to fractured bones

Botanical Name: *Rauvolfia serpentine* (Linn.) Betch. exhuvz.

Trade Name: Serpentine root, Sarpagandhi

Local Name: Paatala garuda

Plant Characters: A small under shrub, about 60 cm height, leaves in whorls of three crowded at the top of the branches, elliptic lanceolate, acuminate, shinning and pale below. Flowers in terminal many flowered, Flowers white with scarlet calyx, corolla white, fruits purplish-black when ripe.

Parts Used: Roots

Uses: The roots are used in the treatment of high blood pressure and are sedative. The root has febrifuge and as an antidote to the bites of poisonous reptiles, snakes and scorpion stings. The root is supposed to cause uterine contraction and promote the expulsion of the foetus.

Botanical Name: *Tylophora indica* (Burm.f.) Merr.

Trade Name: Indian Ipsecacuanha, Emetic swallow-wort

Local Name: Asthma teega or verri paala

Plant Characters: Perennial twining climber with lance-shaped leaves and white flowers producing many flat seeds.

Parts Used: Roots, leaves.

Uses: The roots possess stimulant, emetic, cathartic, expectorant, stomachic and diaphoretic properties. Leaves are used in treatment of asthma.

Botanical Name: *Litsea chinensis* Lam.

Local Name: Naramamidi

Plant Characters: A perennial tree growing to a height of 5 mts. Leaves elliptic ovate or oblong lanceolate, flowers yellow in colour.

Parts Used: Roots, Stem bark.

Uses: Root is sweetish bitter, astringent, tonic, expectorant, aphrodisiac, useful in joint pains, thirst, throat troubles, diseases of the spleen and paralysis. Stem bark is useful in treatment of diarrhoea and dysentery.

Chapter 45

Kinetic Studies of Dye Effluent Degradation by *Aspergillus terreus*

N. Rajamohan*

Environmental Engineering Laboratory, Department of Chemical Engineering, Annamalai University, Annamalai Nagar – 608 002, Tamil Nadu, India

ABSTRACT

In the broad range of industrial waste waters, dye wastes are one of the toughest to treat because of the presence of complex synthetic non-biodegradable dyes that are highly resistant. In this work, *Aspergillus terreus*, was employed for the treatment of dyehouse effluent. The time courses of COD reduction and percentage COD removal were studied. A biokinetic model was formulated which relates the COD removal rate to initial effluent COD arid biomass concentration. The COD removal rate showed a first order dependency with respect to both initial COD and biomass concentration. The specific COD removal rate was evaluated as

$$k = 0.0352 \frac{1}{gDCMh} \text{ at } 32°\text{C}$$

Keywords: Dyehouse effluent, Aspergillus terreus, Kinetics.

Introduction

Synthetic dyestuffs are extensively used in textile and dyeing industries. As a result of which, the effluents from these industries are highly coloured and the disposal of these wastes into receiving

* E-mail: natrajmohan@yahoo.com

waters causes significant damages by affecting the photosynthetic activity. Unfortunately, many synthetic dyes are designed to be resistant to conventional effluent treatment. Hence, a need to develop a novel efficient treatment technology for degrading dye house effluent arises. Bioremediation, a technology involving the exploitation of microbial activities to degrade the wastes, is a very good alternative for the above said purpose. Microbial pollution abatement offers significant advantages like simplicity in design and low operating cost. Many of the studies on metabolism of synthetic dyes by fungal and bacterial strains are of little help in the development of biological processes for the treatment of dye house effluent, as they were carried out under conditions, which does not resemble with an industrial effluent environment, for example, using only a single non-commercial synthetic dye (Nigam *et al.*, 1996).

In the light of the aforementioned studies, this study was designed:

1. To study the kinetic characteristics of degradation of dye house effluent by *Aspergillus terreus*.
2. To determine the order of biodegradation reactions with respect to initial COD of the effluent and Biomass.

Materials and Methods

Culture and Medium

The fungi used for the dye house effluent treatment was isolated from the soil sludge obtained from the point of discharge of dyeing unit wastewater. The isolated fungus was identified as *Aspergillus terreus* by Indian Type Culture Collection, Division of Plant Pathology, IARI, New Delhi. The culture was grown in a growth medium composition of 40 gl^{-1} Glucose, 5.0 gl^{-1} peptone from casein and 5.0 gl^{-1} peptone from meat.

Dye House Effluent

The dye house effluent was collected from a dyeing unit situated in Tirupur region (Tamil Nadu, India). The characteristics of the dye house effluent were chemical oxygen demand (COD)–3600 mgl^{-1}, Biochemical Oxygen Demand (BOD)–1220 $mg.l^{-1}$ and pH–10.1±0.1 (APHA, 1992). The colour of the effluent was dark red due to the presence of dyes. The effluent was refrigerated at 4°C and used without any preliminary treatment.

Analysis

Samples were withdrawn from the conical flasks and centrifuged at 13,000 rpm for 5 minutes to separate mycelia from the media. With the clear supernatant obtained, Chemical Oxygen Demand (COD) was evaluated (APHA, 1992). Biomass quantification was done by separating the mycelial pellet through centrifugation. Then the centrifuged pellet was dried for 5 hours at 80°C in an oven and dry weight was found out. The averages of the last three steady-state measurements were used in the data analysis with the standard deviations less than 5 per cent.

Results and Discussion

Bio-Kinetic Modeling

A Michaelis Menten model type equation (equation-1) has been widely applied to study the kinetic pattern of substrate conversion by enzymes or live biomass under isothermal conditions.

The two model parameters (V_m and K_m) in the nonlinear rate expression include enzyme activity (E_0) and specific rates (k_1, k_{-1} and k_2) and are temperature dependent (Yu *et al.*, 2001).

$$v = -\frac{d[Cs]}{dt} = \frac{V_m[Cs]}{K_m + [Cs]} = \frac{k_2[E_0][Cs]}{k_{-1} + \frac{k_2}{k_1} + [Cs]} \qquad \text{.......(1)}$$

It was reported that equation (2} can be considered as a modified model of Michaelis Menten rate equation (equation-1) when the parameter K_m is not included.

$$\frac{dCs}{dt} = -kM^m Cs^n \qquad \text{.......(2)}$$

where, t is the incubation time(h), M is the biomass concentration (mg DCM l^{-1}), m is the partial reaction order with respect to biomass, n is its partial reaction order with respect to COD value and Cs is the COD value of the effluent (mgl^{-1}). Here, k is the specific COD removal rate with units ($mg^{(1-n)(m+1-1)}$/ $mgDCM^m h$). The relationship between the COD of the effluent and incubation time depending on the partial reaction order (n) is given by:

$$\left(\frac{Cs}{Cs_0}\right)^{(1-n)} = 1 - \frac{(1-n)kM^m}{Cs_0^{(1-n)}}t \quad \text{where } (n \neq 1) \qquad \text{.......(3)}$$

and

$$\left(\frac{Cs}{Cs_0}\right) = \exp(-kM^m t) \quad \text{where } (n=1) \qquad \text{.......(4)}$$

The influence of biomass quantity and the initial dye concentration was studied with the dye house effluent whose initial COD value is 3600 mg l^{-1}. The biomass quantity was varied in the range of 0.2 to 1.0 $gDCMl^{-1}$. When the plot between incubation time and ln (Cs_0/Cs) was made, a high degree of linearity was observed ($R^2 > 0.99$) and it was shown in Figure 45.1. Hence the order of the reaction with respect to initial COD value of the effluent was found to be 1.0

$$\ln Cs = \ln Cs_0 - kM^m t \qquad \text{.......(5)}$$

The effect of initial biomass concentration on effluent COD removal rate was found from the experiments conducted with the different biomass concentrations of 0.20, 0.40, 0.67, 0.80 and 1.0 $gDCMl^{-1}$. The incubation time for these experiments was fixed as 48 h. In this set of experiments, used to establish a relationship between biomass concentration and effluent COD removal rate, the incubation time was fixed as 72 h.

A linear relationship ($R^2 > 0.99$) was obtained between biomass concentration M and kMm that indicates m equals to unity, which was shown clearly in Figure 45.2.

Therefore, the COD removal was a first order reaction with respect to biomass concentration (M). The kinetic model of COD removal by the fungal strain *Aspergillus terreus* was given by:

$$\frac{dCs}{dt} = -kMCs \qquad \text{.......(6)}$$

$$Cs = Cs_0 \exp(-kMt) \qquad \text{.......(7)}$$

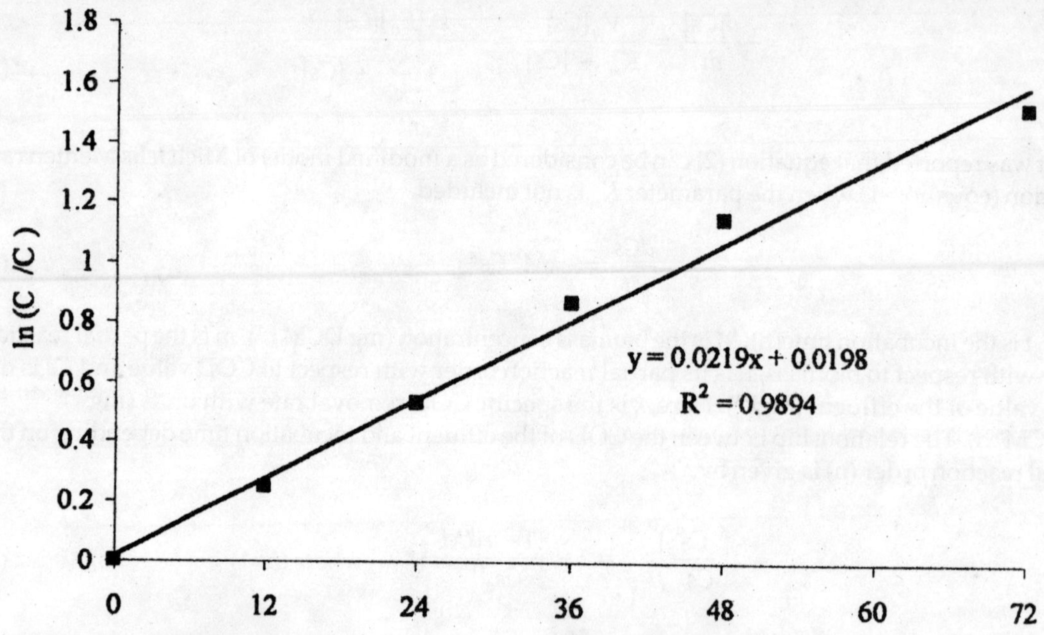

Figure 45.1: Data Fitting and Parameter Estimation of the Kinetic Model for Batch Dye Effluent Degradation by *Aspergillus terreus* (COD = 3600 mg/l, Mo: 0.67 gDCM/l)

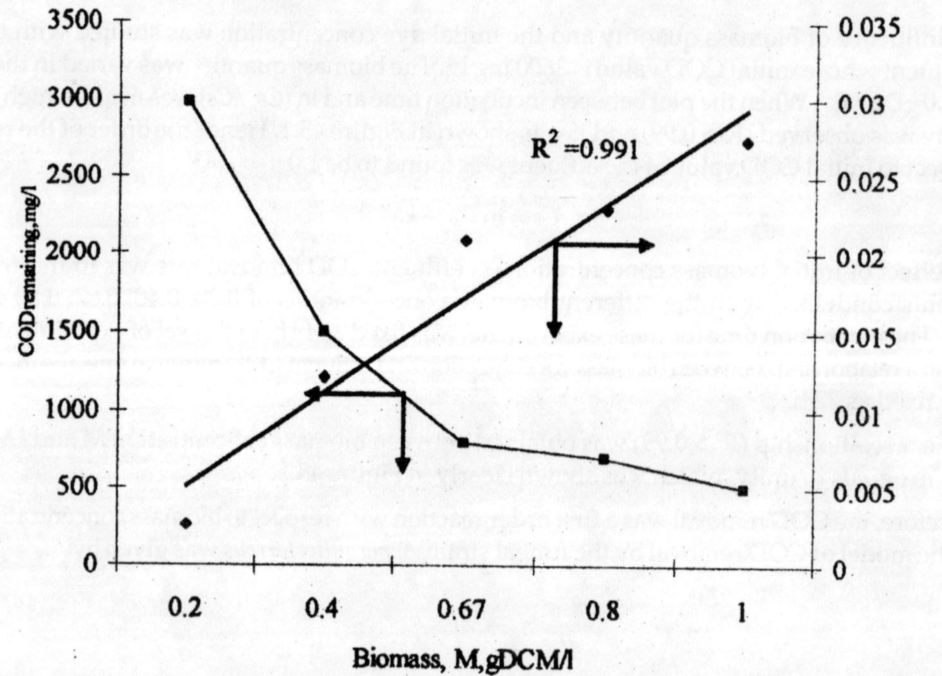

Figure 45.2: Effect of Biomass (*Aspergillus terreus*) Concentration on COD Remaining in Test Effluent with COD: 3600 mg/l for t: 72h

The value of specific COD removal rate, *k*, obtained from the slope of the line in Figure 45.2 was found to be $k = 0.0352 \frac{1}{gDCMh}$ at 32° C.

Conclusion

1. The dye house effluent was effectively degraded by the fungus, *Aspergillus terreus* under aerobic condition.
2. For effluent COD removal, the rate showed a first order dependency on biomass concentration and effluent COD.
3. The specific COD removal rate was found to be $k = 0.0352 \frac{1}{gDCMh}$ at 32° C.

Acknowledgement

The author acknowledges the financial support by the Annamalai University, Annamalai Nagar. Also the author acknowledges the technical support offered by the fellow researchers in the Environmental Engineering Lab, Department of Chemical Engineering, Annamalai University.

References

Banat, I.M., Nigam, P., Datel Singh and Roger Marchant, 1996. Microbial decolorisation of textile dye containing effluents: A review. *Biosource Technology*, 58: 217–227.

Coughlin, M.F., Kinhle, B.K. and Bishop, P.L., 2002. Degradation of acid orange 7 in an aerobic biofilm. *Chemosphere*, 46: 11–19.

Jian Vu, Xiaoweiwang and Pol Lok Yue, 2001. Optimal decolorization and kinetic modelling of synthetic dyes by pseudomonas strains. *Water Research*, 35(15): 3579–3586.

Knapp, J.S. and Newby, P.S., 1999. The decolorisation of a chemical industry effluent by white rot fungi. *Wat. Res.*, 33: 575–577.

Knapp, J.S., Newby, P.S. and Rece, L.P., 1995. The decolorization of dyes by wood-rotting basiomycete fungi. *Enzyme Microbial Biotechnol.*, 17: 664–668.

Lourenco, N.D., Novais, J.M. and Pinheiro, H.M., 2001. Effect of some operational parameters on textile biodegradation in a sequential batch reactor. *J. Biotechnology*, 89: 163–174.

Mou, D.G., Lim, K.K. and Shen, H.P., 1991. Microbial agents for decolorization of dye wastewater. *Biotechnol. Adv.*, 9: 613–622.

O'Neill, C., Hawkes, F.R., Hawkes, D.L., Lourenco, N.D., Pinheiro, H.M. and Delee, W., 1999. Colour in textile effluents: Sources measurement, discharge consents and simulations: A review. *J. Chemical Technol. Biotechnology*, 74: 1009–1018.

Oxspring, D.A., McMullan, G., Smyth, W.F and Marchant, R., 1996. Decolorisation and metabolism of the reactive textile dye, Remazol-Black-B, by an immobilized microbial consortium. *Biotechnol. Lett.*, 18: 527–530.

Poonam Nigam, Geoff Mcmullan, Ibrahim M. Banat and Roger Marchant, 1996. Decolorisation of effluent from the textile industry by microbial consortium. *Biotechnology Letter*, 18(1) 117–120.

Riu, J., Schonsu, I. and Barcelo, D., 1998. Determination of sulfonated azodyes in groundwater and industrial effluents by automated solid phase extraction followed by capillary electrophoresis/ mass spectrometry. *J. Mass Spectrom.*, 33: 653–663.

Shahvali, M., Assadi, M.M. and Rostami, K., 2000. Effect of environmental parameters on decolorization of textile wastewater using phanerochaete chrysosporium. *Bioprocess Engineering*, 23: 721–726.

Rajamohan, N. and Karthikeyan, C., 2005. Optimal degradation and kinetic characteristics of dye effluent treatment by *Galactomyces Geotrichum. Pollution Research*, 24: 55–60.

Rajamohan, N. and Karthikeyan, C., 2004. Treatment of simulated dye house effluent in a rotating perforated tubes biofilm reactor. *Journal of Industrial Pollution Control*, 20(2): 339–344.

APHA, 1992. *Standard Methods for the Examinations of Water and Wastewater*, 18th edition. American Public Health Association, Washington, D.C.

Sumathi and Manju, B.S., 2000. Uptake of reactive textile dyes by *Aspergillus foetidus. Enzyme and Microbial Technology*, 27: 347–355.

Yuzhu, Fu and Viraraghavan, T., 2000. Removal of a dye from an aqueous solution by the fungus *Aspergillus niger. Water Qual. Res.*, 35: 95–111.

Zollinger, H., 1987. *Colour Chemistry: Synthesis, Properties and Application of Organic Dyes and Pigments*, VCH Publishers, New York, pp. 92–102.

Chapter 46

Impact of Open Cast Coal Mining on Soil Environment and its Management

Indu Gupta[1], R.S. Singh[2] and A.C. Gorai[3]

[1]Department of Zoology, S.S.L.N.T. Mahila Mahavidyalaya, Dhanbad, Jharkhand

[2]Scientist, Central Mining Research Institute, Dhanbad, Jharkhand

[3]Reader, R.S.P. College, Jharia, Jharkhand

ABSTRACT

Opencast mining causes land degradation, land fragmentation, soil disruption, soil contamination, erosion and soil quality degradation. The contribution of opencast mining to total coal production around 350 million tones in India during 2003–04 was more than 75 per cent. Mine spoils created due to open-cast mining not fit for plant growth due to their high concentration of heavy metals and destruction of flora and fauna. The study describes the procedures to improve the damage to the soil and improve the ability to support plant growth in long term. It also includes the practices of mine rehabilitation with emphasis on re-vegetation. The same can also be adapted in rejuvenation of mined out areas.

Introduction

In India, there are over 500 coal mines spread over an area of 16,650 sq.km. Mining activity till date are confined to 36 million hectares including the state of Assam, West Bengal, M.P., A.P., Maharashtra and Orissa. Major coal reserves are confined in eastern states of India, known as Jharkhand is full of shrubs, forests and green covers. The entire coal reserves in Indian Sub-continent is riverine in nature. The river valleys in being lust green and a store of biodiversity (Singh, 1995).

Table 46.1: Distribution of Riverine Coal Deposits in India (Singh, 1995)

River Basin	Percentage of Reserve
Damodar river	64.85
Sone and its tributaries	11.60
Mahanadi and tributaries	8.80
Godavari and tributaries	10.29
Others	4.00

With the nationalization of coal industry in the year 1971/72, the opencast mining became extremely popular. Initially, it was introduced as a packet mining. In 1994 the share of opencast mining was 68 per cent of the total coal production. The nature increased to 75 per cent by 2004, however, opencast mining is responsible for extensive damage to the environment and ecology which has caused havoc in the fragile ecosystem.

Table 46.2: Pre-dominantly Opencast Mining Belt (Singh, 1995)

Coal-Field	Coal in mt/sq.km.	Percentage Contribution to Total Surface Production
North Karanpura	11.16	7.30 High concentration
Singrauli	4.95	22.45
I.B. Valley	14.74	6.10 High concentration
Korba	9.69	13.00 High concentration
Wardha	1.23	10.20
Talcher	14.43	8.70 High concentration

Due to opencast mining nearly 60 sq.km. areas per year was estimated to be directly upturned and an additional 150 sq.km. area was likely to be degraded every year due to overburden dumping. Over 70 per cent of coal blocks earmarked for opencast mining during the 8th and 9th Five Year Plans were within the declared forest land. The irrigated land also affected due to opencast mining. The average human population was below 100 sq.km., while the biodiversity was extremely rich with rare species and dense forest.

According to plan of Eastern Coalfields Limited (Anan, 1988), at least 14,000 ha of land in Raniganj Coal field would be damaged due to mining by that time. In Jharia Coal-field, the total of 426.16 ha was identified as degraded area (CMPDI Report, 1986).

During opencast mining, the overlying soil is removed and the fragmented rock is heaped in the form of overburden. This overburden becomes mine spoil. Mining disrupts soil components such as soil horizon and structure, soil microbial population and nutrient cycles that are crucial to sustaining a healthy ecosystem. This disrupted mine spoil are very reluctant and are generally not favorable for soil microbes, invertebrates and vegetational growth.

Mine Spoil

Mine Spoils refers to any overburden material not specifically salvaged in the pre-mining phase (Jha *et al.*, 1992). These are acidic in nature with many scarce organic matter contents. Mine spoils are generally not fit for plant growth. They possess very compacted texture resulting into poor drainage. These are unfit for microbial population and microbes as well as plant growth. Mine spoils are devoid of soil microbes, organic matter, soil moisture and plant available nutrients. They consist of excess amount of heavy metal with little amount of macro nutrients. (Singh *et al.*, 2004)

Chemical Aspects of Mine Spoil

Mine spoils vary in their total heavy metal concentration, pH, cation exchange capacity, organic carbon content and plant available nutrient concentration. Metals in soil solution can be present in the form of free ions; soluble complexes with organic or inorganic may be associated with mobile colloidal materials. Several mine spoils contain toxic levels of soluble elements such as Fe, Zn, Al, Mn, and Cu and inadequate supply of mg, Ca, N and P (Singh *et al.*, 2004). An integrated biotechnological approach has been studied by reclamation of manganese mine spoil dumps (Juwarkar *et al.*, 2002).

Mine spoil vary in their heavy metal concentration, pH, cat-ion exchange capacity, organic carbon and plant available nutrient concentration (Singh *et al.*, 2004). It is generally devoid of macronutrients with substantial amount of micronutrients and heavy metals which are toxic to the microbial population.

Microbial Aspects of Mine Spoil

The high concentration of heavy metal in contaminated mine spoils may inhibit microbial activity (Pierzynski *et al.*, 1994). Soil microbe make the soil structure stable, release required nutrient in inorganic forms by mineralization, produce growth, regulating substance and immobilize heavy metal in soil. Microbial number and microbially mediated soil such as nitrification, denitrification and decomposition of organic matter (Chang *et al.*, 1981; Nordgren, 1988; Bollag *et al.*, 1984; Doelman *et al.*, 1979) are also adversely affected, microbial immobilization of heavy metals in the root zone results into reduced availability and uptake by plants (Pierzynski *et al.*, 1994; Jha, Singh, 1992; Pederson *et al.*, 1980) found that spoils have high bulk density as low porosity. Most of the spoils retained only 25 per cent as much plant available water as the water of natural soil.

Effect of Mining on Soil Environment

1. The clearance of forests for mining results in the loss of genetic resources (Hussain, 1998). Reduce the productivity of soil and create a dust nuisance. Erosion from the overburden and tailing dumps create silt blanketing over agricultural lands and destroy cultivation.
2. Flora-Mining cause disappearance of vegetation and regionally rare plant species.
3. Fauna-on the soil Annelids like earthworm, (*Pheretima posthuma, Genus-Lumbricus*). Arthropods likes termites, white ants, centipede, flies, beetles, julus are very common on soil. Some are useful for the vegetation. These fauna are destroyed due to mining. Loss of fauna including direct losses through clearing and the indirect effects on species through reduction and fragmentation of habitat.
4. Release of toxic elements from overburden: From the environmental point of view the tailings have an impact over the quality of air, water and flora and fauna from the aesthetic aspect. Without doubt the tailings of a processing plant when in current disposed pollute the adjacent area and give prejudice and these terrains for use as agricultural or pastoral destination. (Gergorio *et al.*, 1975).
5. Damage to soils including salination, acidification, pollution and loss of structure.
6. Damage to heritage site: Damaged soil by mining can improve by the following method so that vegetation can grow.

Techniques for Re-vegetation on Mine Spoil

Physical Techniques, Landform Design and Reconstruction

Essential aspects of rehabilitation is reshaping and grading of overburden dump and mined out areas. Following factors should be considered for the shaping of land.

1. *Stability*: The erosion potential of the different materials on site needs to be assessed and a geotechnical investigation may be required. Terraced land form with short steep (angle of repose) stopes and gently sloping terraces (C5 per cent) may be stable and have a higher land capability than a conventional land form of around 15–18 per cent slope.
2. *Drainage Density*: Gradient of slopes increase the drainage density and can change the nature of surface material.
3. *Erosion Control*: Before vegetation, it is important that soil should be controlled by erosion. Soil particle can be lost by three ways *i.e.* blown away by winds, washed away, whole surface may slip away or slump.

Soil having size change from 01–0.05 per cent are effected by wind and 40 per cent are not effected.

Wind erosion can be controlled by three means:

1. Protect the soil surface with a mulch of natural or manufactured material.
2. Maintain the soil in erosion resistant condition *i.e.* must, or with a compacted surface.
3. Reduce the velocity of wind by wind breaks.

On disturbed area water erosion can be controlled by:

1. Slow down of the water flow across the soil surface.
2. Reduce the impact of rain drops on soil surface.
3. Maintain the soil in an erosion resistant condition.

Mulches can be used to protect the soil from raindrops impact. Suitable material for mulches are brush matting, stubble mulch, hay mulch, saw mill wastes and other chemical stabilizers. These mulches are effective only on a small area.

Top soil is the important part for the growth of plant. If the Top soil contains large number of seeds of undesirable species, then it may be better to use the sub-soil as a substrate for rehabilitation. The Top soil contains the majority of the seeds and other plant propagules (such as rhizomes, lignotubers, roots etc), soil microorganism, organic matter, plant nutrients.

The top soil is the upper layer and usually darker than the under lying soils because of the accumulation of organic matter. A void stripping deeper soil horizons which may have poor structure or high clay contents. When the A1 horizon is not obvious, the top 100–300 mm of soil should be recovered. When the aim is to restore the native flora, because most of the seeds of flora stored in top-layer and its removal and return as a thin layer on the surface will maximize the contribution of these seeds to the restoration of micro flora and fauna.

Timing of Top Soil Stripping

Soil should not be stripped or replaced when they are too wet or too dry, as this can lead to compaction, loss of structure and loss of viability of seeds and mycorrhizal innoculum for handling the soil, moisture content vary in different soil local language and experience will help for handling the soil without damage. After setting of seeds clearing and soil stripping should take place. Maximise the store of seed in the soil after setting of seed.

If we want the option is to store other plant (*e.g.* corms bulbs, rhizomes and roots) then the optimum time to strip the soil is when the soil is cool and wet.

Soil Microorganism

Symbiotic microorganisms: Plants form beneficial symbiotic association with a number of soil micro-organisms including fungi, bacteria and actinomycites (single cell plants found usually in soil).

Mycorhiza are a natural component of the ecosystem. Majority of plant species form association with vesicular mycorhiza (VAM) and extomycorrhial fungi. These fungi are effective in increasing the uptake of phosphorous by plants growing in phosphorous deficient soil.

To conserve mycorhizal inoculum Top soil should be:

1. Direct return, wherever possible
2. The pile should be low and revegetated as soon as possible.

Nirtrogen fixation by legumes forms a symbiotic association between the plant and bacteria, known as Rhizobia. This Rhizobia is more tolerant to disturbance and stockpiling than mycorrhizal fungi.

An actinomycetes of the genus Frankia form nodules with the capacity to fix nitrogen on the roots of species from the genera Casuarina and Allocasuarina. These genera are often included to revegetation of Frankia to soil disturbance and stockpiling.

Plant Pathogen: When the soil is infected strictly separate infected soil from uninfected control the movement of infected soil into uninfected area and use of fungicide.

Alternate to Top Soil

If the top soil contains high level of weed soil or plant pathogen then it is unsuitable for rehabilitation, then sub-soil, waste rock or similar material must be used as a substrate for revegetation. To increase chemical properties of such soil, maintain the pH, the physical and chemical properties of proposed substrate should be thoroughly investigated before use in rehabilitation.

The following techniques can improve the plant growth in the long term:

1. Application of animal manure, sewage sludge or other waste.
2. Use of gypsum to reduce the pH of higher alkaline substances.
3. Use of lime to raise the pH of acid substances.
4. Inorganic fertilizer
5. Growing green manure
6. To grow nitrogen fixing species such as legumes.
7. Applying mulch.

Re-vegetation

To restore a native ecosystem, the initial re-vegetation effort is identical to original vegetation. The initial vegetation effort must be establish the building blocks for a self sustaining system so that desired vegetation can grow.

Best time for vegetation is the rainy season. All the preparatory works are completed before the seeds are put in the soil, so that seeds can generate properly.

Species Selection

The species selection depends upon the future land use of the area and the soil condition of the area. If the objective is to restore the native vegetation and fauna then the species are pre-determined. Some species can not grow in the soil where the conditions are different after mining. In this case species should be from outside the mining area. One of the major approaches in terms of establishment of the native ecosystem include search of the local areas for natural analogues of post mining landscape and mine soil which may be used as a model for the past mining ecosystem.

Establishment

Plant species can be established on rehabilitated areas from:

1. Propagules (seeds, lignotubers, corm, bulb, rhizomes and roots) stored in the top soil.
2. Sowing seeds.
3. Spreading harvested plants with bradysporous seeds (seed retained on the plant in persistent woody capsules) on to areas being rehabilitated.
4. Planting nursery-raised seedling.
5. Transplants of individuals from natural areas.
6. Habitat transfer–the transfer of substantial amounts (around 1 m^2 or more in area and 200–300 mm depth) of relatively undisturbed soil with its vegetation intact from natural areas.
7. Invasion from surrounding areas through vectors including birds, animals and wind.

Top Soil

Top soil play an important role in the establishment of native species. It is most important aspects of restoring the full suit of pre-mining species to rehabilitated areas.

Seedling

Sowing seed is an economical and reliable method for establishing some species. Seeding results in a more random distribution of plants than planting the seedlings which leads to more natural looking vegetation. The best species are those which have high germination and survival rates in the field. (Singh *et al.*, 1997) Re-vegetated the overburden mine soil with native species *viz.*, are very successful–Sisum (*Dalbergia Sisoo*), Neem (*Azadirchta indica*), Siris (*Albizia procera*), Gulmohar (*Delonix regia*), Babool (*Acacia nilotica*) these are found to be useful as early colonizer plant species.

Planted Seedling

It is more economical than to seeding. Plant taken from the nursery and nursery raised seeding is more appropriate than to seeding in the top soil. Seedling should always be well adapted before being planted in the field.

Transplanting and Habitat Transfer

Direct transplanting of species is possible by transferring slices or front end loader bucket of soil with the vegetation intact. It is an expensive method. Transplanting is best carried out in wet seasons so that there is maximum survival of plant species.

Seed Collection and Treatment

Seed should be cleaned before storage. Drying the capsules or pods in the sun or in an oven fleshy fruits should be soaked in water to release the seeds.

Clean seeds should be stored in dry containers. The containers should be clearly labeled with details of the species, date collected and collection location. Storage areas should be fumigated regularly. Seeds will be treated with insecticide and fungicide. Seeds must be stored at low humidity (<10 per cent relative humidity) and low temperature. Temperature varies for temperate and tropical plant species.

Seeds of some species require pre-sowing treatment. Germination of most native legumes and other species is enhanced by heat treatment. These species are commonly immersed in boiling water for 30 seconds to five minutes before sowing.

Seed Bed Preparation

Seed bed is a suitable place where seed can germinate. In seed bed soil must be well aerated and seed must be in good contact with soil. In the soil, around the seed, soil must be loose, so that root and seed can grow properly. Seed bed should be free from weeds. In the soil moisture level should be adequate.

Seeding Method

The seeding technique chosen will depend on local factors such as topography, size of the area being rehabilitated and the type of seed various mechanics have been developed for sowing seeds but by hand is still the best options. Care must be taken not to bury the seeds too deep for successful establishment.

Timing of seeding is an important factor for successful revegetation. Seed should be sown immediately after reliable rains or after the break of season. Native seeds require specific moisture and temperature conditions for germination. So that they establish at the optimum time of the year for survival. When seeds are sown early, then they may be eaten by ants, birds, and small mammals, but however, in native vegetation this may be prevented by disturbing the fauna.

Weed Control

Spread of weed is an important consideration in rehabilitation. Care must be taken that weeds are not introduced to the area in manure or as contaminated in seed of desirable species.

Herbicides can be applied selectively using wicks application in some cases, when the weeds are much taller than the desirable species.

Fertilizers and Soil Amendments

The type of fertilizers and the application rate will vary according to the site, soil type and post mining land use. Instead of fertilizer mycorrhizal fungi can be used for the growth of plant. Care should be taken when applying fertilizers to plants. In India, high rates of nitrogen and phosphorous are more favourable for plant growth. If nitrogen is close to roots, then it damages them. High rates of phosphorous and low rates of nitrogen fertilizer will favour the growth of legumes (EPA, 1995).

Calcium carbonate increase the pH of acid soil. Application of lime are in the range of 2–5 tonnes/ha but will vary according to soil type, initial pH and source of lime. Gypsum improve the structure of poorly structured sodic soil. If exchangeable Na is above 6 per cent then soil at around 5–10 tonnes/ha.

Various organic wastes (*e.g.* Animal manures, sewage, sludge and blood and bone) can have value as both fertilizer and soil amendments. However, it is difficult to spread for large-scale rehabilitation.

Nutrient Accumulation

During the mining, vegetation are removed from the soil which lead to the loss of plant nutrients. It is necessary to improve the nutrient in the soil before vegetation. Best method to increase the nitrogen contents in the soil is by establishment of nitrogen fixing plants like legumes, which increase the nitrogen level in ecosystem.

Fauna

To restore the natural ecosystem, the fundamental part is to restore the fauna at the place of mining. Some invertebrates species are introduced if fresh Top soil is placed on the area, but most of the fauna re-colonise from the surrounding area. The total population of the mesofauna (0.1 to 10 mm) varies greatly between sites but can exceed 2000 million per hectare in old grassland soils (Alan Wild, 1996). Some fauna spreads quickly colonise if the areas is fit for their food, shelter and breeding places. If in the surrounding area, population of fauna species is.low, then a captive breeding and release program or re-introduction programme may be appropriate. Animals, like invertebrates are important in many ecological processes such as nutrient cycling, litter decomposition, soil aeration, seed dispersal, seed predation and pollination. Invertebrates also form the food of other fauna group. The abundance and diversity of invertebrates evaluate the success of restoration program.

Maintenance

It includes:

1. Replanting failed or unsatisfactory areas.
2. Repairing of any erosion problem.
3. Watering plants in dry areas especially at the establishment phase.
4. Fire management.
5. Pest and weed control.
6. Control of feral and native population, including fencing.
7. Fertilizer Application.
8. Application of lime or gypsum to control pH and improve soil structure.

Site Monitoring

It is important that after rehabilitation its success should be monitored to assess the extent programme is successful. The success can be defining as to what constitutes a viable ecosystem. What is an acceptable level of species diversity, how indistinguishing should be rehabilitated areas from neighbouring untouched areas and so on.

According components of success criteria could include (EPA, 1995).

1. Physical (stability, resistance to erosion, re-establishment of drainage).
2.

   Biological (species richness, plant density, seed production, fauna return, weed control, productivity, establishment of nutrient cycle).

3. Water quality standards for drainage water and
4. Public safety issue.

Conclusion

Rehabilitation of mined out land is not a operation but it is a part of effective environmental management through all phases of resource development, from exploration to construction, operation and closure and soil is considered to be the medium in which plant life sustain.

Restoration of ecosystem is multi-disciplinary science with complicated life forms ecology even though human have been disturbing the land for many centuries. In biodiversity–there is a technique by which mine site can be developed in to economic improvement of the society. They can achieve this by careful attention to all aspects of rehabilitation and revegetation from initial planning. Cleaning, soil removal, storage and replacement through species selection and re-establishment of vegetation with its associated organism to maintenance of soil micro flora and microfauna for the sustainable development of soil structure and fertility status of mining industry.

References

Chang, F.H. and Broadbent, F.E., 1981. *Soi.*, 132: 461–421.

CMPDI, 1986. Environmental restoration of mined out areas broad estimates: A report, Ranchi.

Doelman, P. and Haanstra, L., 1979. *Soi. Biol. Biochem.*, 11: 487–491.

EPA, 1995. *Overview of Best Practice Environment Management in Mining*. The Environmental Protection Agency of the Australian Environmental Department, Commonwealth of Australia, pp. 30.

EPA, 1995. *Rehabilitation and Revegetation: Best Practice Environmental Management in Mining*. The Environmental Protection Agency of the Australian Environmental Department, Commonwealth of Australia, pp. 30.

Hussain, A., 1998. *Surface Mine Environment and Reclamation*. Samya, Ambikapur, India.

Jha, A.K. and Singh, J.S., 1992. *Rehabilitation of Mine Spoils: Restoration of Degraded of Land Concepts of Strategies*, pp. 211–253.

Juwarkar, A., Khobragade, R., Nimje, M., Dubey, K., Singh, S.K., Gandhi, P.K. and Singh, R.N., 2002. Restoration of biodiversity in coal mine dump. In: *Proceeding of National Seminar on Indian Mining: Problems and Prospects*, 20–21st December. Mining Engineers Association of India, Nagpur, pp. 177–188.

Loveson, V.J. and Dhar, B.B., 1995. *Satellite Remote Sensing Survey Over Some Environmentally Critical Areas in Coalfields of Damodar Basin, India*, pp. 228.

Pierzynski, G.M., Schnoor, J.L., Banks, M.K., Tracie, J.C., Litch, L.A. and Erickson, L.E., 1994. *Vegetative Remediation at Superfund Site, Mining and its Environmental Impact*. Royal Society of Chemistry, pp. 49–69.

Singh, R.S. and Tripathi, N., Singh, S.K., Paul, D. and Gupta, A., 2004. Bioreclamation of contaminated mine spoils. In: *Proceedings International Conference on Soil and Groundwater Contamination Enviro.*, India, pp. 149– 150.

Singh, R.S., Tewary, B.K. and Dhar, B.B., 1997. Reclamation of coal mine overburden dumps in India. In: *Engineering Geology and Environment*, (Eds.) Marinos, Koukis, Tsiambos and Stumaras, Balkema, Rotterdam.

Singh, T.N., 1995. Impact of coal mining on environment and ecology. In: *Proceeding of First World Mining Environmental Congress*, New Delhi, December, pp. 70–72.

Wild, Alan, 1995. Soils and the environment: An introduction. Ch. 5: *Organisms and Soil Processes*, pp. 68–88.

Chapter 47

Bioaccumulation of Trace Metals on the Clam *Meretrix casta*, Before and After Tsunami in Tharangampadi Estuary, Southeast Coast of India

P. Martin Deva Prasath*, R. Vijayakumar and S. Kulathooran

Department of Chemistry, TBML College, Porayar – 609 307, Nagapattinam District, Tamil Nadu, South India

ABSTRACT

Accumulation of trace of metals (Zn, Cu, Fe and Mn) in water, sediments and clam (*Meretrix casta*) at Tharangampadi estuary in Nagapattinam District of Tamil Nadu, lying along the southeast coast of India were studied before and after Tsunami attack by using atomic absorption spectrophotometer. Accumulation of trace metals was observed in the order of sediment > Clam > water. In water, the order was found to be Mn > Fe > Zn > Cu; in sediment it was Mn > Fe > Cu > Zn and in bi-valve it was Fe > Zn > Mn > Cu. In water and bi-valve uniformally the metals recorded higher values after tsunami. In sediments, Zn and Cu recorded higher values whereas Fe and Mn registered lower values after tsunami. Consequently, significant variations were observed in all the four trace metals studied after the tsunami.

Keywords: Tsunami, Clam, Meretrix casta, Trace metals, Tharangampadi, India.

* Corresponding Author: Reader in Chemistry; E-mail: martinprasath@rediffmail.com.

Introduction

Molluscs form an important group of animals noted for their ability to take up large amount of metal ions from solutions. Trace metal accumulations by estuarine mollusks have been studied by Bonadonna *et al.* (1990); Rajathy (1997); Senthilnathan *et al.* (1998) and Kumaresan *et al.* (1998). Bivalves have also been considered as indicators of marine pollution for non-radioactive trace metals (Krishnan Nambisan *et al.*, 1997). Estuary is an important part of marine environment and owing to increased industrial activities and failure to take sufficient safety measures; it is becoming increasingly polluted by manmade activities and sometimes due to natural disasters like the recent tsunami of the 26th December 2004.

A study on distribution of heavy metals in seawater is very important to understand the role in various biogeochemical process of the sea. Significant contributions have been made with reference to oceanic and coastal distribution of various heavy metals (Govindasamy and Azariah, 1999). Further studies reveal that heavy metals are biologically non-degradable and through food chain they may finally pass on to man (Thomas and Jaquet, 1976). Further, the tsunami that struck the Indian coasts on 26th December 2004 with the speed of more than 700 km/h has caused considerable changes in the marine ecosystem by polluting the marine environment with high turbidity and also breaking the coastal vegetations (Kannan *et al.*, 2005). Considering all the above mentioned facts, the present study is aimed to determine the accumulation of trace metals in water, sediment and a clam–*Meretrix casta* in Tharangampadi estuary of the Bay of Bengal, south east coast of India and to compare these values with the pre-tsunami data.

Description of the Study Area

The study area was Tharangampadi, (coastal town station–Lat. 11°2′N; Long 79° 49′ E) one of the severely affected stations of the 26th December 2004 tsunami in Nagapattinam District of Tamil Nadu located along the southeast coast of India (Figure 47.1).

Materials and Methods

Samples of water, sediments and bivalve *Meretrix casta* were collected in Tharangampadi estuary in Nagapattinam District, India on 28th November 2004 (Pre tsunami) and on 26th January 2005 (Post tsunami) and were analyzed.

Surface water samples were collected in precleaned and acid washed polypropylene bottles and they were filtered in Millipore filter paper (mesh size 0.45μ). The samples were preconcentrated with APDC-MIBK extraction procedure (Brooks *et al.*, 1967). The resulting solution was aspirated to the flame atomic absorption Spectrophotometer (Perkin-Elmer Model 373) for the determination of Zn, Cu, Fe, and Mn.

Sediment samples were collected in precleaned, acid washed PVC corer, transferred to clean poly bags and transported to the laboratory. These samples were then washed with metal free double distilled water and dried in hot air oven at 110 °C for 5 to 6 hours and ground to powder in a glass mortar and stored in precleaned polythene bags. 500 mg of the sample was taken and digested with a mixture of 1 ml of concentration H_2SO_4, 5 ml of concentration HNO_3 and 2 ml of $HClO_4$. A few drops of HF was added to achieve complete digestion and the sample was filtered and made up to 25 ml with metal free double distilled water for the estimation of Zn, Cu, Fe, and Mn using flame atomic absorption spectrophotometer (Chester and Hughes, 1967).

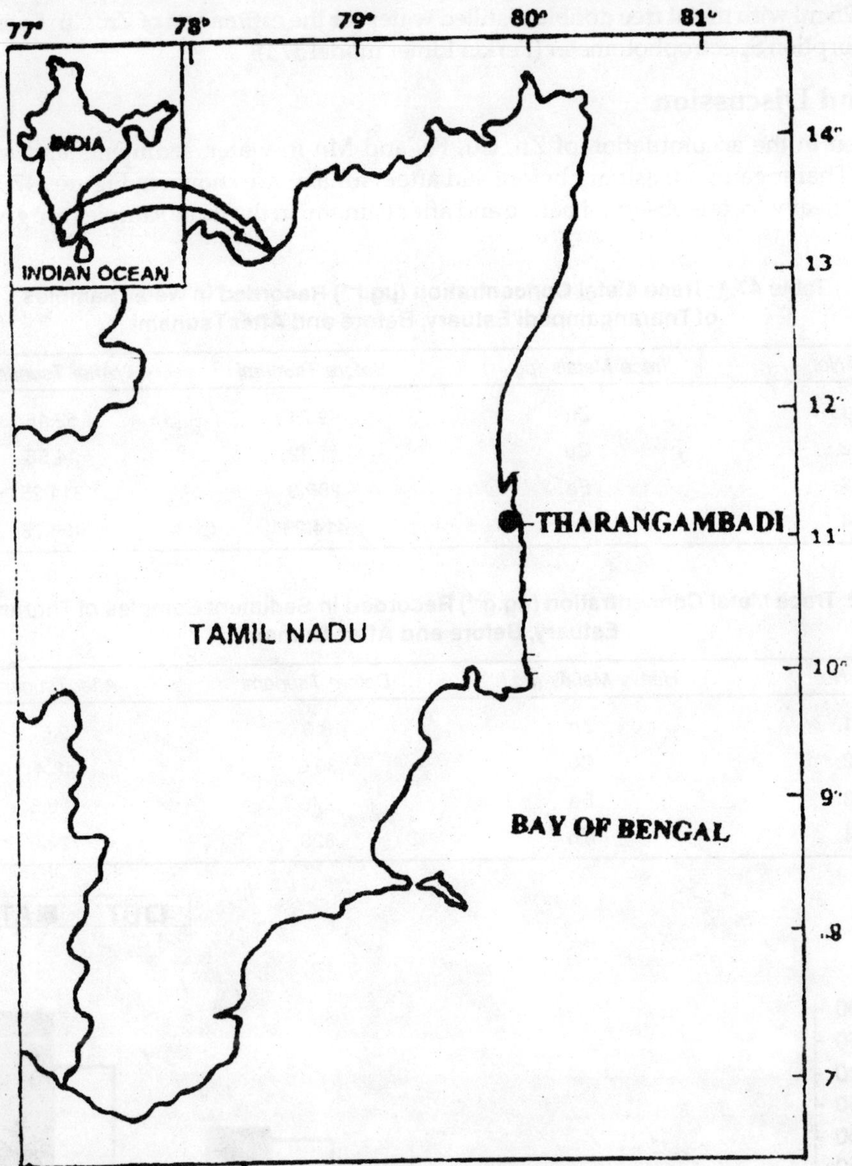

Figure 47.1: Study Area

The method explained by Topping, 1973 was adopted for the analyses of trace-metals in the clam *Meretrix casta.* Ten individuals of the clam of the similar size (6 + 1 cm) were collected and were washed and kept in filtered seawater system to purge the gut content. Whole body tissue of animals were removed, washed with deionized water and brought to the laboratory in frozen condition and stored at –20°C for further analysis. At the time of analysis, the dissected portion of the body muscles were washed and dried in an oven for 24 hours at 110°C to constant weight and then it was powdered. Exactly 500 mg of the powdered sample was digested with 3.1 ratio of nitric acid and perchloric acid in a hot plate for 4 to 6 hours until the clear solution was obtained. The digested samples were diluted to make up 25 ml with metal free double distilled water for the estimation of Zn, Cu, Fe and Mn using Atomic Absorption Spectrophotometer (Perkin Elmer model 373).

Results and Discussion

Variation in the accumulation of Zn, Cu, Fe, and Mn in water, sediments and *Meretrix casta* recorded at Tharangampadi estuary before and after tsunami are shown in Figures 47.2 to 47.4 and the values of heavy metals observed before and after tsunami in the three samples are given in Tables 47.1 to 47.3.

Table 47.1: Trace Metal Concentration (µg.l⁻¹) Recorded in Water Samples of Tharangampadi Estuary, Before and After Tsunami

Sl.No.	*Trace Metals (µg.l⁻¹)*	*Before Tsunami*	*After Tsunami*
1.	Zn	49.21	52.48
2.	Cu	12.42	14.56
3.	Fe	296.3	314.25
4.	Mu	414.23	496.78

Table 47.2: Trace Metal Concentration (µg.g⁻¹) Recorded in Sediment Samples of Tharangampadi Estuary, Before and After Tsunami

Sl.No.	*Heavy Metals (µg.l⁻¹)*	*Before Tsunami*	*After Tsunami*
1.	Zn	8.0	9.6
2.	Cu	30.0	48.4
3.	Fe	270	128.5
4.	Mn	820	724.9

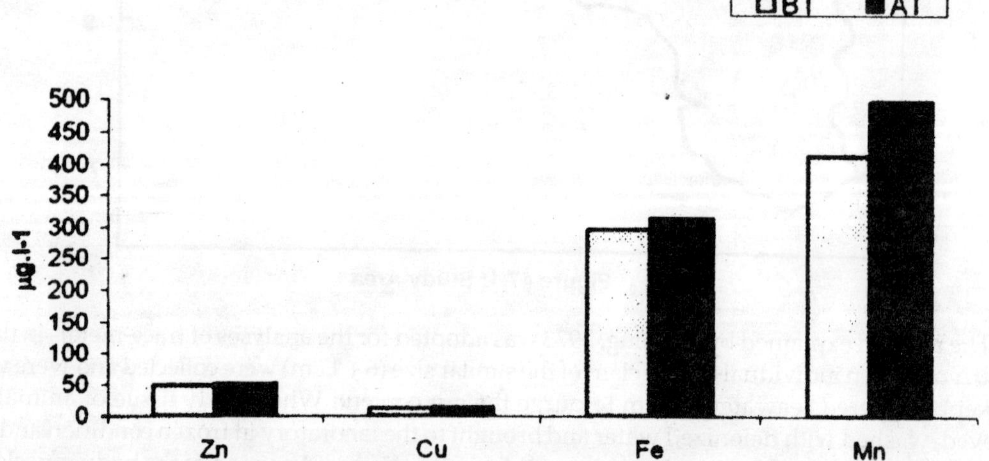

Figure 47.2: Trace Metal Concentration (µg.l⁻¹) Recorded in Water Samples of Tharangampadi Estuary, Before and After Tsunami

BT: Before Tsunami–28th November 2004; AT: After Tsunami–26th January 2005

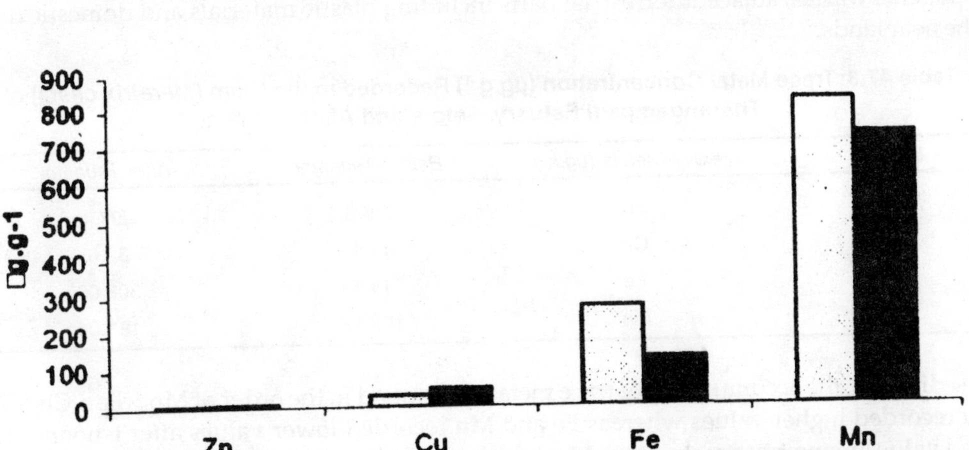

Figure 47.3: Trace Metal Concentration (µg.g^{-1}) Recorded in Sediment Samples of Tharangampadi Estuary, Before and After Tsunami

BT: Before Tsunami–28th November 2004; AT: After Tsunami–26th January 2005

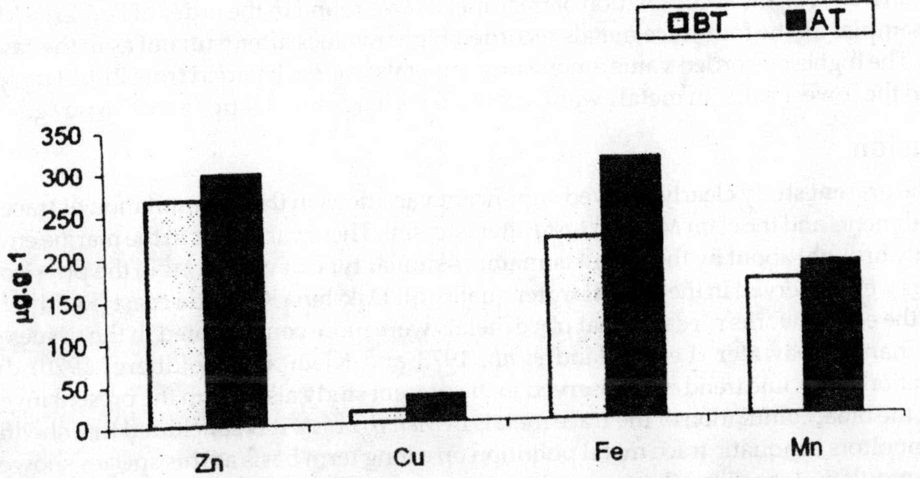

Figure 47.4: Trace Metal Concentration (µg.g^{-1}) Recorded in the Clam (*Meretrix casta*) of Tharangampadi Estuary, Before and After Tsunami

BT: Before Tsunami–28th November 2004; AT: After Tsunami–26th January 2005

In water, the accumulation of trace metals was observed in the order of Mn > Fe > Zn > Cu. Highest recorded value among trace metals was Mn it varied from 414.23 µg/l to 496.78 µg/l and the lowest recorded value among heavy metals was Cu, which varied from 12.42 µg/l to 14.56 µg/l. In the present study, in water samples invariably all the trace metals recorded higher values after tsunami. This variation in metal concentrations is due to the presence of major sources of metal pollution, intensive human activity and discharge of municipal waste and industrial effluents (Govindasamy and Azariah, 1999). Also, the variation in the metal concentration is due to the impact of tsunami that caused large scale seawater inundation and the receding tidal waves carried into the sea, debris, anthropogenic wastes, adjacent terrestrial parts including plastic materials and domestic disposals from the near lands.

Table 47.3: Trace Metal Concentration (µg.g^{-1}) Recorded in the Clam (*Meretrix casta*) of Tharangampadi Estuary, Before and After Tsunami

Sl.No.	Heavy Metals (µg.l^{-1})	Before Tsunami	After Tsunami
1.	Zn	266.8	297.4
2.	Cu	14.4	32.6
3.	Fe	214.14	308.06
4.	Mn	162.12	181.04

In sediments, the accumulation of trace metals was found in the order of Mn > Fe > Cu > Zn > Zn and Cu recorded higher values whereas Fe and Mn recorded lower values after tsunami. Highest recorded value among trace metals was Mn, which varied from 820 µg/g to 724.9 µg/g. The lowest recorded value among metals was Zn, which varied from 8.0 µg/g to 9.6 µg/g. This variation in the heavy metal concentrations was due to the continuous stirring of sediments by the high speed of tsunami tidal actions.

In *Meretrix casta,* the accumulation of trace metals was found in the order of Fe > Zn > Mn > Cu. In bivalve samples all the four trace metals recorded higher values after tsunami as in the case of water samples. The highest recorded value among heavy metals was Fe. It varied from 214.14 µg/g to 308.06 µg/g and the lowest value in metals was Cu, which varied from 14.4 µg/g to 32.6 µg/g.

Conclusion

In the present study clearly showed significant variations in the accumulation of trace metals in water, sediments and the clam *Meretrix casta* after tsunami. These variations in the marine environment is certainly brought about by the recent tsunami as similar type of variations in the physico chemical characters were observed in the coastal water quality off Dakshina Kannada coast (Reddy *et al.,* 2005). Many of the earlier works revealed that trace metals were more concentrated in the tissues of marine animals than in seawater (Leatherland *et al.,* 1973 and Klein and Goldberg, 1970) due to bio magnification. The same trend was observed in the present study also. From the present investigation, based on the bioaccumulation of the trace metals in *Meretrix casta* it is concluded that shellfish can be used as monitors of aquatic trace metal pollution on a long term basis as this species showed greater metal accumulation capacity after tsunami.

References

Bonadonna, L., L. Volterra, F.A-Aulicino and L. Mancini, 1990. Accumulation power of some bivalve mollusks. *Mar. Pollut. Bull.,* 21(1): 81–84.

Brooks, R.R., B.J. Presley and I.R. Kalplan, 1967. APDC-MIBK extraction system for the determination of trace metal in saline water by atomic absorption spectroscopy, *Talanta,* 14: 809–816.

Chester, R. and M.J. Hughes, 1967. A chemical technique for the separation of ferro manganese materials, carbonate minerals and absorbed trace elements from pelagic sediments. *Chem. Geol.,* 2: 249–263.

Govindasamy, C. and J. Azariah, 1999. Seasonal variation of heavy metals in coastal water of the Coromandal coast, Bay of Bengal, India. *Indian J. Mar. Sci.*, 28: 249–256.

Kannan, L.T. Thangaradjou and A. Duraisamy, 2005. Effects of tsunami on the coastal and marine biodiversity of India. *Proceedings of the Brain Storming Session on "Tsunami Mitigation Strategies"* held at Bharathidasan University, 25–26th February.

Klein, D. and E.D. Goldberg, 1970. Mercury in the marine environment. *Environ. Sci. Technol.*, 4: 765–768.

Kumaresan, S., N.V. Vinith Kumar, T. Balasubramanian and A.N. Subramanian, 1998. Trace metals (Fe, Mn, Zn and Cu) in sediments from the Gulf of Mannar Region, southeast of India. *Indian J. Mar. Sci.*, 27: 256–258.

Krishnan Nambisan, P.N., P.T. Lakshmanan and K.Y. Mohammed Salih, 1997. On the uptake of Copper (II) *Meretrix casta* (Cheminitz) and indicator species of metal pollution. *Curr. Sci.*, 45(13): 437–440.

Leatherland, P.M., J.D-Burton, F. Culkin, M.J. McCartney and R.J. Morris, 1973. Concentration of some trace metals in pelagic organisms and of mercury in North Atlantic ocean. *Deep Sea Res.*, 20: 679–685.

Rajathy, S., 1997. Mercury in water, sediment and in some estuarine organisms of the Ennore estuary, Madras, *Tamil Nadu Journal of the Mar. Biol. Ass., India*, 39(1 and 2): 174–178.

Reddy, H.R.V., Katti, R.J., Raveesha, K.P., Chandrashekara, Vikas, S.J., Nagendra Babu, R.S. and Shivakumar, K.S., 2005. Coastal water quality of Dakshina Kannada before and after tsunami. *Current Science*, 88(7): 1026–1027.

Senthilnathan, S., T. Balasubramanian and V.K. Venugopalan, 1998. Metal concentration in mussel *Perna viradis* (Bivalvia/anisomyaria) and Oyster *Carassostrea madrasensis* (Bivalvia/anisomyaria) from some parts in south east coast of India. *Indian J. Mar. Sci.*, 27: 206–210.

Thomas, R.L. and Jacquet, J.M., 1976. Distribution of heavy metals on the fish mugil cephalus. *Fish Res. Board, Canada*, 33: 404.

Topping, G., 1973. Heavy metals in fish from Scottish waters. *Aquaculture*, 1: 373–377.

Chapter 48

Geochemical Study of Fluoride in Groundwater of Sanganer Tehsil (Villages Code Nos. 101–150) of Jaipur District, Rajasthan

Deepika Sohu, J.D. Sharma* and Parul Jain
Reproductive Physiology and Environmental Toxicology Lab, Department of Zoology, University of Rajasthan, Jaipur – 302 004

ABSTRACT

The occurrence of fluoride in groundwater has now become one of the most important health related geo environmental issues in many countries of world. A study has been under taken to determine of fluoride in drinking water collected from 50 villages of Sanganer Tehsil. Various other parameters such as pH, Electrical Conductivity, Total Dissolved Solids, Total hardness, Chloride and Nitrate Concentrations were also measured. The analytical results indicated considerable variations among the analyzed samples with respect to their chemical composition. Majority of samples do not comply with Indian as well as WHO guidelines for most of the water quality parameters measured. The fluoride content of the groundwater of the villages varied from 0.2 to 6.4 ppm causing various ill effects in the inhabitants of the villages. In the study area, 46 per cent villages were having fluoride higher than the permissible limit while 34 per cent villages were found to have fluoride less than the permissible limits and only 20 per cent villages were found to comply the WHO water standards for fluoride.

Keywords: *Fluoride, Groundwater, Fluorosis, Sanganer Tehsil.*

* E-mail: jaishree_ajay@yahoo.co.in.

Introduction

The problem of excessive fluoride in groundwater in India was first reported in 1937 in the State of Andhra Pradesh (Shrott, 1937). In Rajasthan, the first case of skeletal fluorosis was observed at Jobner near Jaipur City (Kalsiwal and Soloman, 1959). Naturally fluoride exists fairly abundantly in the earth's crust and enters the groundwater by natural process. Hydrochemical process controls occurrence of fluoride in groundwater.

High fluoride in groundwater is a considerable health problem in several regions of the world. In India, a considerable part has fairly good distribution of fluoride in groundwater. Excess fluoride problem prevails in 29 countries of world. In India, 20 States are endemic for fluorosis (Susheela, 2001). Millions of men, women and children in many states in India are affected by water borne diseases (Fluorosis) and several lacs of them are in crippling stage leading vegetative life (Vaish, 2004).

However, the fluoride is an essential nutritional trace element in the formation of bones and teeth of all animals (Voznaya, 1981). Excessive ingestion of fluoride through drinking water causes debilitating condition referred to as fluorosis (Mckay, 1925; Teotia and Teotia, 1984; Choubisa *et al.*, 1996 and Kartikeyan *et al.*, 1996).

In Rajasthan, all the 32 districts have been declared as fluoride prone zones. Jaipur district is also one of the worst affected areas. Therefore, in the present study, an attempt has been made for systematic water quality assessment especially with respect to fluoride level in drinking water in rural area of Sanganer Tehsil (Village code Nos. 101–150).

Materials and Methods

Study Area

Sanganer Tehsil is attached with the main city of Jaipur. It lies between 26°49′ to 26°51′ N latitude and 75°46′ to 65°31′E longitude. It covers an area of 636.5 sq.km.

Water Sampling

Drinking water samples were collected in precleaned sterilized plastic containers from different water sources (*i.e.* Hand Pumps, Open Wells, Tube Wells and PHED Water Supply) from the villages during post monsoon season (October–November) 2003. The water samples were analyzed for various, physico-chemical parameters using standard techniques (APHA, AWWA and WPCF, 1976). The parameters studied were pH, Electrical Conductivity, Total Dissolved Solids, Total Alkalinity, Fluoride, Nitrate, Total Hardness as well as Chloride.

Results and Discussion

The results revealed that the drinking water samples had no colour or odour but the taste of water was slightly brackish at most of the locations.

The pH of all water samples was slightly alkaline, which ranged from 7.65±0.03 to 9.3±0.06 Minimum 7.65±0.03 and maximum 9.3±0.06 values of pH were observed from Jotdawala and Dehlawas villages respectively (Table 48.1). However, the permissible limit for pH is 6.9 to 9.2 according to WHO (1992). Therefore, all water samples were within limit for pH except only in Dehlawas where the pH value was found to be out of limit (Table 48.2).

Table 48.1: Analysis of Various Parameters in Drinking Water Samples from Villages of Sanganer (Code Nos. 101–150)

Code No.	*Village Name*	*pH*	*F*	*EC*	*TDS*	*TH*	*Cl⁻*	*TA*	NO_3^-
101	Mahapura	7.85 ±0.03	0.6 ±0.12	3180 ±60.62	1500 ±173.21	640 ±37.85	824 ±43.75	597 ±6.93	38 ±1.15
102	Kishorepura	8.6 ±0.06	2.06 ±0.15	1400 ±42.03	730 ±28.05	435 ±31.23	355 ±24.82	487 ±18.33	44 ±4.0
103	Pawaliya	8.2 ±0.06	1.2 ±0.12	2445 ±200.77	1200 ±30.44	510 ±35.12	653 ±17.79	621 ±31.19	28 ±1.15
104	Roopwas	8.2 ±0.03	1.2 ±0.12	2400 ±63.31	1175 ±68.79	510 ±35.12	653 ±91.72	755 ±18.77	24 ±5.77
105	Manpur Teela	0.1 ±0.06	4.4 ±0.11	750 ±62.92	430 ±64.29	205 ±38.19	262 ±11.13	366 ±13.86	14 ±1.54
106	Charanwas	8.5 ±0.06	2.3 ±0.06	1650 ±26.46	825 ±15.87	480 ±23.09	426 ±40.99	316 ±18.45	19 ±2.89
107	Balawala	8.5 ±0.15	2.1 ±0.20	1650 ±70.23	849 ±175.29	460 ±20.00	510 ±82.03	390 ±42.66	62 ±8.08
108	Tejawala	8.0 ±0.06	0,8 ±0.20	2600 ±52.91	1.300 ±134.19	650 ±30.55	824 ±86.88	559 ±7.23	102 ±27.06
109	Achrawala	8.0 ±0.06	1.2 ±0.30	2580 ±98.49	1200 ±100.00	510 ±15.28	653 ±99.72	925 ±73.76	28 ±1.15
110	Abhaipura	7.85 ±0.03	0.6 ±0.12	2970 ±303.48	1500 ±201.02	640 ±41.63	824 ±46.69	779 ±37.20	78 ±4.16
111	Keshyawala	8.5 ±0.06	2.0 ±0.12	1710 ±18.63	855 ±16.44	480 ±23.09	528 ±16.04	548 ±46.01	38 ±1.15
112	Jagannathpura	8.3 ±0.15	1.8 ±0.12	1800 ±113.58	910 ±102.75	510 ±20.00	605 ±31.56	791 ±69.76	28 ±1.50
113	Madrampura	8.6 ±0.06	2.8 ±0.06	1343 ±23.64	874 ±52.73	480 ±23.09	347 ±38.89	853 ±67.24	34 ±2.31
114	Saipura	7.8 ±0.08	0.6 ±0.12	2940 ±20.02	1440 ±33.29	640 ±32.14	824 ±16.74	791 ±49.65	98 ±3.05
115	Ratalya	8.75 ±0.03	3.8 ±0.06	969 ±25.52	487 ±12.34	320 ±17.32	262 ±30.86	671 ±25.54	78 ±4.16
116	Shri Rampura	8.3 ±0.06	1.2 ±0.12	2350 ±104.05	1200 ±65.5.7	510 ±47.25	653 ±28.50	364 ±12.00	66 ±2.31
117	Lakhana	8.4 ±0.15	1.4 ±0.12	2000 ±87.18	1100 ±52.91	510 ±15.28	605 ±29.26	950 ±31.19	18 ±1.54
118	Jatwala	8.3 ±0.06	1.4 ±0.12	2100 ±23.09	1100 ±46.18	510 ±54.85	624 ±16.74	548 ±20.50	66 ±2.31
119	Bhater	8.75 ±0.03	3.5 ±0.06	1220 ±145.84	630 ±72.24	210 ±17.32	291 ±11.84	841 ±16.16	34 ±1.15
120	Sitarampura	7.8 ±0.06	0.4 ±0.12	4575 ±565.10	2295 ±422.69	970 ±115.90	865 ±196.41	36 ±6.93	78 ±4.16
121	Manyoriyawala	7.85 ±0.03	0.8 ±0.06	2550 ±50.00	1290 ±142.51	640 ±65.06	824 ±14.51	694 ±25.00	34 ±2.31
122	Kokawas	8.25 ±0.03	12. ±0.12	23.10 ±104.40	1200 ±28.86	510 ±20.00	640 ±51.32	706 ±6.93	12 ±3.46
123	Badanpura	8.8 ±0.03	3.2 ±0.12	1240 ±10.00	630 ±30.55	320 ±15.28	320 ±42.19	609 ±24.98	66 ±1.15
124	Dadiya	8.3 ±0.03	1.8 ±0.12	1815 ±17.56	915 ±105.39	510 ±36.06	605 ±14.29	499 ±44.85	54 ±3.05
125	Ganwar Jatan	8.2 ±0.06	1.5 ±0.06	2100 ±49.07	1155 ±131.94	510 ±20.00	624 ±7.02	828 ±57.66	98 ±3.06

Contd...

Table 48.1–Contd...

Code No.	Village Name	pH	F	EC	TDS	TH	Cl^-	TA	NO_3^-
126	Watka	8.75 ±0.31	3.6 ±1.1	1093 ±202.07	548 ±98.96	320 ±30.55	284 ±47.28	500 ±38.19	156 ±20.78
127	Chakwatikawali	7.85 ±0.03	0.4 ±0.12	3240 ±386.04	1580 ±81.45	670 ±78.10	865 ±21.73	719 ±12.00	78 ±4.16
128	Chak amjhar	7.85 ±0.03	0.4 ±0.12	3180 ±39.69	1506 ±60.00	360 ±32.15	824 ±31.02	475 ±23.46	18 ±1.15
129	Dayalpura	8.25 ±0.03	1.6 ±0.12	1900 ±64.29	1065 ±30.00	510 ±20.00	605 ±27.79	1231 ±45.21	30 ±1.15
130	Bhatawala	9.1 ±0.06	4.6 ±0.12	696 ±33.04	348 ±48.23	260 ±30.55	190 ±36.12	488 ±39.15	21 ±1.73
131	Surajpura	8.5 ±0.06	2.5 ±0.10	1425 ±43.66	716 ±18.45	435 ±22.92	355 ±8.08	1268 ±128.89	50 ±9.24
132	Kheri Gokulpura	7.85 ±0.03	1.6 ±0.06	4900 ±125.03	2820 ±55.07	510 ±11.55	892 ±120.68	585 ±44.86	44 ±4.00
133	Ganwar Brahman	8.5 ±0.06	2.3 ±0.10	1498 ±75.51	752 ±36.56	435 ±22.55	355 ±7.02	1109 ±49.17	48 ±1.15
134	Baksawala	8.0 ±0.06	0.8 ±0.12	2486 ±40.10	1269 ±27.39	650 ±15.28	824 ±59.25	768 ±57.66	4 ±0.003
135	Shri Govindpura	7.85 ±0.03	0.6 ±0.12	2826 ±61.10	1300 ±184.14	650 ±56.86	824 ±8.09	341 ±12.01	22 ±2.00
136	Shikarpura	7.85 ±0.03	0.6 ±0.12	2585 ±36.86	1310 ±108.28	640 ±75.49	824 ±11.23	352 ±18.33	38 ±1.15
137	Jotdawala	7.65 ±0.03	0.4 ±0.12	6240 ±326.54	3435 ±271.29	670 ±55.08	1099 ±81.51	280 ±16.65	46 ±6.11
138	Ramsinghpura	7.85 ±0.03	0.4 ±0.12	±4700 ±125.83	2730 ±85.05	970 ±36.06	1099 ±136.63	1049 ±43.09	18 ±1.15
139	Bhambala	7.8 ±0.03	0.4 ±0.12	4700 ±52.91	2700 ±60.00	640 ±30.55	865 ±15.59	426 ±43.19	66 ±30.5
140	Shyopur	8.2 ±0.06	1.4 ±0.12	2150 ±23.63	1171 ±8.02	510 ±22.91	640 ±6.56	1023 ±30.62	34 ±1.15
141	Dehlawas	9.3 ±0.06	6.4 ±0.l2	556 ±62.93	286 ±50.00	60 ±11.55	24 ±2.31	781 ±46.56	70 ±5.03
142	Laxmi Damodarpura	8.4 ±0.06	1.6 ±0.12	2115 ±115.37	1155 ±83.62	510 ±17.56	624 ±19.47	2170 ±46.56	24 ±5.03
143	Chak Sawai Getor	8.8 ±0.06	3.0 ±0.20	1263 ±36.23	633 ±11.36	340 ±10.00	340 ±28.00	890 ±37.64	42 ±2.31
144	Manoharpura	8.5 ±0.06	2.2 ±0.20	1670 ±28.87	870 ±37.51	435 ±12.58	426 ±50.16	634 ±36.66	22 ±1.15
145	Jagatappura	9.2 ±0.06	6.2 ±0.12	580 ±49.39	380 ±76.37	85 ±5.00	43 ±11.24	902 ±18.45	14 ±1.15
146	Barh Teelawala	8.0 ±0.06	0.8 ±0.12	2480 ±84.11	1200 ±34.03	640 ±25.16	824 ±84.91	853 ±39.15	46 ±6.11
147	Teelawala	7.7 ±1.2	0.2 ±0.06	6735 ±368.34	3.705 ±355.70	970 ±55.68	1099 ±53.26	1030 ±25.71	78 ±4.16
148	Mahal	8.2 ±0.06	1.6 ±1.15	1845 ±45.37	930 ±105.63	512 ±11.14	605 ±5.19	987 ±37.21	50 ±9.24
149	Ramnagaria	7.85 ±0.03	0.60 ±0.12	2650 ±80.98	1378 ±97.15	240 ±37.86	824 ±7.50	560 ±18.88	94 ±54.72
150	Barh Karol	8.2 ±0.06	1.2 ±0.12	2115 ±30.14	1155 ±20.03	510 ±17.56	624 ±4.61	1450 ±189.04	34 ±1.15

Table 48.2: Showing Permissible Limits and Level of Water Quality Parameters of Villages

Parameters	*Permissible Limit*	Villages (%)		
		Below Limit	*Optimum Limit*	*Out of Limit*
pH	6.9–9.2	–	98	2
EC	300 μmhos/cm	–	–	100
TDS	500–1500 mg/l	8	76	16
TH	100–500 mg/l	4	–	96
Cl^-	200–600 mg/l	4	28	46
TA	200 mg/l	2	–	98
F	1.0–1.5 ppm	34	20	46
NO_3^-	40–50 mg/l	52	16	32

Fluoride in groundwater is mainly due to leaching from rocks. The minimum value of fluoride (0.2±0.06 mg/l) was observed in Teelawala village and maximum level (6.4±0.12 mg/l) was observed in Dehlawas village (Table 48.1). According to WHO (1992), the permissible limit for fluoride is 1.0–1.5 mg/l. Fluoride content of 20 per cent water samples was found to be with in the limit. The water samples from 17 locations (34 per cent) were having fluoride below the limit while 46 per cent water samples were having fluoride higher than the permissible limit (Table 48.2). The fluoride content exceeding 2 mg/l causes fluorosis with resultant skeletal damage in animals. However, presence of less than 0.8 mg/l fluoride in water causes dental caries in children. Fluoride showed a positive and significant correlation with pH (r = 0.9516) (Teotia *et al.*, 1981, Trivedi, 1988 and Sharma *et al.*, 2005) (Table 48.3).

Table 48.3: Correlation Coefficient "r" Among Various Water Quality Parameters of Drinking Water of Sanganer Tehsil (Code Nos 101–150)

Parameters	*pH*	*EC*	*TDS*	*TH*	*Cl^-*	*TA*	*F^-*	*NO_3^-*
pH	1	–0.8206	–0.7719	–0.8019	–0.9712	+0.0813	+0.9516	–0.0618
EC		1	+0.9908	+0.7771	+0.8131	–0.1362	–0.7126	+0.0991
TDS			1	+0.7407	+0.8419	–0.0981	–0.6759	+0.0819
TH				1	+0.8319	–0.0135	–0.8186	+0.1363
Cl^-					1	+0.9953	–0.9216	+0.0256
TA						1	+0.0359	–0.1536
F^-							1	–0.0096
NO_3^-								1

EC is the capacity of water to carry ions. A large variation was observed in electrical conductivity. All the samples were having EC more than the permissible limit *i.e.* 300 μmhos/cm (Table 48.2) according to USPH. Maximum (6735±368.34 μmhos/cm) and minimum (556±62.93 μmhos/cm) values of EC were observed at Teelawala and Dehlawas villages respectively (Table 48.1). EC signifies the amount of total dissolved salts which indicate inorganic pollution of water. Here TDS found to be varied from minimum 286±50.00 mg/l from Dehlawas village and maximum 3705±355.70 mg/l

Teelawala village (Table 48.1). Data revealed that 16 per cent drinking water samples were having TDS more than the permissible limit *i.e.* 500–1500 mg/l (WHO, 1992), while 8 per cent drinking water samples were having TDS below the permissible limit. Only 76 per cent drinking water samples were complying the TDS standards according to WHO, 1992 (Table 48.2). In the present study, a positive correlation (r = 0.9908) was observed in EC and TDS which is in accordance of Gupta *et al.* (1994) and Sharma *et al.* (2005) (Table 48.3).

Hardness of water is mainly due to carbonates, bicarbonates, sulphates and chlorides of Ca and Mg. Total hardness varied from 60±11.55 to 970±55.68 mg/l from Teelawala and Dehlawas villages respectively (Table 48.1). Total hardness was found to be either out of limit (96 per cent water samples) or below the limit (4 per cent) (Table 48.2). TH showed a negative and significant correlation with pH (r = –0.8019) and fluoride (r = –0.8186) as earlier reported by Trivedi (1988) and Sharma *et al.* (2005) (Table 48.3).

Chloride is considered to be pollution indicating parameters and it is responsible for the salty taste of water. The maximum chloride content (10.99±53.26 mg/l) was observed from Teelawala, Joldawala and Ramsinghpura Villages and minimum (24±2.31 mg/l) was observed from Dehlawas village (Table 48.1). The water samples of 33 locations (46 per cent) were having chloride content higher than the permissible limit *i.e.* 200–600 mg/l (WHO, 1992), whereas 6 per cent water samples were having chloride below the desirable limit. Only 28 per cent water samples were found to have chloride content within limit (Table 48.2).

Total alkalinity of water is mainly due to the presence of carbonates and bicarbonates of the water. The alkalinity was measured from 36±6.93 mg/l (Sitarampura) to 2170±46.56 mg/l (Laxmi Damodarpura) (Table 48.1). TA of 98 per cent water samples was found to be exceeding the permissible limit (200 mg/l according to WHO, 1992), while 2 per cent samples were having TA below the limit. No water sample was found to meet the WHO standards for Total Alkalinity (Table 48.2). TA showed a positive but non significant correlation with Fluoride (r = 0.0359) as reported by Gupta *et al.* (1994) (Table 48.3).

In the present study, nitrate content ranged from 4±0.003 mg/l from Baksawala to 156±20.08 mg/l from Watka village (Table 48.1). Nitrate, content of 32 per cent water samples was higher than the permissible limit *i.e.* 40-50 mg/l (WHO, 1992), while 52 per cent water samples were having nitrate content lower than the limit and only 16 per cent water samples were having nitrate within the desirable limit (Table 48.2). Higher concentration of nitrate in water is responsible for a disease methanoglobanemia or blue baby syndrome.

Conclusion

Most of the water samples collected from the villages of Sanganer Tehsil do not meet the water quality standards for fluoride and other water quality parameters. Hence, it is not suitable for consumption without any prior treatment because it lead to various physiological disorders such as dental and skeletal fluorosis, rapid aging, growth retardation, gastro, intestinal discomforts, genetic disorders as well as very serious sociological problems. So, the water must be treated well before consumption.

Acknowledgement

Authors are thankful to State Groundwater Development and Head Department of Zoology, University of Rajasthan, Jaipur.

References

APHA, AWWA and WPCF, 1976. *Standard Methods for Examination of Water and Wastewater*, 14th edn., American Public Health Association, Washington D.C.

Choubisa *et al.*, 1996. Prevalence of fluorosis in some villages of Durngarpur district of Rajasthan. *Indian Env. Health*, 38(2): 119–126.

Gupta, S.K., Gupta, R.C. and Seth, A.K., 1994. Reversal of clinical and dental fluorosis. *Indian Pediatrics*, 31(4): 439.

Kalsiwal, R.M. and Solomon, S.K., 1959. Fluorosis in a case report. *J. Asso. Phys., India*, 7: 56–59.

Kartikeyan, G., A. Puis and Apparao, B.V., 1996. Contribution of fluoride in water and food to the prevalence of in areas of Tamil Nadu in South India. *Fluoride*, 7: 177–181.

Mckay, F.S., 1925. Mottled enamel: A fundamental problem of dentistry. *Dental Cosmos*, 67: 847.

Sharma, J.D., Sharma, M.K., Jain P. and Sohu, D., 2005. Quality status of potable water of Tehsil Sanganer, District Jaipur, Rajasthan. *Asian J. Exp. Sci.*, 19(2): 113–118.

Shrott, H.E., McRobert, G.R., Bernard, T.W. and Mannadinayar, A.S., 1937. Endemic fluorosis in the Madras Presidency. *Ind. J. Med. Res.*, 25: 553–561.

Susheela, A.K., 2001. *A Treatise to Fluorosis*. F.R.R.D.F., New Delhi, India.

Teotia, S.P.S., Teotia, M. and Singh, R.K., 1981. Hydrogeo skeletal fluorosis in India: An epidemiological study. *Fluoride*, 14: 69–74.

Teotia, S.P.S. and Teotia, M.C., 1984. Endemic fluorosis: A challenging national health problem, *J.A. Pl.*, 32: 347–352.

Trivedi, P., 1988. Relationship between fluoride, total alkalinity, total hardness in groundwater of Pali district in arid and semi arid region of W. Rajasthan. *Proc. Nat. Acad. Sci., India*, 58(A)(1): 7–11.

Vaish, P., 2004. Control and mitigation of fluoride in drinking water through active community participation in Dungarpur district of Rajasthan: A sustainable approach. *National Workshop on Control and Mitigation of Excess Fluoride in Drinking Water*, 5–7th February, T–3: 30–38.

Voznaya, N.F., 1981. *Chemistry of Water and Microbiology*. Mir Publishers, Moscow.

WHO, 1992. *International Standard for Drinking Water*, World Health Organisation, Geneva, Switzerland.

Chapter 49

Studies on Environmental Degradation of Two Wetlands in and Around Guwahati City, Assam, India

P.K. Das[1] and [2]Aruna Bhattacharyya

[1]Department of Geological Sciences, Gauhati University, Guwahati – 781 014, Assam

[2]Department of Geology, Dimoria College, Khetri, Assam

ABSTRACT

Solabeel and Deepar beel are two notable wetlands of Guwahati city. These two beels have been identified as a large water storage beels. The beels once provided habitat for a large number of flora and fauna and received rain water from nearby hills and plains. Previously the both beels accommodates huge water and could control flood of Guwahati area, but nowadays due to huge encroachment, dumping of solid wastes, the morphology of the beels have changed and reduces in size. The quality of the water and soil of the beels has exceeds tolerable limit destroying the aquatic habitat.

To evaluate the nature of pollution and its effect, a systematic study has been a carried out from different locations of the water and soil samples and the different parameters have been determined.

Huge population explosion, rapid urbanization, unauthorized occupation of the beel areas, lack of social awareness for environmental protection and sort of proper planning, the area of wetlands is gradually decreasing and has changed the ecosystem of the beels. Some remedial measures have been suggested to protect the beels for future generations.

Keywords: Environmental degradation, Wetlands, Guwahati city.

Introduction

The term 'wetland' is generally applied to seasonally or permanently waterlogged areas of land. Wetlands include lakes, rivers, estuaries and marshes. Wetland habitats are vibrant with a diversity of living beings.

Wetlands are transitional areas between aquatic and terrestrial ecosystems where the water table is usually at or near the surface. Wetlands are among the most productive life support systems in the world and are of immense socio-economic and ecological importance to mankind.

Wetlands play critical role in the improvement of water quality through the airest pollutants (organic and inorganic) from direct release to the natural water bodies and by removal of sediments, production of oxygen, recycling of nutrients and treatment of wastewater by removing nitrogen phosphorus, bacteria and viruses. The wetlands act as habitat for fish and shellfish and for birds, mammals, reptiles and amphibians. Unfortunately, biodiversity and the ecological functions of wetlands are not widely recognized.

Guwahati, the premier city of North-east India, has covered an area of more than 300 sq.km. of the Brahmaputra valley and occurs in the northern fringe of Assam and Meghalaya plateau. The morpho-tectonic features under the influence of fluvio-denudational process of erosion has given rise to the various land forms like hills, valleys, beels, swamps, and marshy lands. The beels, swamps and marshy lands are the main wetlands of the area.

There are few wetlands in Guwahati city. But Sola beel and Deepar beel play a significant role for restoring the ecosystem and environmental hazards mainly flood problem of the city. Therefore, this type of study is confined within these two beels.

Sola beel, one of the notable wetlands of Guwahati city, is situated near Paltanbazar, which has been identified as a large water storage beel. The beel once provided habitat for a large number of aquatic flora and fauna and receives rain water from nearest hills and plains. Previously the beel accommodates huge water and could control flood of Guwahati area, but nowadays due to huge encroachment by the people, dumping of solid waste and Municipality waste, the morphology of the beel has changed and reduces in size. The quality of the beel water has exceeds tolerable limit destroying the aquatic habitat.

Deepar beel is situated on the southern bank of the river Brahmaputra and west Guwahati area. A permanent freshwater lake on an abandoned channel of the Brahmaputra, Deepar beel is a large natural wetland of immense biological and environmental importance and major storm water storage basin for the Guwahati city. Moreover wetland is rich in biological diversity. Large scale encroachment, heavy siltation from the surrounding hills, accumulation of toxic wastes, construction of railway line along the southern boundary of the beel, industrial development within its periphery, hunting and trapping of wild birds, unregulated fishing practice, invasion of undesirable aquatic weeds, earth-filling and soil cutting etc. disturb the ecosystem and waging a grim battle for survival. Once spreading over an area of 41 sq.km., the beel now barely covers 5 sq.km. and its depth has been reduced to 2.25 metres from 5 metres.

Materials and Methods

To evaluate the nature of environmental degradation and pollution and their effects, a systematic study has been taken up from different locations. Various physical and chemical parameters were undertaken for water and soil samples followed after Tribedi and Saxena (1987), Trivedi and Raj

(1996) and ISI handbook (1986). The different parameters like colour, turbidity, pH, BOD, COD, alkalinity etc. were studied.

Results and Discussion

The colour of water of the Sola beel is almost dark and faulty smelling. The dark colour of the water is due to solid waste deposited there and not easily biodegradable and decrease the photosynthetic activity and other self natural purification of the beel water (Dara, 1995). pH of the water ranges from 6.2–6.5 and it is slightly acidic in nature. Biological activity remains usually restricted to a narrow pH range of 6.5 to 8.5 (Goel, 1996). The hardness ranges from 3.6–5.4. The Dissolved Oxygen (DO), Biological Oxygen Demand (BOD), Chemical Oxygen Demand (COD) shows maximum pollution has taken place (Table 49.1). The soil colour of the beel is almost dark brown. The result of the chemical analysis of soil is shown in Table 49.2.

Table 49.1: Water Analysis of Sola Beel

Sample	*Colour*	*pH*	*Total Alkalinity*	*Total Hardness mg/l*	*DO mg/l*	*BOD mg/1*	*COD mg/l*	*Chloride mg/l*
1	Black	6.5	5	0.6	Nil	40.0	83.6	1.4
2	Greenish with black	6.5	6.5	3.6	Nil	38.0	53.2	1.0
3	Green dark	6.4	6.4	4.8	Nil	38.0	68.4	1.1
4	Greenish	6.5	6.5	4.8	Nil	40.0	60.6	1.2
5	Normal	6.2	6.2	5.4	Nil	36.0	33.3	1.1

Table 49.2: Chemical Analysis of Soil Near Sola Beel

Sample	*Color*	*pH*	*Water Holding Capacity*
1. Northern part of Beel	Dark brown	6.6	0.23 per cent
2. Near Fish Market	Dark brown	7.5	0.23 per cent
3. Eastern side of Beel	Dark brown	6.9	0.24 per cent

The results of chemical analysis of water and soil of Deepar beel were shown in Tables 49.3 and 49.4. The results indicate that the water and soil are less polluted than the water and soil of the Sola beel. Moreover, environmental degradation is also less than the Sola beel. Still, it losses its unique ecosystem balance gradually.

Table 49.3: Water Analysis of Deepar Beel

Sample	*Colour*	*pH*	*Conductance μmho km*	*Turbidity*	*Total Alkalinity mg/l*	*Total Hardness mg/l*	*DO mg/l*	*BOD mg/l*	*COD mg/l*	*Chloride mg/l*
1	Colurless	6.3	217	Nil	1.7	3.0	6.9	20.8	68.0	
2	Colurless	6.5	226	Nil	1.5	2.6	5.7	29.0	49.0	
3	Colurless	6.0	247	Nil	1.9	4.1	6.2	32.5	53.0	
4	Colurless	6.1	212	Nil	2.5	4.8	6.5	26.9	63.0	
5	Colurless	6.2	229	Nil	2.2	2.4	6.6	36.0	69.0	1.1

Table 49.4: Chemical Analysis of Soil Near Deepar Beel

Sample	*Color*	*pH*	*Water Holding Capacity*
1. Near Borhola Beel	Dark brown	6.6	0.23 per cent
2. Azara side	Dark brown	7.5	0.23 per cent

Remedial Measures

1. Scientific management of waste-utilize sewage waste as substitute of wood, cattle feed, fertilizer and fuel production *e.g.* CH_4 gas released due to bacterial growth can be used as a fuel, use of sewage water for fish farming etc. These type of practices should be made to conserve Sola beel and Deepar beel.
2. Clearance of the water by hyacinth and reclamation of beel water should be done. The Government of Assam has already taken this step in 1995 according to the decision of the "State Wetland Committee".
3. Sewage and water treatment plant should be established by Municipal body.
4. Social awareness will be helpful for saving the beels.
5. Pollution Control Board must enforce laws to save the beels.
6. Fencing around the beel should be completed immediately.
7. New settlement and further encroachment should be stopped.
8. Plantation can be done around the beels.
9. Recreation place can be made around the beels.
10. The beels may converted to a Government Fishery which will help state's economic development.
11. NGO's should come forward to restore the ecosystem of the beels.

References

Dara, S.S., 1995. *A Textbook for Environmental Chemistry and Pollution Control*, New Delhi.

Dey, S.K., 1996. *Environmental Chemistry*, New Delhi.

Goel, P.V., 1996. *Water Pollution Effects and Control.*

Goswami, I.D. and Barman D.K., 1998. Geo-environmental impact on the geohydro-morphological setting of Guwahati city and its environment. In: *International Conference on Disaster Management*, Guwahati, Assam, India, April 23–26.

I.S.I., 1996. *Methods of Sampling and Test (Physical and Chemical) for Water used in Industry*. Manak Bhawan, Bahadur Shah Jafar Marg, New Delhi.

Saxena, M.M., 1987. *Environmental Analysis (Water, Soil and Air)*. Agri-Botanical Publication.

Tribedi, P.K., 1987. *Practical Methods in Ecology and Environmental Science*. Environmental Publication.

Chapter 50

Effect of Tamoxifen on Human Erythrocytes: An *in vitro* Study

K. Vijayalakshmi*, S. Seema and S.P. Sasipriya
Post Graduate Department of Biochemistry,
Bharathi Women's College, Chennai – 108, T.N., India

ABSTRACT

Tamoxifen, an anti-estrogen drug is used widely to treat breast cancer. Tamoxifen has many adverse side effects. Tamoxifen irritates the walls of veins causing bleeding. Hence it was planned to study the effect of tamoxifen on erythrocytes. The total nitrite content during haemolysis was studied to evaluate the cause for membrane damage. Various concentrations of tamoxifen was taken and percentage haemolysis was studied. Tamoxifen leads to haemolysis in a dose-dependent manner. 0.4 mM tamoxifen caused maximum haemolysis. The protective effect of coenzyme Q_{10} was also studied. Coenzyme Q_{10} reduced the percentage haemolysis induced by tamoxifen. Total nitrite content increased during haemolysis caused by tamoxifen. Total nitrite content decreased in the presence of coenzyme Q_{10}.

Keywords: Tamoxifen, Coenzyme Q_{10}, Nitrite, Haemolysis, Erythrocytes.

Introduction

Tamoxifen has been the gold standard hormonal agent used for the treatment of breast cancer for more than 20 years (Fisher *et al.*, 1998). Tamoxifen citrate has potent anti-estrogenic properties and

* Corresponding Author: Dr. K. Vijayalakshmi, Reader in Biochemistry, PG Department of Biochemistry, Bharathi Women's College (Autonomous), Chennai – 600 108, Tamil Nadu, India; E-mail: viji42research@yahoo.com.

competes with estradiol for estrogen receptor proteins. Tamoxifen is extensively metabolised after oral administration. A single oral dose of 20 mg tamoxifen gives an average peak plasma concentration of about 40 ng/ml after 5 hours. The decline in plasma concentrations of tamoxifen is biphasic with a terminal elimination and half-life of about 5 to 7 days. N-desmethyl tamoxifen was the major metabolite found in patient's plasma. The oral administration of tamoxifen enters the blood stream, and links with hormones, nutrients, oxygen and other molecules and circulates through the tissues of the body. If breast cancer cells have estrogen receptor tamoxifen slips into the receptor, "locks" the secretion of wore estrogen and thereby prevents breast cancer to occur.

Tamoxifen has adverse side effects. Tamoxifen causes damage to retina. It has been reported that tamoxifen causes endothelial cell damage in blood vessels. The constant irritation and inflammation weakens the veins causing bleeding.

There is growing evidence that nitric oxide (NO^o) has an unpaired electron and its derivatives produced by activated phagocytes play a role in the multistage carcinogenic process (Hakler *et al.*, 2001). It has been demonstrated that oxygen radicals and nitric oxide derivatives such as peroxynitrite and NO_2 can effectively damage DNA despite the presence of multiple antioxidant defense and repair systems (Maeda, 1998 and Tamir, 1996).

Antioxidants counteracts with free radicals and protect the tissues. There are a number of antioxidants like betacarotene, coenzyme Q_{10}, vitamin C and vitamin E. According to Mortensen *et al.* (1985), Coenzyme Q_{10} has remarkable antioxidant properties. He has also reported that CoQ_{10} regulates the flow of oxygen. Coenzyme Q_{10} stimulates the immune system to produce more IgG antibodies and activates the macrophages. Coenzyme Q_{10} has been used to treat breast cancer patients as reported by Folkers (1996). It functions as an important fat soluble antioxidant stabilising the membrane. Hence it was planned to study the effect of tamoxifen on erythrocytes with and without coenzyme Q_{10} supplements to see if it could protect the erythrocytes from free radical damage.

Materials and Methods

Chemicals

Tamoxifen citrate, Coenzyme Q_{10} were purchased from Sigma Chemicals U.S.A. Naphthyl ethylene diamine dihydrochloride and Sulphanilamide were obtained from Qualigens. All other chemicals used were of analytical grade.

Red Cell Preparation

Adult human blood was drawn into a tube containing small amount of 3.8 per cent Sodium Citrate solution. After centrifugation plasma and white cells were removed and the red cells were washed 3 times with PBS (9 parts of 0.9 per cent Sodium Chloride to 1 part of 0.1M potassium dihydrogen phosphate/dipotassium hydrogen phosphate, pH 7.4) at room temperature and used for *in vitro* work. The number of RBCs were enumerated.

Haemolysis

Serial concentrations of the drug tamoxifen citrate (0.05, 0.1, 0.2, 0.3 and 0.4 mM) was tested for per cent haemolysis. Haemolysis was studied at different time intervals of 10, 20, 30, 40, 50 and 60 minutes.

An aliquot was taken immediately and the total nitrite was determined by the method of Green *et al.* (1982). The dosage of 0.4 mM tamoxifen gave the maximum percentage of haemolysis and this was chosen for further study. Keeping this concentration of tamoxifen as constant, various concentrations of coenzyme Q_{10} (1 mM, 2.5 mM. 5 mM, 7.5 mM and 10 mM) were tried and haemolysis was studied. An aliquot was taken immediately and the total nitrite content was determined as earlier.

Results and Discussion

Figure 50.1 shows the effect of tamoxifen on haemolysis. Various concentrations of tamoxifen proportionately increased the percentage of haemolysis. It is seen from the figure that 0.4 mM concentration of tamoxifen after one hour incubation with RBC cells produced about 47 per cent haemolysis. Tamoxifen at a concentration of 20 mg/day results in 40 ng/ml of plasma. Hence this concentration of 0.4 mM was chosen for *in vitro* work.

Figure 50.2 shows the total nitrite content during haemolysis induced by tamoxifen. From the Figure 50.2 it is observed that with increasing tamoxifen concentration the total nitrite content increased. The nitrite content was analysed and 0.4mM concentration of tamoxifen produced 450nmoles of total nitrite during one hour period.

Tamoxifen increased serum nitrate and nitrite levels as reported by Coskun *et al.* (2004). This increase in nitrite production would penetrate through the RBC membrane and undergo degradation. The free radical NO° can be degraded in a variety of reactions but one of the most important is through reacting with heme protein. The reaction of nitric oxide with oxy haemoglobin occur at a very high rate to form nitrosoyl haemoglobin (HbNO and meth Hb) as reported by Cassoly *et al.* (1974) and Smith *et al.* (1996).

The haemolysis induced by tamoxifen could be due to the increased production of nitrite. The increased production of nitrite in turn affect haemoglobin molecule resulting in haemolysis.

To alleviate the toxic effect of tamoxifen, chemoprotective agent like coenzyme Q_{10} was studied. Coenzyme Q_{10} plays a critical role in generating energy. It is a potent antioxidant that counteracts with the free radicals. It also functions as an important fat soluble antioxidant and proved to be a membrane stabiliser as reported by Ernster and Dallner (1995) and Crane (2000).

Hence we planned to study the effect of coenzyme Q_{10} on tamoxifen induced haemolysis. Figure 50.3 shows the effect of coenzyme Q_{10} on haemolysis induced by 0.4 mM concentration of tamoxifen. Coenzyme Q_{10} is very effective in controlling haemolysis induced by tamoxifen in a dose-dependent manner. 10 mM concentration of coenzyme Q_{10} is very effective in controlling haemolysis induced by tamoxifen.

Figure 50.4 shows the total nitrite content produced during haemolysis induced by 0.4 mM tamoxifen in the presence of coenzyme Q_{10}. At 10 mM concentration of coenzyme Q_{10} minimum amount of nitrite was produced.

Conclusion

Tamoxifen at different concentrations lead to the disruption of RBC membrane resulting in haemolysis. During the process of haemolysis, nitrite was produced. Tamoxifen may lead to the production of nitrite which causes RBC membrane damage. To alleviate this toxic effect of tamoxifen, coenzyme Q_{10} was tried to counteract this toxic effect. Coenzyme Q_{10} prevents the production of nitrite

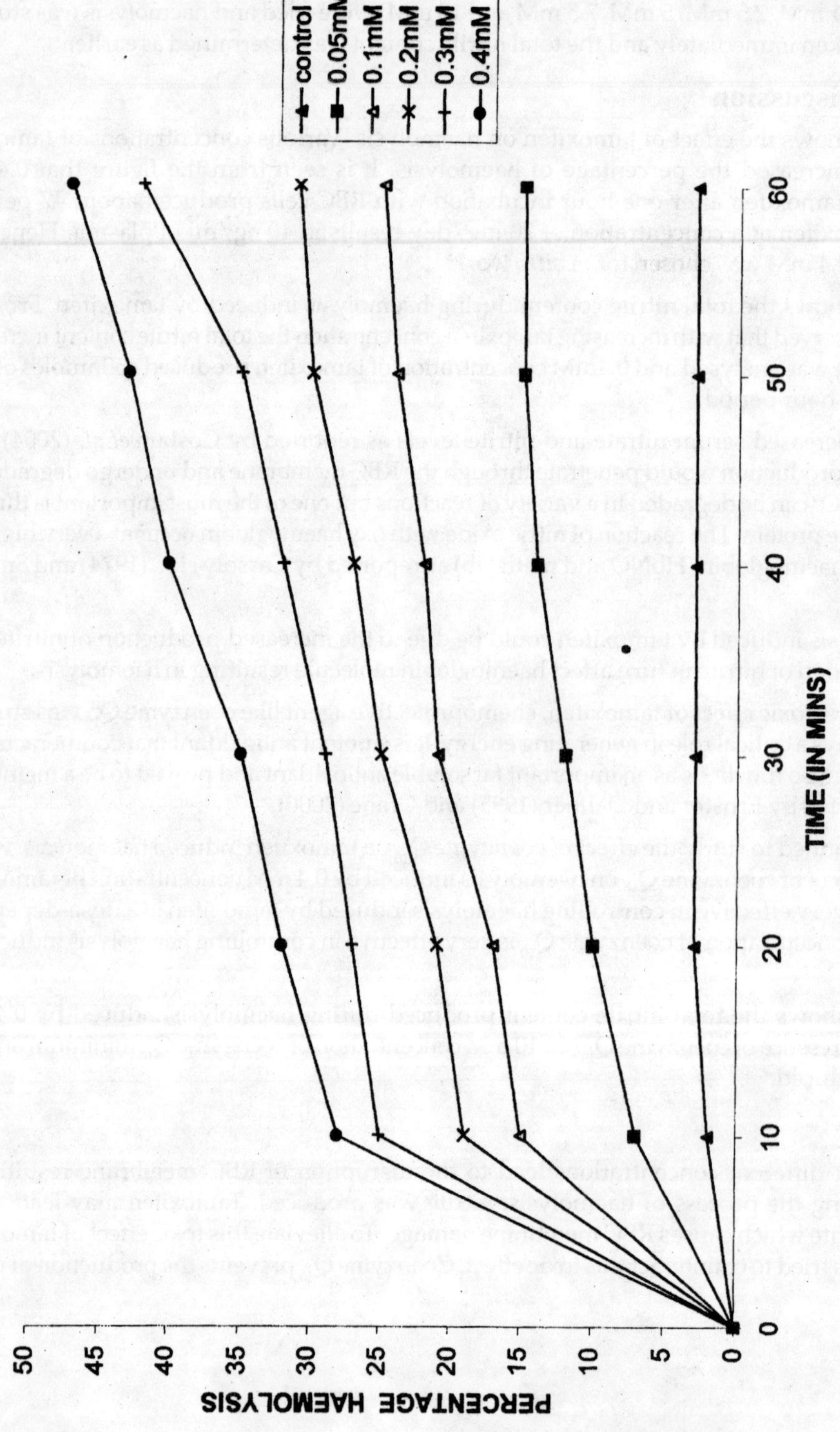

Figure 50.1: The Effect of Tamoxifen on Haemolysis

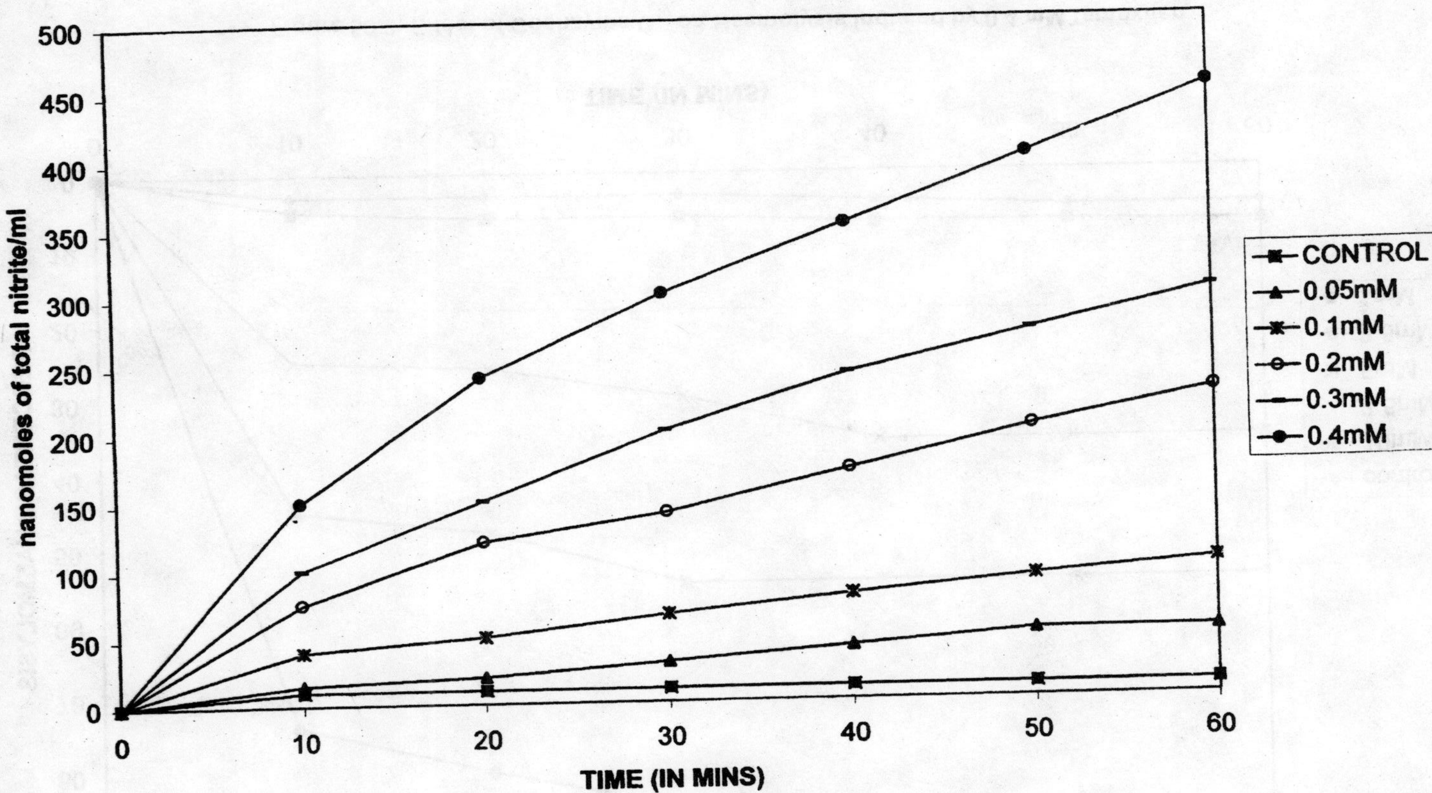

Figure 50.2: Total Nitrite Content During Tamoxifen Induced Haemolysis

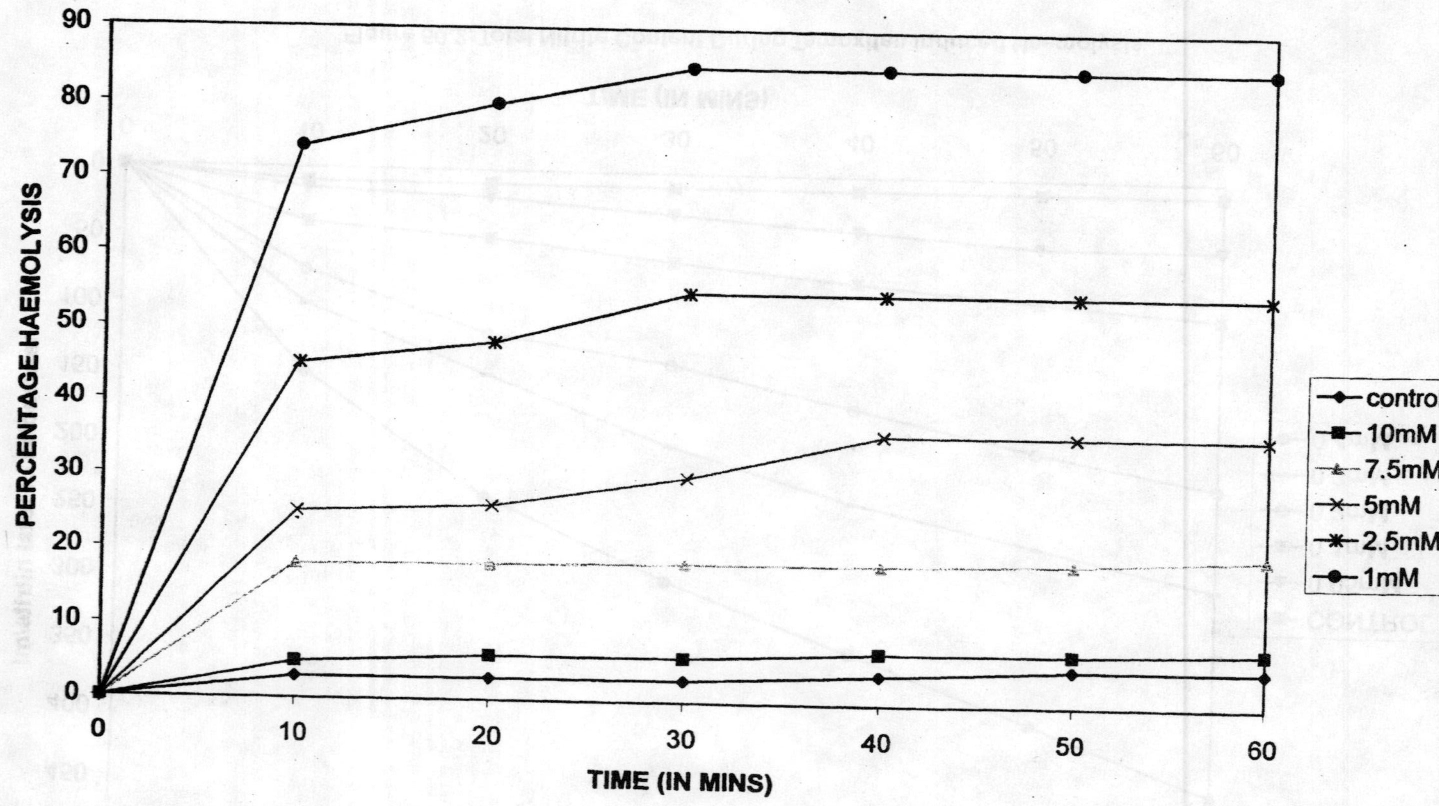

Figure 50.3: Effect of Coenzyme Q_{10} on Haemolysis Induced by 0.4 mM Tamoxifen

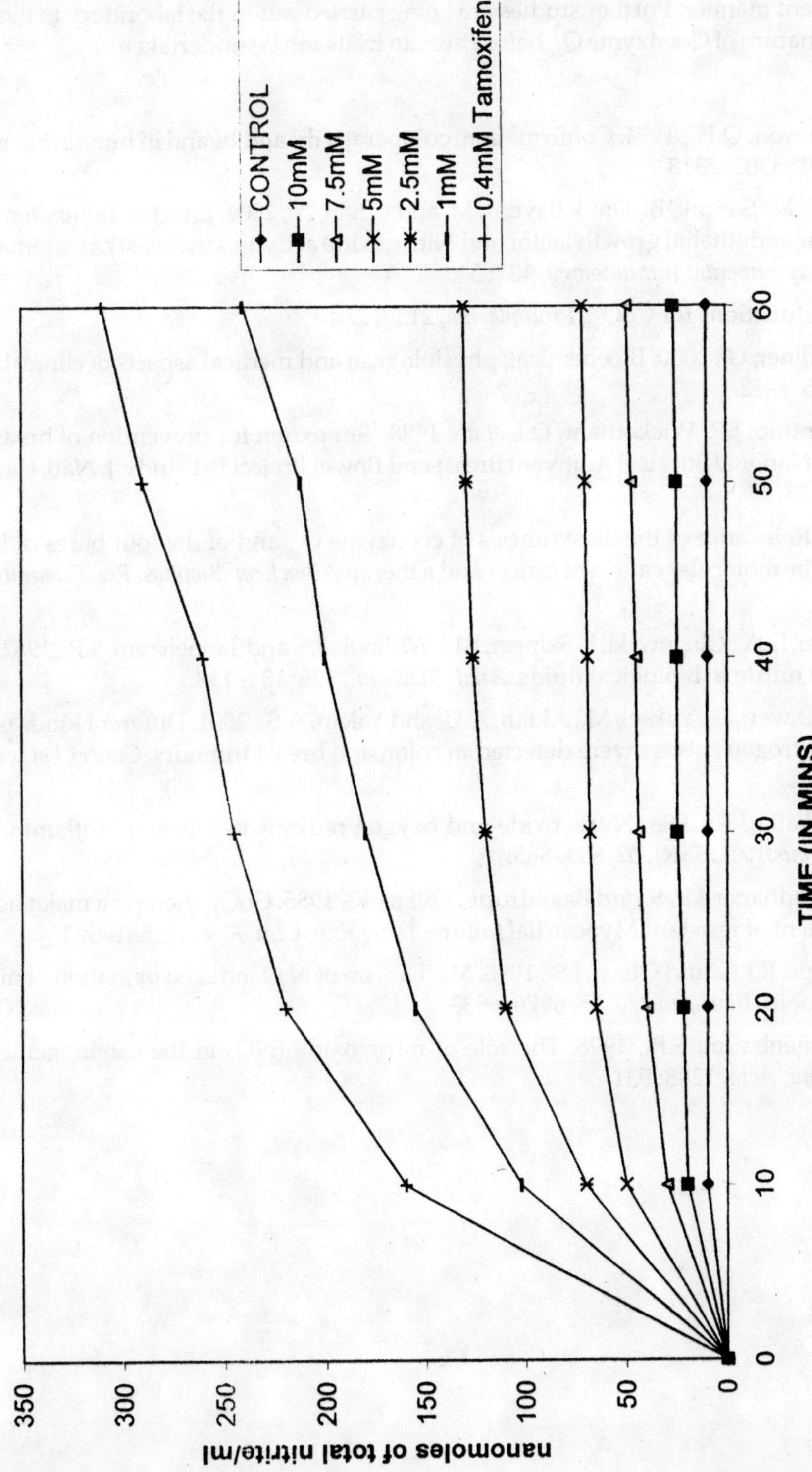

Figure 50.4: Total Nitrite Content in the presence of 0.4 mM Tamoxifen and Coenzyme Q_{10}

in a dose-dependent manner. Further studies are being carried out in the laboratory to evaluate the chemoprotective nature of Coenzyme Q_{10} before human trials can be undertaken.

References

Cassoly, R. and Gibson, Q.H., 1974. Confirmation, co-operativity and ligand in human haemoglobin. *J. Mil. Biol.*, 91: 3301–3313.

Coskun, U., Gunel, N., Sancak, B., Onuk Bayran, M. and Cihan, A., 2004. Effect of Tamoxifen on serum IL-18 Vascular endothelial growth factor and Nitric oxide activities in breast carcinoma patients. *Clinical and Experimental Immunology*, 137: 546.

Crane, 2000. New functions for CoQ_{10}. *Protoplasma*, 213: 127.

Ernster, L. and Dallner, G., 2000. Biochemical, physiological and medical aspects of clinical trials. *US Pharmacist*, 25: 1–12.

Fisher, B., Constantino, J.P., Wickerham, D.L *et al.*, 1998. Tamoxifen for prevention of breast cancer: Report of the National Surgical Adjuvant Breast and Bowel Project P-1 study. J. Natl. Cancer Inst., 90(18): 137–188.

Folkers, K., 1996. Relevance of the biosynthesis of coenzyme Q_{10} and of the four bases of DNA as a rationale for the molecular causes of cancer and a therapy. *Biochem. Biophys. Res. Commun.*, 224(2): 385–391.

Green, L.C., Wagner, D.A., Ologowski, J., Skipper, P.L., Wishnok, J.S. and Tannehaum, S.R., 1982. Analysis of nitrate and nitrite in biological fluids. *Anal. Biochem.*, 126: 131–138.

Hakler, G., Sayin-Ozveri, E., Yuksel, M., Aktan, A.U. and Yalein, A.S., 2001. Different kinds of reactive oxygen and nitrogen species were detected in colon and breast tumours. *Cancer Lett.*, 165: 219–224.

Maeda, H. and Akaike, T., 1998. Nitric oxide and oxygen radicals in infection, inflammation and cancer. *Biochemistry (USSR)*, 63: 854–865.

Mortensen, S.A., Vadhanarikit, S. and Baandrupu, Folker, K., 1985. CoQ_{10} therapy a major advance in the management of resistant Myocardial failure. *Drug Exp. Clin. Res.*, 11: 581–593.

Smith, K.A., Phillips, R.I.G. and Olson, J.S., 1996. Mechanism of N_2O induced oxidation of myoglobin and haemoglobin. *Biochemistry*, 35: 6976–6983.

Tamir, S. and Tannenbaum, S.R., 1996. The role of nitric oxide (NO) in the carcinogenic process. *Biochem. Biophy. Acta*, 1288: F31.

Chapter 51

Natural Radioactivity Levels in Sediments Along the East Coast of Region of Potronovo-Pondicherry Coast, East Coast of Tamil Nadu, India

E. Manikandan[1], B. Rajamannan[1*], R. Ravisankar[2], R. Sakthivel[3], M. Arumugam[3], V. Gajendiran[4] and V. Meenakshisundaram[4]

[1]Department of Physics, [3]Department of Earth Science, Annamalai University, Annamalainagar – 608 102, Tamil Nadu, India

[2]Department of Physics, SSN College of Engineering, Chennai – 603 110, Tamil Nadu, India

[4]Radiological Safety Division, IGCAR, Kalpakkam – 603 102, Tamil Nadu, India

ABSTRACT

The activity concentration of primordial radionuclides in beach sediment samples from South East Coast of region of Porto-Novo-Pondicherry Coast of Tamil Nadu, India has been estimated using gamma-ray spectrometer. The specific concentration for ^{232}Th, ^{238}U and ^{40}K are ranged from (7.78–3687.74 Bq.kg^{-1}), (BDL –469.21 Bq.kg^{-1}) and (BDL-377.11 Bq.kg^{-1}) respectively. The measured activity concentrations for these radionuclides were compared to world average activity of soil. These results indicate no radiological anomaly. The data presented in this study will serve as a base line survey for primordial radionuclides concentration in the study area.

***Keywords:** Radionuclides, Beach sediment, Specific activity and Gamma ray.*

* *Corresponding Author: E-mail: rajamannan@rediffmail.com.

Introduction

Radiation is present in every environment of the earth's surface and beneath the earth and in the atmosphere. The earth is bombarded by cosmic rays from the space and all matter contains some traces of radioactive substance. Mankind has all long lived in natural radiation environment present due to the various natural sources of terrestrial and extra terrestrial origin. The extra terrestrial sources of radiation are cosmic rays and other radioactive elements produced due to the interaction of cosmic rays with the atmosphere. The sources of terrestrial radiation are radionuclides present in the earth's crust or in the atmosphere. Most of the terrestrial radionuclide originates from ^{238}U and ^{232}Th series and the singly occurring primordial radionuclide ^{40}K. This radionuclide may enter into our food chain and cause internal exposure. The distribution of both the natural and anthropogenic radionuclides is an essential prerequisite for the evaluation and control of public exposures.

Natural radiation level may be enhanced as a result of industrial and technological activities such as phosphate industry, thermal power plants, oil refineries etc. The objective of the work is to collect beach sediment samples from Porto-Novo to Pondicherry, East Coast of Tamil Nadu, India and to study the distribution of these radionuclides in the study area.

Sample Collection and Preparation

Twenty-two sediment samples were collected during low tide. The samples were collected 10 to 20 m away from the high tide when it makes towards the coast. At each sample site, covering a sampling area of 1 m^2, five wet samples were collected, each weighing about one kilogram. Among these five samples from a site, four were from the corners and one from the center. The sediment samples were collected at a depth of 5 cm. The samples were placed in plastic pouches and transported to the laboratory. The study area and sample collection locations are indicated in Figure 51.1.

The five samples collected from a sample site were mixed and weighed after removing stones and other unwanted materials. They were packed in sample containers as detailed in references IAEA, 1989. The samples are stored in the sealed containers for one-month period to attain equilibrium between ^{226}Rn and its daughter products (Mollah *et al.*, 1986). The samples were subjected to gamma spectral analysis with a counting time of 20,000 seconds. A 3″ × 3″ NaI (Tl) detector was employed with adequate lead shielding which reduced the background by a factor of about 95. IAEA standard sources were used to estimate the efficiency factors for the required geometry. The concentrations of various nuclides of interest are determined in $Bq.kg^{-1}$ using the count spectra. The peaks corresponding to 1.46 MeV (^{40}K), 1.76 MeV (^{214}Bi) and 2.614 MeV (^{208}Tl) were considered for arriving at the activity levels. The activity concentrations Cu, C_{Th} and C_K for ^{238}U, ^{232}Th and ^{40}K respectively (in $Bq.kg^{-1}$) were deduced from the spectral data by noting the counts in the respective energy for each location regions of interest (ROI).

Results and Discussion

The distribution of natural radionuclides in the beach sediment samples at 22 sampling sites and the activity ratio are presented in Table 51.1. From the table it is seen that the activities of these vary from 7.78 to 3687.74 $Bq.kg^{-1}$ and the arithmetic mean is 451.26 $Bq.kg^{-1}$. The activity concentration of ^{238}U varies from BDL to 469.21 $Bq.kg^{-1}$ and the arithmetic mean of 62.03 $Bq.kg^{-1}$. Similarly the activity concentration of ^{40}K varies from BDL to 377.11 $Bq.kg^{-1}$ and the arithmetic mean of 157.68 $Bq.kg^{-1}$ these are illustrated in Figure 51.2. The revised world wide median activity counteraction of ^{238}Th, ^{238}U and ^{40}K are 30, 35 and 400 $Bq.kg^{-1}$ is compiled by UNSCEAR (2000).

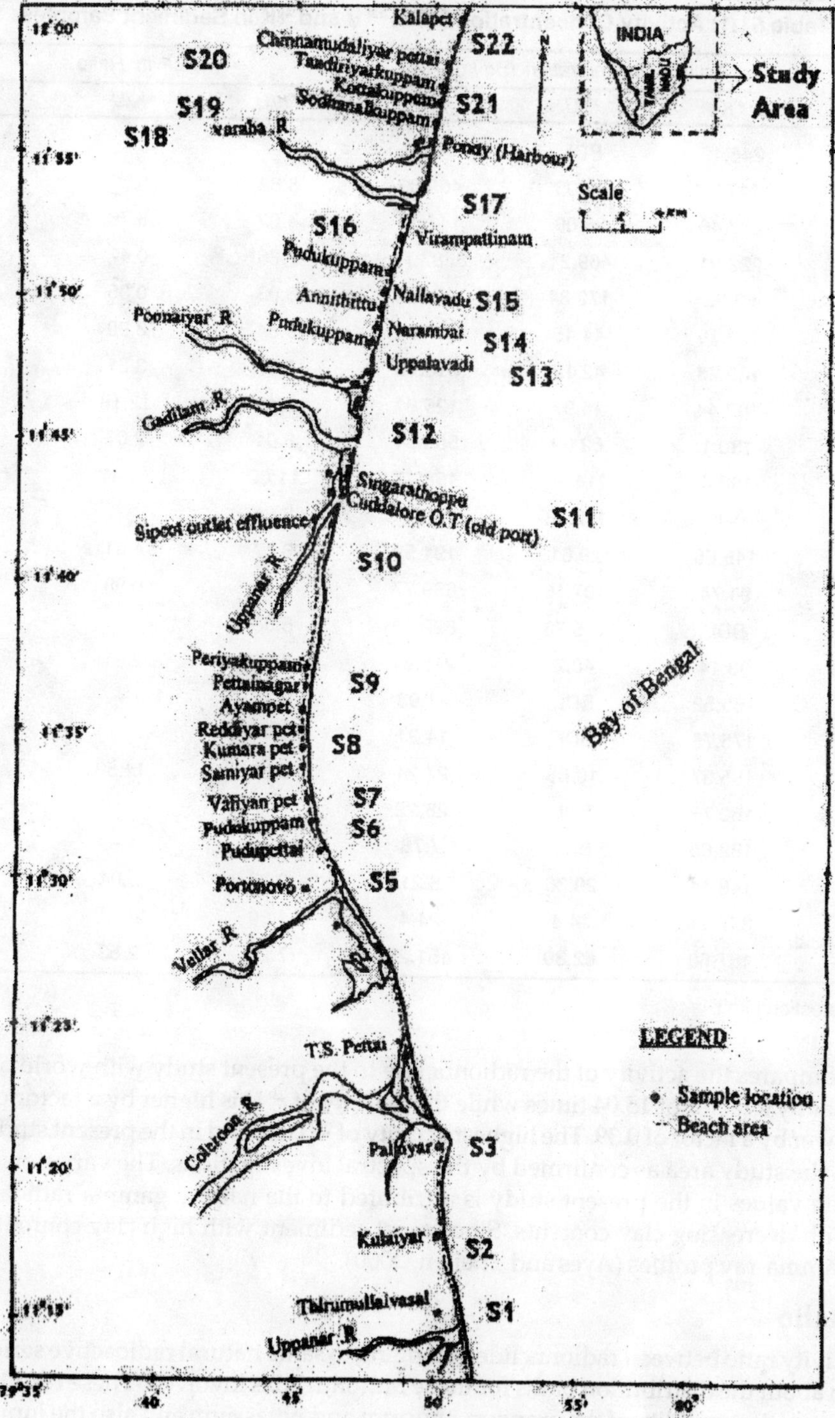

Figure 51.1: Sampling Location Map

Table 51.1: Activity Concentration ^{232}Th, ^{238}U and ^{40}K in Sediment Samples

Sl.No.	Activity Concentration (Bq.kg^{-1})			Activity Ratio's		
	^{40}K	^{238}U	^{232}Th	Th/U	K/U	K/Th
1.	246.12	BDL	58.09	–	–	4.23
2.	174	46.05	409.03	8.88	3.77	0.42
3.	237.46	35.08	141.11	4.02	6.76	1.68
4.	222.91	469.21	3687.74	7.85	0.47	0.06
5.	63.32	172.34	1039.27	6.03	0.36	0.06
6.	173.79	44.45	437.54	9.84	3.90	0.39
7.	139.78	42.12	316.13	7.50	3.31	0.44
8.	182.44	14.97	125.81	8.40	12.18	1.45
9.	130.1	62.69	558.84	8.91	2.07	0.23
10.	193.4	14.4	158.47	11.00	13.43	1.22
11.	BDL	132.33	1144	8.64	–	–
12.	145.66	29.61	191.59	6.47	4.91	0.76
13.	94.74	107.15	689.77	6.43	0.88	0.13
14.	BDL	105.73	622.63	5.88	–	–
15.	93.14	40.2	215.81	5.36	2.31	0.43
16.	165.52	BDL	21.93	–	–	7.54
17.	175.75	BDL	14.21	–	–	12.36
18.	155.37	10.65	27.24	2.55	14.58	5.70
19.	160.75	BDL	28.22	–	–	5.69
20.	189.65	BDL	7.78	–	–	24.37
21.	148.14	29.36	8.21	0.27	5.04	18.04
22.	377.11	14.4	24.4	1.69	26.18	15.45
Mean	157.68	62.30	451.26	7.24	2.53	0.34

BDL: Below Detection Limit.

- If one compares the activity of the radionuclide to the present study with world average value, ^{232}Th is higher by a factor of 15.04 times while the activity of ^{238}U is higher by a factor of 1.77 and ^{40}K activity is lower by a factor of 0.39. The highest activity of ^{232}Th found in the present study is due to the monazite in the study area as confirmed by the spectral investigations. The variation in the activity concentration values in the present study is attributed to the natural gamma radioactivity values increases with decreasing clay contents. Since most sediment with high clay content is associated with high gamma-ray profiles (Ayes and Theilen, 2000).

Activity Ratio

The activity ratio between radionuclide belonging to same natural radioactive series can provide information about the distribution paths for those radionuclides involved. These results were used to establish the relative mobility of the uranium, thorium and potassium and also the input routes of the sediments. The measurement of activity ratio in sediments may provide clues to their source or supplies

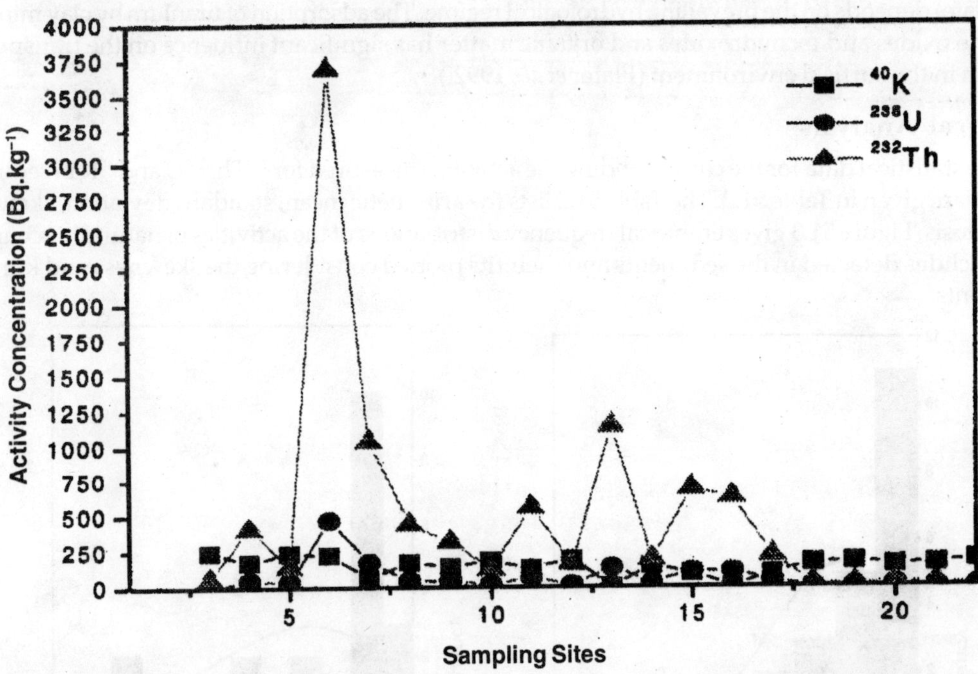

Figure 51.2: Sampling Sites vs Activity Concentration (Bq.kg^{-1})

as well as their pathways in addition to scavenging by particulars (Chung and Chung, 1996). Hence there is a demand to calculate the activity ratio in the present study. The mean activity (Th/U) obtained from the study is found to 7.24.

The theoretically expected Th/U ratio for normal continental crust is about 3.0 while the corresponding value obtained for this ratio is 7.24 which higher than to the expected value. This is due to the different chemical behaviors of Th and U may result in concentration difference another possible reason may be thorium content (and therefore Th/U ratio) of the sediments is much greater in fine size fractions. This concentration of thorium in the fine fraction may reflect a higher content of thorium rich in resistant minerals or relatively a higher content of thorium rich resistant minerals or relatively larger amounts of thorium fixed by clays. The presences of clay minerals in the study area from the spectral (FTIR and XRD) investigations sustain the above statements.

The activity ratios of (Th/U) in almost all the samples are greater than unity, which indicates either a net loss of ^{238}U or an addition of ^{232}Th by weathering processes (Scott, 1968). Further the higher value of the activity ratio (Th/U) supports Th enrichment during weather and transport processes (Balakrishna, 2001).

The Th/U, K/U and K/Th ratio may provide an indication whether relative depletion or enrichment of radioisotopes had occurred. From the Table 51.1, the average values are also calculated and are equal to 7.24, 2.53 and 0.34 respectively. The crystal o rocks have the following ratio Th/U = 3.5, K/U = 1×10^4 and K/Th = 3×10^3, the ratio values in the present are compared to crystal rocks; it is found that, the values of K/Th and K/U is lower. This might reflect a quite different enrichment/ depletion of the radionuclides.

This is may be due to the relative mobility of uranium (largely dissolved) and Thorium (largely particulate) depends on the prevailing hydrological regime. The adsorption of uranium by clay minerals insoluble oxides and ox hydroxides and organic matter has significant influence on the transport of uranium in the surfical environment (Plater *et al.*, 1992.)

Statistical Analysis

The statistical data for the corresponding the activities measured for ^{232}Th, ^{238}U and ^{40}K in sediment samples are given in Table 51.2. The Table 51.2 lists the arithmetic mean, standard deviation, skewness and kurtosis. Figure 51.3 gives empirical frequency distributions of the activities of naturally occurring radionuclides detected in the sediments and their fits plotted considering the skewness and kurtosis coefficients.

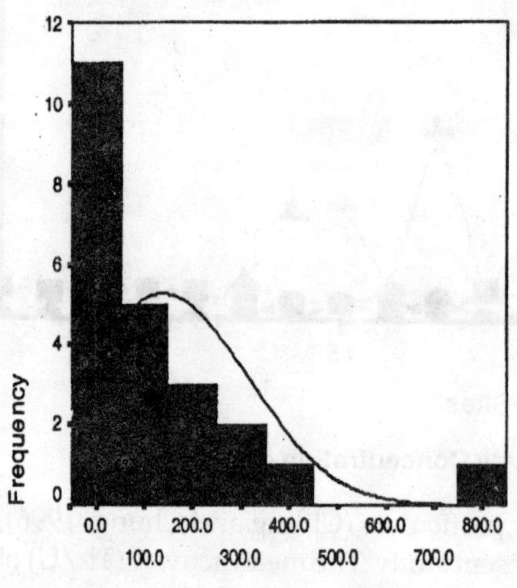

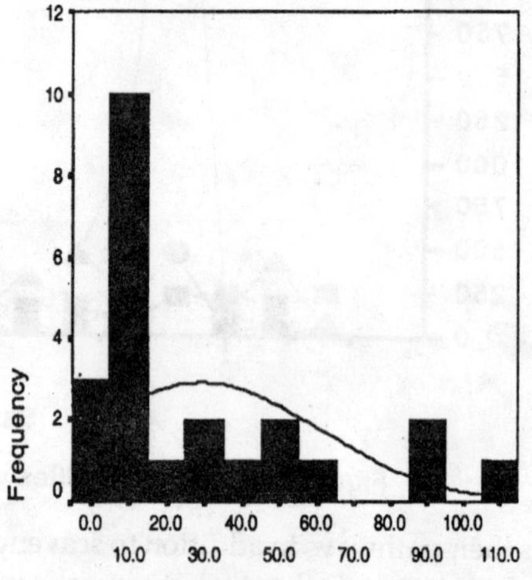

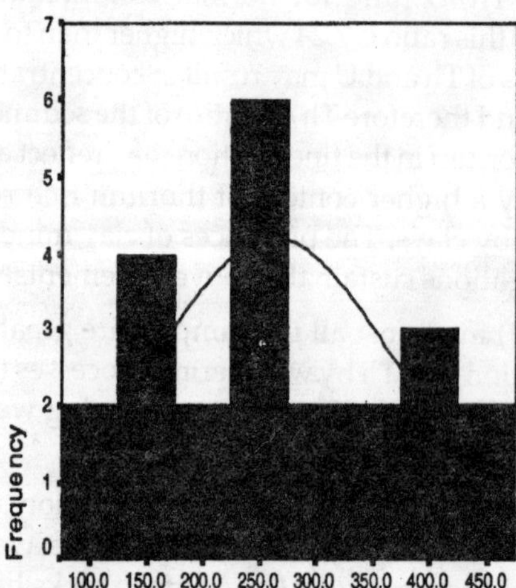

Figure 51.3: Frequency Distribution of Activity Concentration in Bq.kg^{-1}

Table 51.2: Statistical Data Analysis

Functions	^{40}K	^{238}U	^{232}Th
Mean	157.68	62.31	451.26
Median	163.13	32.34	175.03
Std. Dev.	81.21	102.29	795.62
Skewness	0.28	3.29	3.51
Kurtosis	2.04	12.55	13.98
Freq. Dist.	Normal	Normal	Normal

The positive values of the skewness coefficient indicate that these distributions are asymmetric with right tail longer than left. The very low kurtosis coefficients of ^{40}K suggest that the distribution is close to normal. The positive values of these coefficients for the three radionuclides indicate that the distributions are asymmetrical about the mean and are higher and narrower than would expected for normal distributions. In the correlation between ^{232}Th and ^{238}U, the best fitting relation is of linear type with a correlation coefficient of 0.954, it's illustrated in Figure 51.4.

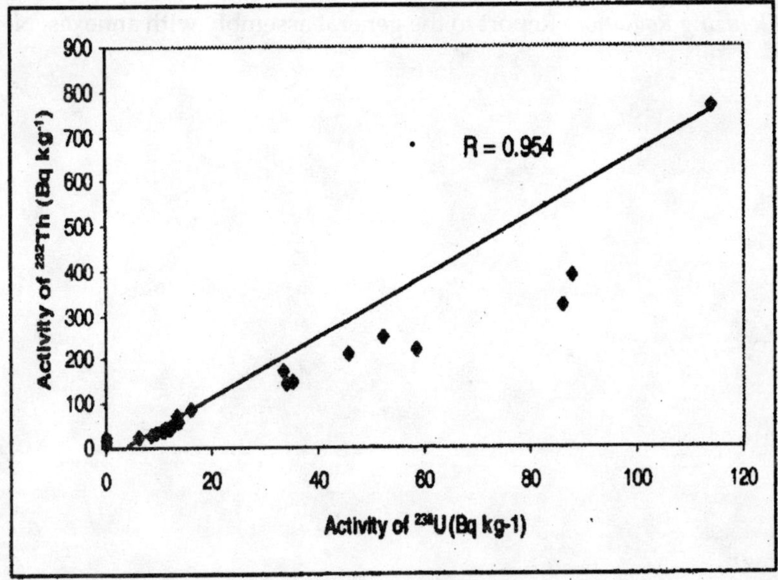

Figure 51.4: Correlation Between ^{238}U and ^{232}Th

Conclusion

The analytical results proved that the samples of particular area (S_4) significantly contain three radioactive isotopes (^{232}Th, ^{238}U and ^{40}K) all of which is originated due to the monazite content in this place. This is due to weathering of rocks in Western Ghats and Niligri Hills and also due to the transportation of thorium bearing minerals by long shore current. The activity ratio (Th/U) in the present study is greater than unity in all the samples indicate that either a net loss of ^{238}U or ^{230}Th by weathering processes. From the environmental protection point of view, the study area is free from any radiological hazard.

References

Ayres, A. and Theilen, 2001. Natural gamma-ray activity compared to geotechnical and environmental characteristics of near surface marine sediments. *J. App. Geophysics.*, 48: 1.

Balakrishna, K., Shankar, R., Sarin, M.M and Manjunatha, 2001. Distribution of U-Th, nuclides in the riverine and coastal sediments of the tropical South West Coast of India. *J. Env. Radio.*, 57: 21.

Chung, Y. and Chang, W.C., 1996. Uranium and thorium isotopes in marine sediments off Northeastern, Taiwan. *Mar. Geo.*, 133: 89.

IAEA, 1989. Measurements of radionuclides in food and the environment. *Technical Report*, Series No. 295.

Mollah, A.S., Ahmed, G.U., Husain, S.R. and Rahman, M.M., 1968. The natural radioactivity of some building materials used in Bangladesh. *Health Physics*, 50: 849.

Plater, A.J. Ivanovich, M. and Dugdale, R.E., 1992. Uranium series disequilibrium in river sediments and waters; the significance of anamalous activity ratios. *App. Geo.*, 7: 101.

Scott, R., 1968. *Earth and Planet Sci. Letts.*, 4: 245.

United National Scientific Committee on the Effects of Atomic Radiation (UNSCEAR), 2000. *Sources and Risks of Ionizing Radiation*. Report to the general assembly with annexes, New York, United Nations.

Chapter 52

Performance of *Mentha piperita* against *T. castaneum* Herbst (Coleoptera : Tenebrionidae)

Sudhakar Gupta
Department of Animal Science, M.J.P. Rohilkhand University, Bareilly – 243 006, Uttar Pradesh, India

ABSTRACT

The mortality rate of eggs was recorded 22.00 and 22.34 per cent in 1 and 2 per cent diet respectively, as compared to control (15.89 per cent). The rate of larval mortality increased from 6.0 per cent (L_1) to 12.19 per cent (L_3) in 2 per cent diet which was identical to those reared on diet mixed 1 per cent *Mentha piperita* leaf powder. In case of 2 per cent mixture of *Mentha piperita* leaf powder, the highest mortality was recorded in L_3 stage 12.19 per cent. The maximum mortality of pupae was recorded 2.54 per cent in 1 per cent treatment. The fresh biomass of untreated larvae of *T. casteneum* increased rapidly from L_1 = 0.06 to L_6 = 1.9 mg and declined in pupae (1.8 mg). Among the larvae given access to 2 per cent *Mentha piperita* leaf powder, the fresh weights of individuals from L_4 to adult stage (L_4 = 0.91, L_5 = 1.24, L_6 = 1.47, pupae = 1.37 and adult = 1.27 mg) were significantly lower (P/_ 0.01) than those of controls (L_4 = 1.27, L_5 = 1.64, L_6 = 1.91, pupae = 1.80, and adult = 1.70 mg). The dry weights of some of the developmental stages treated with 2 per cent leaf powder of *Mentha piperita* (L_5 = 0.46, pupae = 0.51 and adult = 0.44 mg) were significantly lower (P/_ 0.05 and P/_ 0.01) than those of controls. The absolute growth rates at L_4 stage (control = 0.072, 1 per cent treatment = 0.069, and 2 per cent treatment = 0.053 mg/day) were maximal but afterward it declined negatively in pupae (control = – 0.010, 1 per cent treatment = – 0.011 and 2 per cent treatment = – 0.0216 mg/day) and adults (controls = – 0.011, 1 per cent treatment = – 0.0082 and 2 per cent treatment = – 0.014 mg/day). The egg production by untreated adults of *T. castaneum* started on the 4th day after emergence of adults and an average of 6.6 eggs were laid on this day. Subsequently, the eggs laid were recorded on the 6th day (19.3), 8th day (23.3), 10th day (24.0), 12th day (26.0), 14th day (26.6), 16th day (27.6), 18th day (26.6) and 20th day (26.0). At 2 per cent treatment, however, the mean daily egg production was lower (4th day = 03.0, 6th day = 8.0, 8th day = 12.0, 10th day = 13.6, 12th day = 14.3, 14th day = 17.0, 16th day = 14.6, 18th day = 15.3 and 20th day = 14.0) than those of controls. These observations indicated that dietary incorporation of *Mentha piperita* leaf powder reduced the number of eggs to 81.09 per cent at 1 per cent treatment and 54.28 per cent at 2 per cent treatment in comparison to those of controls on the 20th day.

Keywords: *T. castaneum, Mentha piperita, Growth-rate, Oviposition.*

Introduction

T. castaneum (Herbst) is the most common pest of stored grain which causes severe losses in almost all parts of the world. The biology and certain aspects of the growth-rates of this red flour beetle on its natural diets have been variously reported by Klekowski *et al.* (1967) and Petrusewicz and Macfadyen, (1970). The effects of dietary incorporation of *Mentha piperita* leaf powder on the absolute daily growth life-table characteristics and oviposition of *T. castaneum* have been investigated in the present study.

Materials and Methods

The stock culture of *T. castaneum* was maintained on refined wheat flour in BOD incubator at 30±2°C and 70±5 per cent RH. In case of the treatments, *Mentha piperita* leaf powder was mixed in wheat flour in two different doses of 1 and 2 per cent (w/w) and the insects were reared on these diets. Desired number of eggs were collected by allowing the adults of similar age-group to lay eggs in refined wheat flour diets and sifting the eggs out with a 70 mesh sieve. Batches of 100 eggs each were sorted out and kept separately in petridishes, (6″ diameter) containing these dietary formulations, in BOD incubator for rearing. Five replicates of each experiment were run along with an appropriate control set-up and biomasses (fresh and dry) of the developmental stages were determined gravimetrically (Singh *et al.*, 1976) in order to estimate the growth rates of the immature stages of *T. castaneum.* For this purpose the samples of developmental stages were dried in oven at 60°C for 24 hrs. and the life table characteristics of *T. castaneum* were determined (Odum, 1983 and Krebs, 1985). Accordingly, the following parameter's were determined.

Equations

$$I_x = n_x/n_o \qquad \text{.......1}$$

$$d_x = n_x - (n_x + 1) \qquad \text{.......2}$$

$$q_x = d_x/n_x \qquad \text{.......3}$$

$$T_x = \sum_{x}^{\infty} L_x \qquad \text{.......4}$$

$$e_x = T_x/n_x \qquad \text{.......5}$$

where,

X: Age interval

n_o: Number of individuals at the beginning of the experiment.

n_x: Observed number of alive individuals at the start of age interval x.

I_x: Proportion surviving to start of age interval x.

d_x: Number of organisms dying within age interval x to x +1.

q_x: Rate of mortality during age interval x to x+1.

T_x: Units of individuals times time unit.

e_x: Mean expectation of life for organisms alive at the start of age x.

L_x: Number of individuals alive on the average during the age interval x to x +1.

The absolute growth-rates were estimated using the following equation (Klekowski *et al.*, 1967, Petrusewicz and Macfadyen, 1970 and Campbell *et al.*, 1976).

Absolute daily growth (mg/day) = $W_2 - W_1/t_2 - t_1$

where,

W_1 and W_2: The mean biomasses of the individuals at times t_1 and t_2 respectively.

Results and Discussion

The life-table characteristics of the immature and adults stages of *T. castaneum* reared on refined wheat flour (Control) containing 1 and 2 per cent *Mentha piperita* leaf powder are presented in the (Table 52.1) revealed a total mortality in about 32 days of the life-table of *T. castaneum*. The maximum rate of mortality of eggs was recorded 22.00 per cent in 1 per cent and 22.34 per cent in 2 per cent diet respectively where as control showed only 15.89 per cent. At 1 per cent treatment of M. *piperita* leaf powder the larval mortality was the highest (9.13 per cent). At 2 per cent treatment of M. *piperita* leaf powder the mortality at L_3 stage (12.19 per cent) was highest than that of L_6 (7.96 per cent) stages. Pupae showed a fairly high mortality (2.54 per cent) in 1 per cent diet as compared to that of control (2.04 per cent). The observations indicated that incorporation of *Mentha piperita* leaf powder adversely effected the survival of larvae and promoted survival of pupae of *T. castaneum* contact presumably on account of its repellency or contact toxicity specific to larvae while pupae experience better survival (Tripathi *et al.*, 2000 and Varma and Duby, 2001).

The mean biomasses (fresh weights) of the untreated individuals of *T. castaneum* increased sharply from: Egg = 0.031, L_1 = 0.06, L_2 = 0.139, L_3 = 0.39, L_4 = 1.27, L_5 = 1.64, L_6 = 1.91 mg followed by a decline in pupae (1.8 mg) and adults (1.7 mg). At 1 per cent treatment with *Mentha piperita* leaf powder, however, the fresh body weight of pupa (1.67 mg) was significantly (P/_ 0.05) lower than that of control although, other stages were not affected (Table 52.2). At 2 per cent treatment, however, the fresh weight all the developmental stages from L_4 stage onwards (L_4 = 0.91, L_5 = 1.24, L_6 = 1.47, pupae = 1.37 and adults = 1.27 mg) were significantly lower (P/_ 0.01) than those of the controls (Table 52.2).

The mean biomasses (dry weights) of the individuals at various developmental stages (Table 52.3) revealed that the standing crop biomass of the earlier larval instars (L_1 to L_3) did not change on account of *Mentha piperita* leaf powder. The biomasses of remaining stages, such as, L_5 (0.46 mg), pupae (0.51 mg) and adult (0.44 mg) reared at 2 per cent treatment of *Mentha piperita* leaf powder, were significantly lower (P/_ 0.05 and P/_ 0.01) in comparison to those of control L_5 (0.58 mg), pupae (0.78 mg) and adult (0.72 mg) (Table 52.3).

Table 52.1 : Life Table of *T. castaneum* Reared at 30±2°C. (L_1–L_6 = Larval instars)

Life Stage	*Duration (Days)*	n_x	l_x	d_x	q_x	L_x	T_x	e_x
Egg	3.5	100	1.00	20.667	0.2066	89.666	587.307	5.8730
L_1	1.5	79.333	0.7933	03.00	0.0378	77.833	497.641	6.2728
L_2	2.5	76.333	0.7633	03.00	0.0393	74.831	419.808	5.4996
L_3	4.5	73.33	0.7333	4.997	0.0681	70.831	344.977	4.7044
L_4	4.5	68.333	0.6833	5.003	0.0732	65.831	274.146	4.0119
L_5	4.0	63.33	0.6333	02.67	0.0421	61.995	208.315	3.2893
L_6	5.5	60.66	0.6066	2.000	0.0329	59.66	146.32	2.4121
Pupa	6.0	58.66	0.5866	01.33	0.0226	57.995	86.66	1.4773
Adult	5.0	57.33	0.5733	0.00	0.00	28.665	28.665	0.5
1 per cent leaf powder of *M. piperita*								
Egg	3.5	100	1.00	22.00	0.220	89.000	551.980	5.519
L_1	1.5	78.000	0.780	03.670	0.047	76.165	462.980	5.935
L_2	2.5	74.330	0.743	05.000	0.067	71.830	386.810	5.203
L_3	4.5	69.330	0.693	06.330	0.091	66.165	314.980	4.543
L_4	4.5	63.000	0.630	05.000	0.079	60.500	248.820	3.949
L_5	4.0	58.000	0.580	02.000	0.034	57.000	188.320	3.246
L_6	5.5	56.000	0.560	03.670	0.065	54.160	131.320	2.345
Pupa	6.0	52.330	0.523	1.330	0.025	51.660	77.160	1.474
Adult	5.0	51.000	0.510	0.000	0.000	25.500	25.500	0.500
2 per cent leaf powder of *M. piperita*								
Egg	3.5	100	1.00	22.34	0.223	88.830	539.960	5.399
L_1	1.5	77.660	0.776	04.660	0.060	75.330	451.130	5.800
L_2	2.5	73.000	0.730	04.670	0.0639	70.660	375.800	5.147
L_3	4.5	68.330	0.683	08.330	0.121	64.160	305.14	4.465
L_4	4.5	60.000	0.600	03.000	0.050	58.500	240.980	4.016
L_5	4.0	57.000	0.570	02.670	0.046	55.660	182.480	3.201
L_6	5.5	54.330	0.543	04.330	0.0796	6.160	126.820	2.330
Pupa	6.0	50.000	0.500	00.340	0.006	49.830	74.660	1.490
Adult	5.0	49.660	0.496	0.00	0.00	24.830	24.830	0.500

These observations revealed that *Mentha piperita* leaf powder affected the biomass accumulation at different developmental stage of *T. castaneum* at 2 per cent dose only perhaps by acting as a powerful repellent adversely affecting the feeding of the larvae as reported by Tripathi *et al.* (2000) and Gupta (2006) and against *Trogoderma granarium* by Gupta and Singh (2005).

The observation on the absolute growth of the various developmental stages of *T. castaneum* revealed an almost identical rate of earlier immature stages (L_1 to L_3) in all the experiments. Beyond L_3 stage, however, the growth rates of developmental stages reared on 1 per cent leaf powder (L_4 = 0.069, L_5 = 0.02, L_6 = 0.050, pupa = – 0.011 and adult = – 0.008 mg/day) and those reared on 2 per cent treatment with *Mentha piperita* leaf powder (L_4 = 0.053, L_5 = 0.025, L_6 = 0.032, pupae = – 0.021 and adult

= – 0.0114 mg/day) were lower than those of controls (L_4 = 0.072, L_5 = 0.029, L_6 = 0.048, pupae = – 0.010 and adult = – 0.011 mg/day) (Figure 52.1). Obviously, a negative growth rate in pupae and adults occurred on account of the fact that pupae did not feed and use their body nutrients for metabolism (Campbell *et al.*, 1976 and Singh *et al.*, 1976). The fact that *Mentha piperita* leaf powder brought about a reduction in growth-rate of *T. castaneum* indicated its adverse effects reducing the growth of body tissues in the immature stages of *T. castaneum* as reported by Tripathi *et al.* (2000) and Kokate *et al.* (2000).

Table 52.2: The Fresh Weights of Immature Adult Stages of *T. castaneum* Reared on 1 and 2 per cent Leaf Powder of *M. piperita* at 30±2°C

Life Stage / *Diet Mixture*	*Egg*	L_1	L_2	L_3	L_4	L_5	L_6	*Pupa*	*Adult*
Refined wheat flour (mg±SE)	0.0313 ±0.00067	0.0663 ±0.00263	0.1395 ±0.01024	0.397 ±0.02384	1.277 ±0.06092	1.648 ±0.02555	1.919 ±0.06880	1.806 ±0.00702	1.706 ±0.04910
1% *M. piperita* Leaf Powder in wheat flour (mg±SE)	0.03 ±0.00115	0.0645 ±0.00235	0.1332 ±0.00246	0.363 ±0.0470	1.23 ±0.0881	1.57 ±0.03511	1.73 ±0.16623	1.67* ±0.03929	1.63 ±0.0185
2% *M. piperita* Leaf powder in wheat flour (mg±SE)	0.03 ±0.00066	0.0623 ±0.00033	0.1228 ±0.02535	0.328 ±0.1080	0.914** ±0.0062	1.24** ±0.0782	1.47** ±0.1181	1.37** ±0.0504	1.27** ±0.0145
Critical difference at 0.01 level	0.0049	0.01040	0.08310	0.16274	0.32482	0.27097	0.65112	0.19461	0.16477
Critical difference at 0.05 level	0.0032	0.00686	0.05485	0.10742	0.21441	0.17886	0.42981	0.12846	0.10876

Significant at *p/_ 0.05; **p/_ 0.01.

Table 52.3: The Dry weights of Immature Adult Stages of *T. castaneum* Reared on 1 and 2 per cent Leaf Powder of *M. piperita* at 30±2°C

Life Stage / *Diet Mixture*	*Egg*	L_1	L_2	L_3	L_4	L_5	L_6	*Pupa*	*Adult*
Refined wheat flour (mg±SE)	0.0166 ±.00067	0.042 ±0.00393	0.0606 ±0.01020	0.136 ±0.01850	0.464 ±0.04663	0.581 ±0.01912	0.846 ±0.05174	0.7833 ±0.01202	0.7266 ±0.04055
1 per cent *M. piperita* leaf powder in wheat flour (mg±SE)	0.016 ±0.00033	0.0408 ±0.00475	0.0612 ±0.00553	0.1333 ±0.0225	0.447 ±0.05910	0.527 ±0.02811	0.803 ±0.0964	0.736 ±0.00881	0.695 ±0.03329
2 per cent *M. piperita* leaf powder in wheat flour (mg±SE)	0.0164 ±0.00083	0.040 ±0.00120	.0.059 ±0.0051	0.122 ±0.0018	0.363 ±0.0470	0.464* ±0.0466	0.646 ±0.03711	0.516** ±0.02027	0445** ±0.01365
Critical difference at 0.01 level	0.00338	0.01902	0.03840	0.08812	0.26854	0.17459	0.34954	0.07613	0.11640
Critical difference at 0.05 level	0.00223	0.01255	0.02535	0.05817	0.17726	0.11525	0.23073	0.05025	0.10826

Significant at *p/_ 0.05; **p/_ 0.01.

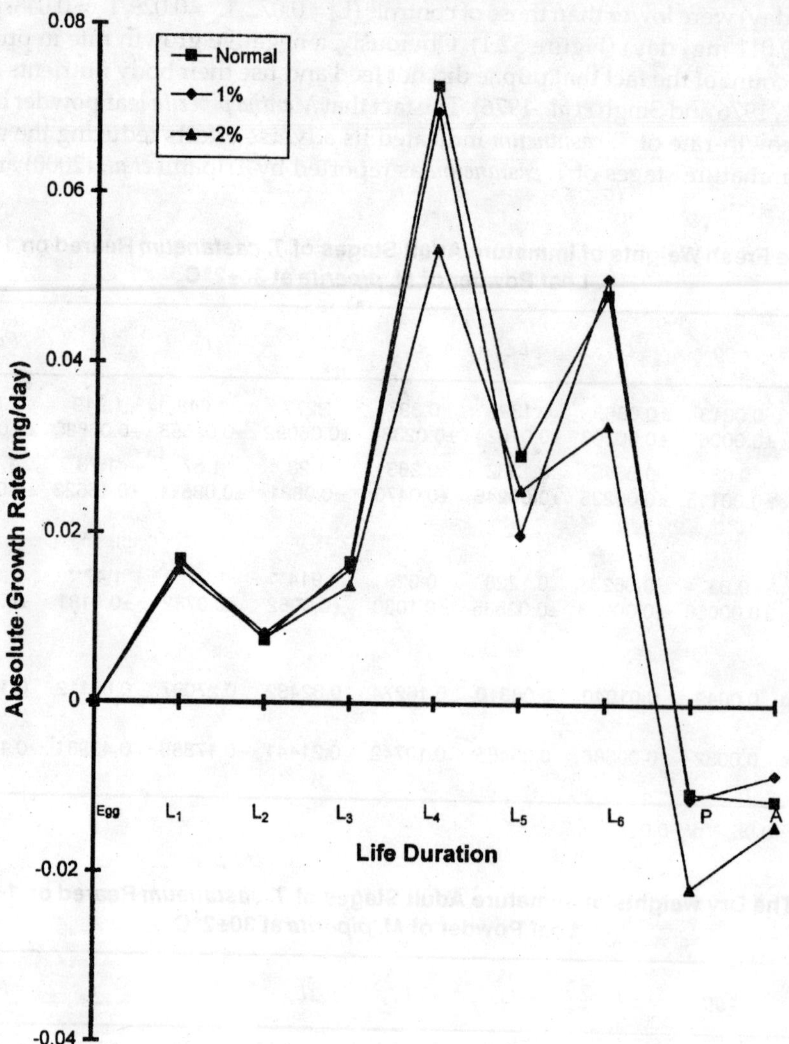

Figure 52.1: Absolute Growth Rates of Immature and Mature Stages of *T. castaneum* Reared on 1 and 2 per cent Leaf Powder of *Mentha piperita* (L_1–L_6 = Larva, P = Pupa, A = Adult)

The egg production by untreated adults of *T. castaneum* only started on the 4th day (6.66 eggs per pairs) after its emergence. Subsequently, the average number of total eggs laid were recorded on 6th day (19.33) 8th day (23.33), 10th day (24.00), 12th day (26.00), 14th day (26.66), 16th day (27.00), 18th day (26.66) and 20th day (26.00) (Table 52.4). At 1 per cent treatment with *Mentha piperita* leaf powder, however, the mean daily egg laying was lower (4th day = 05.66, 6th day = 12.33, 8th day = 19.00, 10th day = 21.66, 12th day = 22.33, 14th day = 22.66, 16th day = 21.33, 18th day = 22.00 and 20th day = 20.33) than those of controls (Table 52.4). At 2 per cent treatment, however, there were further decline in the number of eggs laid (4th day = 3.00, 6th = 8.00, 8th day = 12.00, 10th day = 13.66, 12th day = 14.33, 14th day = 17.00, 16th day = 14.66, 18th day = 15.33 and 20th day = 14.00) in comparison to controls. The egg laying capacity of *T. castaneum* females treated with 1 and 2 per cent *M. piperita* leaf powder was reduced to 81.09 and 54.28 per cent, respectively.

Table 52.4: Mean Daily Egg Production in *Tribolium castaneum* Reared on 1 and 2 per cent Leaf Powder of *Mentha piperita*

Number of Days	*Refined Wheat Flour (Control)*	*1% Mentha piperita Leaf Powder in Wheat Flour*	*2% Mentha piperita Leaf Powder in Wheat Flour*
2	–	–	–
4	06.66	05.66	03.00
6	19.33	12.33	08.00
8	23.33	19.00	12.00
10	24.00	21.66	13.66
12	26.00	22.33	14.33
14	26.66	22.66	17.00
16	27.66	21.33	14.66
18	26.66	22.00	15.33
20	26.00	20.33	14.00
Total	206.30	167.30	111.98
Percentage (%)	100	81.09	54.28

The oviposition of *T. castaneum* given access to *Mentha piperta* leaf powder mixed with wheat flour, (Table 52.4) was adversely affected vis-à-vis control both in case of 1 and 2 per cent treatments lowering the total egg production from 206.3 (control) to 111.98 (2 per cent treatment). The fact that 2 per cent *Mentha piperita* leaf powder brought about a reduction in egg laying capacity of *T. castaneum* to the tune of 54.28 per cent than that of control which indicates its effects either as fumigant or repellent for *T. castaneum*, as reported by Tripathi *et al.* (2000) and Varma and Dubey (2001) in case of other insects, or as a potent inhibitor of egg laying.

Acknowledgement

I express my grateful thanks to Dr. N.B. Singh, Professor and Head, Department of Animal Science, M.J.P. Rohilkhand University, Bareilly (U.P.) for providing worthy guidance and necessary facilities for research work. I also thank Dr. R.K. Agarwal, Deputy Director, Indian Grain Storage Management and Research Institute, Hapur (U.P.) for his valuable help and suggestions in this work.

References

Campbell, A., Singh, N.B. and Sinha, R.N., 1976. Bioenergetics of the granary weevil, *Sitophilus granarius* (L.) (Coleoptera : Curculionidae). *Can. J. Zool.*, 54: 786–798.

Gupta Sudhakar, 2006. Effect of Mint (*Mentha arvensis* Linn.) leaf powder on growth and development of *Tribolium castaneum* (Herbst). *J. Food Sci. Tech.*, 43(1): 22–25.

Gupta, S. and Singh, N.B., 2005. Effect of 'menthol mint' leaf powder on growth rate and life table of *Trogoderma granarium*. *J. Appl. Bio. Sci.*, 31(1): 12–16.

Klekowski, R.Z., Prus, T. and Zyromska-Rudzka, H., 1967. Elements of energy budget of *Tribolium castaneum* (Hbst.) in its developmental cycle, pp. 859–79 In: *Secondary Productivity of Terrestrial Ecosystem*, (Ed.) K. Petrusewicz. PWN, Warszawa, Krakow.

Kokate, C.K., Purohit, A.P. and Gokhale, S.B., 2000. *Pharmacognosy*. Nirali Prakashan, 41, Budhwar Peth, Jogeshwari Mandir Lane, Pune, pp. 314–565.

Krebs, C.J., 1985. *Ecology: The Experimental Analysis of Distribution and Abundance*, 3rd Edn. Harper and Row Publ., New York.

Odum, E.P., 1983. *Basic Ecology*. CBS College Publ., Saunders College Publishing, Philadelphia.

Petrusewicz, K. and Macfadyen, A., 1970. *Productivity of Terrestrial Animals: Principles and Methods*. IBP Handbook No. 13, Blackwell Publ., Oxford.

Singh, N.B., Campbell, A. and Sinha, R.N., 1976. An energy budget of *Sitophilus oryzae* (Coleoptera : Curculionidae). *Ann. Ent. Soc. Am.*, 69 (3): 503–511.

Tripathi, A.K., Prajapati, V., Aggarwal, K.K. and Kumar, S., 2000. Effect of volatile oil constituents of 'Mentha' species against the stored grain pests, *Callosobruchus maculatus* and *Tribolium castaneum*. *J. Med. Aro. Plant Sci.*, 22: 549–556.

Varma, J. and Dubey, N.K., 2001. Efficacy of essential oils of *Caesulia axillaris* and *Mentha arvensis* against some storage pests causing biodeterioration of food commodities. *International J. Food Microbiology*, 68: 207–210.

Chapter 53

Effect of Isopod Parasite, *Cymothoa indica* on Pearl Spot, *Etroplus suratensis* (Bloch) from Parangipettai Coastal Waters (Southeast Coast of India)

M. Rajkumar*, P. Perumal, P. Santhanam and N. Veerappan

Centre of Advanced Study in Marine Biology, Annamalai University, Parangipettai – 608 502, Tamil Nadu, India

ABSTRACT

The present study reported for the first time the effect of isopod parasite, *Cymothoa indica* infestation on *Etroplus suratensis*, a culturable and commercially important fish from Parangipettai, South-east Coast of India. The weight loss in host fishes (male 33.23 per cent; female 49.57 per cent) was observed. The weight of uninfected female fish was found to be higher than that of infested one. The calus like thickening developed on the gill arch and gill filaments of host fish due to the constant irritation caused by the appendages of the parasite. The reduction of gill surface area was due to the attachment of the parasites and observed maximum reduction of first gill arch was mainly due to the heavy pressure exerted by the parasite. The swimming capacity of the fish was also found to be affected. Details of gross lesions observed in the branchial chamber, buccal cavity and body surface were enumerated. Heavy infestations of parasitic juveniles have the potential to kill small fingerlings.

Keywords: *Isopod fish parasite, Loss of weight, Gross lesions, Gill surface reduction.*

---

* Corresponding Author: E-mail: arunachalashivamdr@yahoo.com.

Introduction

Several facultative and obligatory parasitic members of the order Isopoda are deleterious to fishes, capable of exhibiting tremendous destructive activity (Trilles, 1979; Maxwell, 1982). Parasites interfere in the biology of fishes as they affect their behaviour, health and distribution (Rohde, 1993). Isopods associated with fish constitute a significant population of crustacean ectoparasites, yet little is known about their biology (Bunkley-Williams and Williams, 1998).

To date, about 450 isopod species have been identified as parasites in freshwater and marine teleosts, as well as elasmobranchs (Kabata, 1985; Moiler and Andress, 1986). Most are cymothoids flabelliferans, a group of isopods with a short free-living planktonic phase. Cymothoids are exclusively parasitic and the presence of a few adults can cause damage to hosts. They are protandric hermaphrodites, living on the skin, in the gill chambers, or in the mouth of the fish the position is thus highly specific (Baer, 1951 and Trilles, 1969). They feed on host blood, plasma in wounds and fish tissue, sometimes resulting in massive tissue damage (Bunkley-Williams and Williams, 1998).

Diseases are a most serious limiting factor in the aquaculture production and knowledge of the fish diseases vis-a-vis parasites is highly essential for a successful aquaculture, particularly in a country like India, which has long and highly productive coastal waters. But, a glance at the relevant literature shows that the tropical coastal and brackish-waters, which likely to contain many unknown parasites and pathogens, are poorly documented.

Isopoda absorb their nourishment directly from the host's body and thus depend upon their hosts for feeding (Pillai, 1958). Segal (1987) observed the death of host fish, *Meniida beryllina* died due to the attack of isopod parasite, *Nerocila acuminate* which lived the physical damage produced by the isopod parasite upon the host has been studied by a few workers. Mini Nair (1987) however, found that the feeding intensity of fishes was unaffected due to the infestation by isopods.

In spite of all these reports, works pertaining to pathological effects of isopod parasites on the physiology of host fishes are scanty particularly from Parangipettai environments. Hence the present attempt was made to study the effect of isopod infestation on (host) fish, with reference to reduction of the respiratory surface area and loss of weight in the pearl spot, *Etroplus suratensis*.

Materials and Methods

The samples of fish, *E. suratensis* were collected from Vellar estuary (Lat. 11°29′N, Long. 79°96′E) (Parangipettai coast) and examined thoroughly for the presence of isopod parasites. *E. suratensis* is an Asian cichlid (Family: Cichlidae) endemic to Southern India and the Island of Sri Lanka and they are popularly called as Pearl spot (Green chromide) (Keshava *et al.*, 1988). It is an oval-shaped fish, laterally compressed fish with a pointed head. The body colour is ereamy olive green to greenish brown. The body is marked with six to eight slightly diagonal transverse bars which may at times, be indistinct. Each scale has a golden spot and the fins are the same colour as the body. The anal fin may have some blue iridescence. The back of the dorsal fin is a reddish-orange colour. Their face lacks the spangling of the body and is instead a metallic apple green, which is very pronounced on the forehead and radiating into their eye. The dorsal fin is long and the base extends to the caudal peduncle. The soft rays are more elongated than the hard front rays of the dorsal fin. The total length of the fish was ranged from 400–460 mm. It is a common species occurring throughout the year (Milts and Lambert, 2004). The site of attachment, orientation of parasites (on the host) and the number of parasites in each location were recorded.

To study the effect of infestation on the pearl spot fish, (*E. suratensis*), the length and weight data were analysed. The fishes were categorised as infested and uninfested. The average length and weight of infested and uninfested fishes were determined. The loss of weight and percentage loss of weight of male and female fishes were calculated. Respiratory surface area of gill arch was measured to find out the influence of parasites on the gill surface area. The gill arches of uninfested and infested fishes were carefully dissected out and blotted to remove the moisture. Then the total surface area of gill arch of both infested and uninfested fish was compared, and the differences were noted and considered as the reduction of respiratory area due to infestation.

Results

Effect on Weight of Host Fish

The effect of *C. indica* infestation on the weight of *E. suratensis* is presented in (Tables 53.1 and 53.2) (Figures 53.2 and 53.3). It is evident from the results that the infestation affected the weight was 6.7 g when compared to that of parasitized fish. Further the weight loss was more pronounced in female (49.57 per cent) than in males (33.23 per cent).

Table 53.1: Effect of *Cymothoa indica* in Relation to the Body Weight (g) of *Etroplus suratensis*

Nature of Infestation	*Mean Body Weight*	*Reduction of Weight*	*Percentage Reduction of Weight*
Uninfected	87.97	18.24	6.72
Infected	66.64	10.40	

Figure 53.1: *Cymothoa indica* Parasite Attach on Mouth of the *Etroplus suratensis*

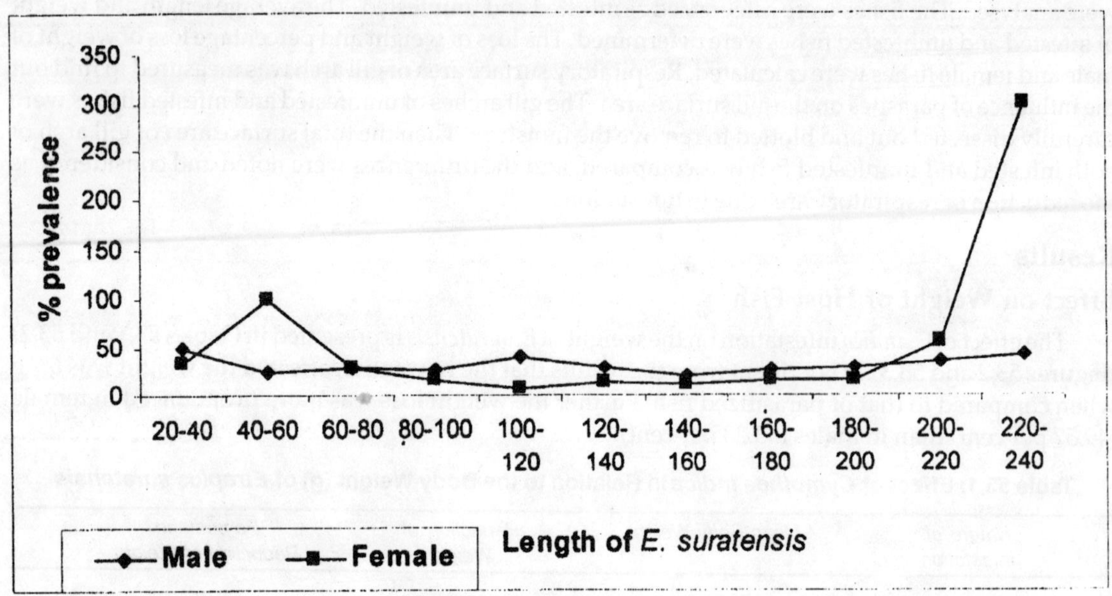

Figure 53.2: The Percentage Prevalence in Relation to Size and Sex of *Etroplus suratensis*

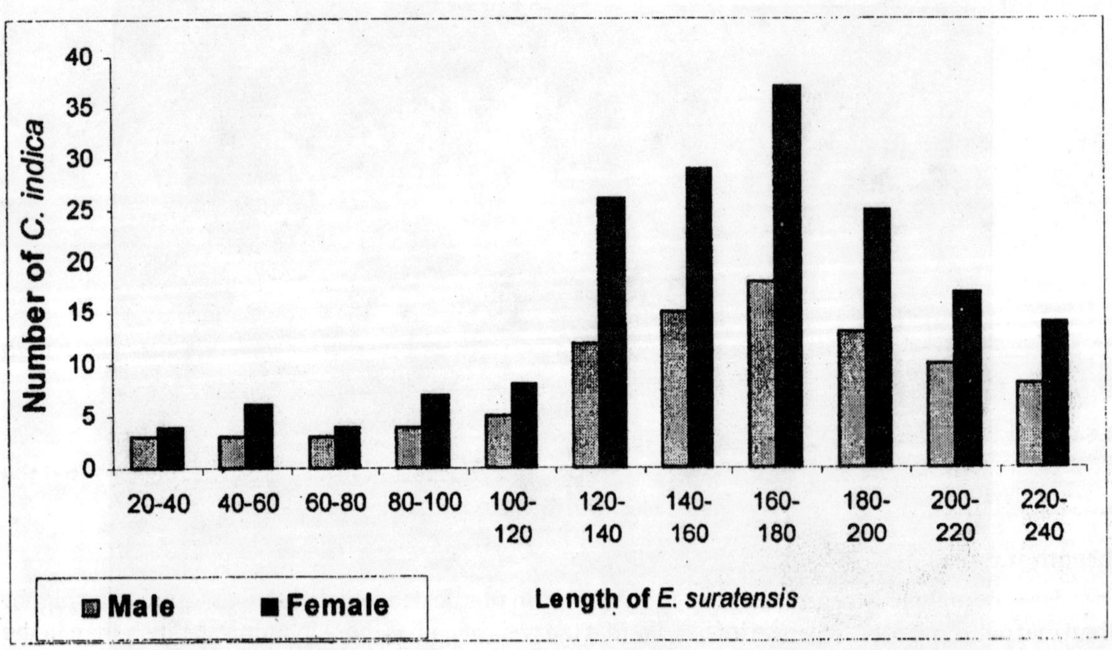

Figure 53.3: The Number of Male and Female *C. indica* in Relation to Length of *Etroplus suratensis*

Table 53.2: Percentage Weight (g) Variation in Male and Female of *Etroplus suratensis*

	Sex of Host	*Uninfected*	*Infected*
Mean Weight	Male	54.72	30.48
	Female	41.60	35.56
Loss of weight	Male	–	10.13
	Female	–	17.63
Percentage loss of weight	Male	–	33.23
	Female	–	49.57

Table 53.3: Infestation Rate of *Cymothoa indica* on *Etroplus suratensis* During Different Months (July 2003–June 2004), from Parangipettai Coast

Months	*No. of Fishes Examined*	*No. of Fishes Infested (% Prevalence)*	*No. of Parasites Collected (Mean Intensity)*
July 2003	41	18 (43.9)	23 (1.2)
August	44	13 (29.5)	16 (1.2)
September	43	16 (37.2)	22 (1.3)
October	58	14 (24.1)	25 (1.7)
November	22	10 (45.4)	12 (1.2)
December	18	7 (38.8)	13 (1.8)
January 2004	21	5 (23.8)	10 (2.0)
February	32	8 (25.0)	10 (1.2)
March	37	11 (29.7)	12 (1.0)
April	20	6 (30.0)	8 (1.3)
May	26	3 (11.5)	5 (1.6)
June	35	9 (25.7)	10 (1.1)
Total	**397**	**120 (30.2)**	**166 (1.3)**

Effect on Length and Weight Relationship

The effects on length and weight relationship of *E. suratensis* by the infestation of *C. indica* in both male and female fishes were compared and the results are as follows:

Length and Weight Relationship in Male Fish

The rate of growth is almost similar in uninfested and infested fishes. It can be inferred that the rate of growth by weight and average size was not affected much by infestation.

Length and Weight Relationship in Female Fish

The rate of increased growth in terms by weight in uninfested female fish was found to be higher than that of the infested. It can be inferred is that the weight gain is more in uninfested fish than in the infested fish.

Table 53.4: Infestation of *Cymothoa indica* on *Etroplus suratensis* in Relation to Sex at Different Months

Months	*Male Fishes*			*Female Fishes*		
	No. of Fishes Examined Prevalence)	*No. of Fishes Infested (% Intensity)*	*No. of Parasites Collected (Mean Prevalence)*	*No. of Fishes Examined Prevalence)*	*No. of Fishes Infested (% Intensity)*	*No. of Parasites Collected (Mean Prevalence)*
July 2003	29	12 (41.3)	16 (1.3)	12	6 (50.0)	7 (1.1)
August	21	8 (38.0)	11 (1.3)	23	5 (21.7)	5 (1.0)
September	18	9 (50.0)	13 (1.4)	25	7 (28.0)	9 (1.2)
October	30	8 (26.6)	12 (1.5)	28	6 (21.4)	13 (2.1)
November	15	6 (40.0)	8 (1.3)	7	4 (57.1)	4 (1.0)
December	11	4 (36.3)	9 (2.2)	7	3 (42.8	4 (1.3)
January 2004	16	(3 (18.7)	6 (2.0)	5	2 (40.0)	4 (2.0)
February	22	6 (27.2)	8 (1.3)	10	2 (20.0)	2 (1.0)
March	20	8 (10.0)	9 (1.1)	17	3 (17.6)	3 (1.0)
April	12	4 (33.3)	6 (1.5)	8	2 (25.0)	2 (1.0)
May	16	2 (12.5)	3 (1.5)	10	1 (10.0)	2 (2.0
June	23	6 (26.0)	7 (1.1)	12	3 (25.0)	3 (1.0)
Total	**233**	**76 (32.6)**	**108 (1.4)**	**164**	**44 (26.8)**	**58 (1.3)**

Effects on Respiratory Surface Area

Variations in the respiratory surface area of *E. suratensis* were due to the infestation of *C. indica*. Maximum reduction in respiratory surface area was noticed in the first gill arch and a minimum in the third gill arch. Thus considerable variation in the respiratory area was observed due to the attachment of the parasite (Figure 53.1).

Gross Lesions

The gross lesions were observed in the branchial chamber, buccal cavity and body surface is reported.

Branchial Chamber

Infested fishes had extremely pale gills indicating severe anemia. Gill rakers were seriously lost, apical edges damaged and out of gill lamellae heavily destroyed. Some secondary gill lamellae were fused or thickened. Between the gill lamellae calus-like thickening were observed. Further, the gill lamellae of the first and second gill arches were found eroded towards posterior position due to *C. indica*. Due to the lodging of parasites at the gill clefts, the gill arches showed lesions with a wide depression. The parasites further stimulated over secretion of the mucus production.

Buccal Cavity

Gross lesions observed in the buccal cavity of infested fishes showed small pin-holes in the cartilage support of the gill arches (developed), through which dactyls of peropods of isopod were penetrated claws are dig into the host tissues. The result is usually a localized destruction of the epidermis and inflammatory response around the attachment area was noticed.

Body Surface

Isopods make frequent shifts in position on the host causing a serious wound. They often move about as they feed. The activities also stimulate mucus secretion, epidermal proliferation and dilation of dermal capillaries. Mucus cells increase in number in the epidermis peripheral to the wound. Due to increased mucus secretion, there are signs of inflammation in the dermis beneath the wound. The digestive secretion apparently causes significant damage to the host's tissues.

Discussion

E. suratensis is commonly distributed and occurring throughout the year in the Parangipettai coastal waters. They are commercially important and contribute to sizeable market landings from the brackish waters of the southeast coast of India. *E. suratensis* being the largest have many desirable features which make them ideal fishes for aquaculture. Its wide salinity tolerance, ability to breed in confined waters, fast rate of growth, good body weight, tasty flesh, highly adaptable feeding habits, robust and sturdy body and good market price are some of the favourable characteristics for selection of this fish as a candidate species for brackish water aquaculture. In some parts of South India, *E. suratensis* is considered a delicacy and priced as much as Rs. 30/- per Kg which is comparable to the market value of other high quality table fishes (Keshava *et al.*, 1988).

Infestation causes serious problems to host animals either directly or indirectly affecting the physiological status of host. Loss of weight has been probably the most common effects on crustacean infestation. The present study also revealed that there was significant weight loss in the infested pearl spot fish, *E. suratensis* thereby proving the fact that infestation causes weight loss in fishes.

Kabata (1984) expressed that the general effects as loss of weight can be attributed to more than one cause and the most obvious one could be the loss of food reserve drawn from the various depots and other tissues to help in coping with the ranges of infestation. Low weight of infested fish might result from failure to grow normally. However, the present study on the length and weight relationship proved that the infestation alter the growth of the fish. It shows that the increase in length of fish is not in accordance with the increase in weight of fish due to infestation. Trilles (1979) found that several species of European cymothoids were capable of slowing down the growth in their hosts, although they did not affect weight-size ratios of the fish. However, Maxwell (1982) reported that the significant difference in the length and weight relationship of infested and uninfested fishes of Jack Mackeral, *Trachurus decliuis* when infested by *Cerotothea impricatus.*

The present study showed significant reduction of respiratory surface area because of the infestation of cymothoids in the branchial chamber was noticed. The dorsal surface of the parasite was always in close contact with the first gill arch, causing more atrophy of the first gill filament. The pressure exerted towards second to fourth gill arches were comparatively lesser than the first gill arch and the damage was also less.

Kroger and Guthrie (1972) stated that male, *Olencira praegustor* damaged the gill ofjuvenile Atlantic Menhaden, *Brevoorita tyrannus.* Stephensen (1976) stated that the gill damage was owing to buccal parasites. Bowmen (1960) found that the gill filaments of Hawaiian Moray eel, *Gymnothorax eurostus* were missing in the anterior or posterior most regions of gill arches due to the attachment of *Lirneca puhi.* Overstreet (1978) reported that the similar males and juveniles of isopod, *Olencira praegustator* parasitic on Gulf Menhaden cause change on the gill filaments through feeding.

In the present observation the parasites (*C. indica*) mostly attach with buccal region and the branchial chamber of the (host-fish) first gill arch. *C. indica* first described from Bangkok, Thailand has been recorded by Chilton (1924) from Chilka Lake and is stated to infest the mouth of *Gobius giuris*, causing deformation of the host's tongue. Subsequently, Panikkar and Aiyar (1937) recorded *C. indica* on *Etroplus macuatus, E. suratensis* and *Glossogobius giuris* from the Adyar estuary, Chennai (India). Evangeline (1963) reported the occurrence of this parasite, in the Adyar fish farm, from a variety of hosts, *viz., E. suratensis, Tilapia mossambica, Macrones gulio, Gobius giuris, Polynemus tetradactylus, Pomadasys hasta* and *Sphyraena obtusata*. It is thus evident that *C. indica* exhibits a wide geographic and host distribution. Jayadev Babu and Sanjeeva Raj (1984) noticed the infestation in several regions like the chin, nape and pectoral fin base and the buccal cavity. But, Ravichandran *et al.* (1999) found that the dorsal surface of the parasite facing the first gill arch where the parasites were attached in *Joryma brachysoma*. The damage of gill filaments thus was not only for feeding but also the pressure exerted by the dorsal side of the parasite.

The gross size and shape of parasites can act as physical irritants, which may be responsible for the presently observed damages of the branchial tissues. The reduction in the surface area was thus due to several factors such as mode of attachment, movement, size and duration of stay of the parasites.

Pale gills damaged in gill rakers and erosion of gill lamellae are the severe gross lesions were observed as a consequence of isopod infestation. Pale gills of infested fishes indicate anemia, which may be due to the loss of blood and the construction of branchial circulation by the attachment of parasite and of the haemophageous nature of the branchial cymothoids (Romestand and Trilles, 1977; Romestand, 1979). The callus like thickening observed on the gill arch and gill filaments may be due to the constant irritation caused by the body and appendages of the parasite. Erosion and thickening are the two unique morphological changes noticed owing to the infestation. These changes are mainly due to the heavy pressure exerted by the parasite and also by their feeding nature. Kabata (1985) observed destruction of host tissues as a result of the pressure exerted by the parasite body, when present in the gill cavity. Longer stay of parasite within the gill chamber may also prevent and obstruct the normal growth of the gill arches. This may be the reason for the erosion of gill arch and fusion of gill lamellae. More mucus secreted in the infested gills may be due to the reaction to irritation created by the parasite. Romestand and Trilles (1977) observed that the secretion of mucus on the ligament surface as a reactive response of the host against infestation. As a consequence of the attachment of peropods host tissues were compressed and eroded at the attachment sites, which were surrounded by an inflammed peripheral welt of peropods.

Lesions associated with reproduction of parasite to the host are related to the direct activity of the parasites. A marked increase in the size of the parasite may be seen with the development of marsupium filled with juvenile isopods. This can significantly increase the pressure atrophy caused by the presence of the parasites. These reproduction related activity in the host increases the chance that the parasite will serve as a vector for microbial parasites such as hematozoans (Smith, 1975). Uninfested fishes were more so active than the infested, and were quite still or swam feebly than the infested fishes, resulting in low rate of activity.

Assessment of the general effects of parasites on the condition of their hosts is beset with numerous difficulties. The main difficulty in garging the extent of damage inflicted on the fish lies the normal condition of the fish. Certain reports suggested that the effects of parasitization might become apparent after a latent period of fairly long duration (Reichenbach-Klinke *et al.*, 1968). The parasitic infections also result in the abnormal behaviour of the host. The parasitic infections change the blood picture and also the parasitic condition of many species leads to the secondary infections by bacteria and fungi. Some reports (Venkataraman and Sreenivasan, 1952 and 1954) indicated that the environmental factors may influence the bacterial flora of the skin and gills.

In general, parasitic infection of fishes mainly depends upon a host of factors such as age, size, sex, maturity, stage, behaviour, feeding and breeding, lifecycle, physico-chemical and particularly environment factors. The negative impact of parasites on host's growth and survival has been demonstrated for several parasite-host systems, both in aquaculture and in natural populations (Sindermann, 1987). However, host-parasite relationships are generally very complex and difficult to clarify. With the exception of cases of mass mortalities caused by outbreaks of parasites, assessment of the effects of parasite infection in natural fish populations is particularly difficult because of the presence of predators or scavengers which rapidly remove the moribund or dead fish.

It appears that *C. indica* occurs in a limited range of teleost fishes and that it takes shelter in the host mainly for the purpose of breeding (Misra and Nandi, 1986). In the present study, the pearl spot fish, *E. suratensis* parasitised by the isopod, *C. indica* showed loss of weight, loss of fat content, changes in the water content of various tissues and reduction in respiratory surface area of the host fish.

References

Baer, J.G., 1951. Ecology of the family Anchuridae (Crustacea : Isopoda) with remarks on certain morphological peculiarities. *J. Linn. Soc. (Zool.)*, 36: 109–160.

Bowmen, T.E., 1960. Description and notes on the biology of *Lironeca puhi*, sp. (Isopoda : Cymothoidae). Parasite of the Hawaiian murrey eel, *Gymnothorax eurostus (Abbott). Crustaceana L.*, pp. 82–91.

Bunkley-Williams, L. and E.H. Jr. Williams. 1998. Isopods associated with fishes: A synopsis and corrections. *J. Parasitol.*, 84(5): 893–896.

Chilton, C., 1924. Fauna of the Chilka Lake II. Tanaidacea and Isopoda. *Mem. Indian Mus.*, 5: 877–878.

Evangeline, G., 1963. Occurrence of the isopod fish parasite, *Cymothoa indica* in the Adyar Fish Farm, Madras. *Madras J. Fisheries*, 1.

Jayadev Babu, S. and S. Sanjeevaraj, 1984. Isopod parasites of fish of Pulicat Lake. In: *Proc. Symp. Coastal Aquaculture (Finfish)* 3: 818–823.

Kabata, Z., 1984. Diseases caused by metazoans: Crustaceans in diseases of marine animals. Vol. IV, Part I. *Introduction Pisces Biologische Anstalt, Helgoland* (Ed.) O. Kinne, pp. 321–399.

Kabata, Z., 1985. *Parasites and Disease of Fish Cultured in Tropics: Isopoda*. Taylor and Francis, London pp. 265–271.

Keshava, P. Santha Joseph and M. Mohan Joseph, 1988. Reproduction of Pearl-spot, *Etroplus suratensis* (Bloch) in the Nethravati-Gurpur estuary, Mangalore. In: *The first Indian Fisheries forum, Proceedings Asian fisheries Society*, Indian Branch, Mangalore, (Ed.) M. Mohan Joseph. pp. 237–241.

Kroger, R.L. and J.F. Guthrie, 1972. Incidence of parasitic Isopods, *Olencira praegustator* in juvenile Atlantic Menhaden. *Copeia*, 2: 374–379.

Maxwell, J.G.H., 1982. Infestation of jackmackeral, *Trachurus dacliuis* (Jengus), with cymothoid isopod, *Ceratothoa imbricatus* (Labricus) in Southeastern Australian waters. *J. Fish. Biol.*, 29(3): 341–349.

Mills, Dick and Lambert Derek, 2004. *The Aquarium Fish Handbook: The Complete Reference from Anemone Fish to Zamora Woodcats*, 1st Edition for the United Kingdom.

Mini Nair, 1987. Studies on the systemics and ecology of isopods of the southwest coast of India. *Ph.D. Thesis*, University of Kerala.

Misra, A. and N.C. Nandi, 1986. A new host record of *Cymothoa indica* Schioedte and Meinert (Crustacea : Isopoda) from Sundarbans, West Bengal. *Indian J. Fish.*, 33: 229–231.

Moller, H. and K. Anderss, 1986. *Diseases and Parasites of Marine Fishes*. Moiler, Kiel.

Overstreet, R.M., 1978. Marine maladies worms germs and other symbionts from the Northern gulf of Mexico Mississippi. *Alabama Sea Grant Consortium*, pp. 140.

Panikkar, N.K. and R.G. Aiyar, 1937. On a cymothoan parasite on some brackishwater fishes from Madras. *Curr. Sci.*, 5(8): 429–430.

Pillai, N. K., 1958. Tanaidocea and isopods of Travancore. *Ph.D. Thesis*, University of Kerala, pp. 152.

Ravichandran, S., A.J.A. Ranjit Singh, N., Veerappan and T. Kannupandi, 1999. Effect of isopod, parasite *Joryma brachysoma* on *Ilisha melastoma* from Parangipettai coastal waters (South east coast of India). *Ecol. Env. and Cons.*, 5(2): 95–101.

Reichenbach-Klinke, W., F. Braun, H. Held and S. Riedmuller, 1968. Vorlaufige Ergebnisse Vergleichender-physiologischer undersuchungen an coregonen verschiedener obserbayerischer seen (Fettgehatt, Blutbild, Fermentspiegel, Parasitisierung). *Arch Fischereiwiss*, 19: 114–130.

Rohde, K., 1993. *Ecology of Marine Parasites*, 2nd Edn. CAB International, Wallingford, pp. 1–298.

Romestand, B., 1979.Etude ecophysiologique des parasitoses a Cymothoadiens. *Ann. Parasitol. Hum. Comp.*, 54(4): 423–448.

Romestand, B. and J.P. Trilles, 1977. Influence decymothoiddicus (Crustacea, Isopoda, Flabellifera) succrtaines con stand hematologiques despoissons hotes. *Zeitschrift fur. Parasitenkunde*, 52: 91–95.

Segal, E., 1987. Behaviours of juvenile *Nerocila acuminata* (Isopoda : Cymothoidae) during attack and attachment and feeding on fish prey. *Bull. Mar. Sci.*, 41(2): 351–360.

Sindermann, C.J., 1987. Effects of parasites on fish populations: Practical considerations. *Int. J. Parasitol.*, 17: 371–382.

Smith, F.G., 1975. *The Pathology of Fishes*. Chap. 6: Crustacean parasites of marine fishes, pp. 189–203.

Trilles, J.P., 1969. Recherches sur les isopodes 'Cymothoidae' des cotes francaises. Apercu general et comparatif sur la bionomie et la sexualite de ces Crustaces. *Bull. Soc. Zool. Fr.*, 94(3): 433–445.

Trilles, J.P., 1979. Les cymothoidae (Isopoda, Flabellifera, parasites de poissons) du Rijksmuseum van Naturlijke Historie de le idem II. Afrique, Amerique et regions indo-ouest-pacifiques. *Zool meded, Leide.*, 54: 245–275.

Venkataraman, R. and A. Sreenivasan, 1952. A preliminary investigation of the bacterial flora of the mackerals of the west coast. *Ind. J. Med. Res.*, 40: 524–533.

Venkataraman, R. and A. Sreenivasan, 1954. Bacteriology of inshore and of mackerals off Telicherry (Malabar). *Proc. Nat. Inst. Sci.*, 20: 651–655.

Index